Regression Models for Categorical, Count, and Related Variables

Regression Models for Categorical, Count, and Related Variables

An Applied Approach

JOHN P. HOFFMANN

UNIVERSITY OF CALIFORNIA PRESS

University of California Press, one of the most distinguished university presses in the United States, enriches lives around the world by advancing scholarship in the humanities, social sciences, and natural sciences. Its activities are supported by the UC Press Foundation and by philanthropic contributions from individuals and institutions. For more information, visit www.ucpress.edu.

University of California Press
Oakland, California

Cataloguing-in-Publication data on file with the Library of Congress.

ISBN 978-0-520-28929-1 (pbk.: alk. paper)
ISBN 978-0-520-96549-2 (ebook)

Stata® is a registered trademark of StataCorp LP, 4905 Lakeway Drive, College Station, TX 77845 USA. Its use herein is for informational and instructional purposes only.

Manufactured in the United Sattes of America

25 24 23 22 21 20 19 18 17 16
10 9 8 7 6 5 4 3 2 1

For Brandon

CONTENTS

PREFACE

About a dozen years ago I published a book titled *Generalized Linear Models: An Applied Approach* (Hoffmann, 2004). Based on a detailed set of lecture notes I had used for several years in a graduate sociology course, its goal was to provide a straightforward introduction to various regression models designed to answer research questions of interest to social and behavioral scientists. Although I did not keep track of how often the book was used in university and college courses, I did receive the occasional email asking me for assistance with the data sets used in the book or with matters of clarification. Moreover, it has garnered several hundred citations—admittedly a modest amount in these days of an ever-increasing number of print and online books and research articles—on Google Scholar, so I assume it has served as a useful guide to at least a few students, instructors, and researchers.

The book relied on the statistical software Stata® for most of its examples. As many readers may know, Stata is user-friendly yet sophisticated software for conducting quantitative research. I have now used Stata for more than 20 years and continue to be impressed by its evolution and depth. It contains a remarkable range of tools for applied statisticians and researchers from many disciplines. Yet its evolution has taken it far beyond what was presented in my earlier book.[1] Moreover, some of the material I provided there has been superseded by innovations in the statistical sciences. Therefore, I opted to entirely revamp the material presented in that book with the material presented in this book. Although there are similarities, this book takes a broader view of the types of models used by social and behavioral scientists. Whereas I was primarily concerned with generalized linear models (GLMs) such as linear, logistic, and Poisson regression in my earlier work, I've found that most researchers need a more general toolkit of approaches to data analyses that involve multiple variables.

---

1. I hope it is clear that, for various pedagogical purposes, I rely on Stata to estimate the models in this book. However, Appendix A provides code in SAS, SPSS, and R that may be used to estimate most of the models presented in the following chapters.

In particular, I continue to see studies in my home discipline and related disciplines that rely on categorical or discrete variables: those that are grouped into nonoverlapping categories that measure some qualitative condition or count some phenomenon. Although most studies recognize the need to use specific models to predict such outcomes—or to use them as explanatory variables—there is no concomitant attention to many of the nuances in using such models. This includes models that fall under the general category of GLMs, but it also involves models, such as those designed for data reduction, that do not fit neatly under this category. Yet, I see the need for a more general treatment of categorical variables than is typically found in GLM and related books.

Therefore, this book provides an overview of a set of models for predicting outcome variables that are measured, primarily, as categories or counts of some phenomenon, as well as the timing of some event or events. At times, though, continuous outcome variables are used to illustrate particular models. In addition, Chapter 9 presents multilevel models, which are included since they have gained popularity in my home fields of sociology and criminology, and offer such an intriguing way of approaching socially relevant research questions. Finally, since data reduction techniques such as factor analysis are, implicitly, regression models designed to predict a set of variables simultaneously with one or more latent variables, the final chapter (Chapter 10) reviews these models, with sections that show how to estimate them with categorical variables. At the conclusion of each chapter, a set of exercises is provided so that readers may gain practice with the models. A solutions manual and the data sets are available on the publisher's website (http://www.ucpress.edu/go/regressionmodels).

I have to admit, though, that I do not address an issue that has become more prominent in the social and behavioral sciences in recent years. Specifically, this book takes a "traditional" approach to modeling by relying on frequentist assumptions as opposed to Bayesian assumptions.[2] Frequentists assume that we may, theoretically, take repeated random samples from some predetermined population and make inferences to that population based on what we find in the samples. In general, they assume that the parameters—quantities representing something about a population—are fixed. Of course, in the social and behavioral sciences, we rarely have the luxury of taking repeated random samples so we usually rely on a single sample to infer characteristics of the population.

Bayesians typically assume that the data are fixed and the parameters may be described probabilistically. Bayesian statistics also differ from frequentist statistics in their focus on using "prior beliefs" to inform one's statistical models. In frequentist statistics, we rely mainly on patterns in the available data to show statistical relationships. In Bayesian approaches, the data are combined with these prior beliefs to provide posterior beliefs that are then used for making inferences about some phenomenon. Frequentist and Bayesian

---

2. See Stone (2013) for a general introduction to Bayesian thinking and Congdon (2014) for a review of Bayesian data analysis. As of version 14 (released in April 2015), Stata now has built-in commands for Bayesian analysis. SAS and R also have Bayesian routines available; and there is specialized software such as WinBUGS and Stan.

statistical frameworks are often described, perhaps unfairly, as relying on *objective* versus *subjective* probability.

Although this presentation, like my earlier book, rests firmly on a frequentist foundation, I hope readers will consider in more detail the promise (some argue advantage) of Bayesian methods for conducting statistical analyses (see Gelman and Hill, 2007, and Wakefield, 2013; both books provide an impressive balance of information on regression modeling from Bayesian and frequentist perspectives). Perhaps, if I feel sufficiently energetic and up-to-date in a few years, I'll undertake a revision of this presentation that will give equal due to both approaches to statistical analyses. But, until that time and with due apologies to my Bayesian friends, let's get to work.

ACKNOWLEDGMENTS

I am indebted to many students and colleagues for their help in preparing this book and other material that presaged it. First, the many students I have taught over the past 20 years have brought wisdom and joy to my life. They have taught me much more than I have ever taught them. Some former students who were especially helpful in preparing this material include Scott Baldwin, Colter Mitchell, Karen Spence, Kristina Beacom, Bryan Johnson, and Liz Warnick. I am especially indebted to Kaylie Carbine, who did a wonderful job writing Appendix A, and to Marie Burt for preparing the index. I would also like to thank Seth Dobrin, Kate Hoffman, Stacy Eisenstark, Lia Tjandra, S. Bhuvaneshwari, and Mansi Gupta for their excellent assistance and careful work in seeing this book to print. Several confidential reviewers provided excellent advice that led to significant improvements in the book. My family has, as always, been immensely supportive of all my work, completed and unfinished, and I owe them more than mere words can impart. This book is dedicated to my son, Brandon, who lights up every life he comes across.

## ONE

Review of Linear Regression Models

As you should know,[1] the linear regression model is normally characterized with the following equation:

$$y_i = \alpha + \beta_1 x_1 + \beta_2 x_2 + \cdots + \beta_k x_k + \varepsilon_i \quad \{\text{or use } \beta_0 \text{ for } \alpha\}.$$

Consider this equation and try to answer the following questions:

- What does the y_i represent? The β? The x? (Which often include subscripts i—do you remember why?) The ε_i?
- How do we judge the size and direction of the β?
- How do we decide which xs are important and which are not? What are some limitations in trying to make this decision?
- Given this equation, what is the difference between prediction and explanation?
- What is this model best suited for?
- What role does the mean of y play in linear regression models?
- Can the model provide causal explanations of social phenomena?
- What are some of its limitations for studying social phenomena and causal processes?

---

1. This book assumes familiarity with linear regression models that are estimated with ordinary least squares (OLS). There are many books that provide excellent overviews of the model and its assumptions (e.g., Fox, 2016; Hoffmann and Shafer, 2015; Weisberg, 2013).

Researchers often use an estimation technique known as *ordinary least squares (OLS)* to estimate this regression model. OLS seeks to minimize the following:

$$\text{SSE} = \sum_{i=1}^{n} \left(y_i - \hat{y}_i \right)^2 .$$

The SSE is the *sum of squared errors*, with the observed y and the predicted y (y-hat) utilized in the equation. In an OLS regression model[2] that includes only one explanatory variable, the slope (β_1) is estimated with the following least squares equation:

$$\hat{\beta}_1 = \frac{\sum \left(x_i - \bar{x} \right)\left(y_i - \bar{y} \right)}{\sum \left(x_i - \bar{x} \right)^2 \big/ \left(n - 1 \right)} .$$

Notice that the variance of x appears in the denominator, whereas the numerator is part of the formula for the covariance ($\text{cov}(x,y)$). Given the slope, the intercept is simply

$$\hat{\alpha} = \bar{y} - \left\{ \hat{\beta}_1 \times \bar{x} \right\}.$$

Estimation is more complicated in a multiple OLS regression model. If you recall matrix notation, you may have seen this model represented as

$$\mathbf{Y} = \mathbf{X}\boldsymbol{\beta} + \boldsymbol{\varepsilon} .$$

The letters are bolded to represent vectors and matrices, with $\mathbf{Y}$ representing a vector of values for the outcome variable, $\mathbf{X}$ indicating a matrix of explanatory variables, and $\boldsymbol{\beta}$ representing a vector of regression coefficients, including the intercept (β_0) and slopes (β_i). The OLS regression coefficients may be estimated with the following equation:

$$\hat{\boldsymbol{\beta}} = \left(\mathbf{X'X} \right)^{-1} \mathbf{X'Y} .$$

A vector of residuals is then given by

$$\boldsymbol{\varepsilon} = \mathbf{Y} - \mathbf{Y}\hat{\boldsymbol{\beta}} .$$

Often, the residuals are represented as e to distinguish them from the errors, ε. You should recall that residuals play an important role in linear regression analysis. Various types of residuals also have a key role throughout this book. Assuming a sample and that the

---

2. The term "OLS regression model" is simply a shorthand way of indicating that the linear regression model is estimated with OLS. As shown in subsequent chapters, another common estimation technique is maximum likelihood estimation (MLE). Thus, an ML regression model refers to a model that is estimated using MLE.

model includes an intercept, some of the properties of the OLS residuals are (a) they sum to zero ($\Sigma \varepsilon_i = 0$), (b) they have a mean of zero ($E[\varepsilon] = 0$), and (c) they are uncorrelated with the predicted values of the outcome variable ($r(\varepsilon, \hat{y}) = 0$).

Analysts often wish to infer something about a target population from the sample. Thus, you may recall that the standard error (SE) of the slope is needed since, in conjunction with the slope, it allows estimation of the t-values and the p-values. These provide the basis for inference in linear regression modeling. The standard error of the slope in a simple OLS regression model is computed as

$$\text{SE}\left(\hat{\beta}_1\right) = \sqrt{\frac{\sum\left(y_i - \hat{y}_i\right)^2 / n - 2}{\sum\left(x_i - \bar{x}\right)^2}} = \sqrt{\frac{\text{SSE}/n - 2}{\text{SS}[x]}}.$$

Assuming we have a multiple OLS regression model, as shown earlier, the standard error formula requires modification:

$$\text{SE}\left(\hat{\beta}_i\right) = \sqrt{\frac{\sum\left(y_i - \hat{y}_i\right)^2}{\sum\left(x_i - \bar{x}\right)^2 \left(1 - R_i^2\right)\left(n - k - 1\right)}}.$$

Consider some of the components in this equation and how they might affect the standard errors. The matrix formulation of the standard errors is based on deriving the variance-covariance matrix of the OLS estimator. A simplified version of its computation is

$$\sigma_\varepsilon^2\left(\mathbf{X'X}\right)^{-1}, \text{ with } \sigma_\varepsilon^2 \text{ estimated by } \hat{\sigma}_\varepsilon^2 = \frac{\varepsilon'\varepsilon}{n - k} = \frac{\sum \varepsilon_i^2}{n - k}.$$

Note that the numerator in the right-hand-side equation is simply the SSE since $\left(y_i - \bar{y}\right) = \varepsilon_i$ or e_i. The right-hand-side equation is called the *residual variance* or the *mean squared error* (MSE). You may recognize that it provides an estimate—albeit biased, but consistent—of the variance of the errors. The square roots of the diagonal elements of the variance–covariance matrix yield the standard errors of the regression coefficients. As reviewed subsequently, several of the assumptions of the OLS regression model are related to the accuracy of the standard errors and thus the inferences that can be made to the target population.

OLS results in the smallest value of the SSE, if some of the specific assumptions of the model discussed later are satisfied. If this is the case, the model is said to result in the best linear unbiased estimators (BLUE) (Weisberg, 2013). It is important to note that this says *best linear*, so we are concerned here with linear estimators (there are also nonlinear estimators). In any event, BLUE implies that the estimators, such as the slopes, from an OLS regression model are unbiased, efficient, and consistent. But what does it mean to say they have these qualities? Unbiasedness refers to whether the mean of the sampling distribution of a statistic equals the parameter it is meant to estimate in the population. For example, is the slope estimated from the sample a good estimate of an analogous slope in the population? Even though

| Unbiased, efficient | Biased, efficient | Unbiased, inefficient | Biased, inefficient |

FIGURE 1.1

we rarely have more than one sample, simulation studies indicate that the mean of the sample slopes from the OLS regression model (if we could take many samples from a population), on average, equals the population slope (see Appendix B). Efficiency refers to how stable a statistic is from one sample to the next. A more efficient statistic has less variability from sample to sample; it is therefore, on average, more precise. Again, if some of the assumptions discussed later are satisfied, OLS-derived estimates are more efficient—they have a smaller sampling variance—than those that might be estimated using other techniques. Finally, consistency refers to whether the statistic converges to the population parameter as the sample size increases. Thus, it combines characteristics of both unbiasedness and efficiency.

A standard way to consider these qualities is with a target from, say, a dartboard. As shown in figure 1.1, estimators from a statistical model can be imagined as trying to hit a target in the population known as a parameter. Estimators can be unbiased and efficient, biased but efficient, unbiased but inefficient, or neither. Hopefully, it is clear why having these properties with OLS regression models is valuable.

You may recall that we wish to assess not just the slopes and standard errors, but also whether the OLS regression model provides a good "fit" to the data. This is one way of asking whether the model does a good job of predicting the outcome variable. Given your knowledge of OLS regression, what are some ways we may judge whether the model is a "good fit"? Recall that we typically examine and evaluate the R^2, adjusted R^2, and root mean squared error (RMSE). How is the R^2 value computed? Why do some analysts prefer the adjusted R^2? What is the RMSE and why is it useful?

A BRIEF INTRODUCTION TO STATA[3]

In this presentation, we use the statistical program Stata to estimate regression models (www.stata.com). Stata is a powerful and user-friendly program that has become quite popular in the social and behavioral sciences. It is more flexible and powerful than SPSS and, in

---

3. There are many excellent introductions and tutorials on how to use Stata. A good place to start is Stata's YouTube channel (https://www.youtube.com/user/statacorp) and the following website: http://www.ats.ucla.edu/stat/stata/. This website also includes links to web books that demonstrate how to estimate OLS regression models with Stata. For a more thorough treatment of

my judgment, much more user-friendly than SAS or R, its major competitors. Stata's default style consists of four windows: a command window where we type the commands; a results window that shows output; a variables window that shows the variables in the data file; and a review window that keeps track of what we have entered in the command window. If we click on a line in the review window, it shows up in the command window (so we don't have to retype commands). If we click on a variable in the variables window, it shows up in the command window, so we do not have to type variable names if we do not want to.

It is always a good idea to save the Stata commands and output by opening a log file. This can be done by clicking the brown icon in the upper left-hand corner (Windows) or the upper middle portion (Mac) of Stata or by typing the following in the command window:

`log using "regression.log"`   *the name is arbitrary*

This saves a log file to the local drive listed at the bottom of the Stata screen. To suspend the log file, type `log off` in the command window; or to close it completely type `log close`.

It is also a good idea to learn to use .*do* files. These are similar to SPSS syntax files or R script files in that we write—and, importantly, save—commands in them and then ask Stata to execute the commands. Stata has a do-editor that is simply a notepad screen for typing commands. The Stata icon that looks like a small pad of paper opens the editor. But we can also use Notepad++, TextEdit, WordPad, Vim, or any other text-editing program that allows us to save text files. I recommend that you use the handle .do when saving these files, though. In the do-editor, clicking the *run* or *do* icon feeds the commands to Stata.

AN OLS REGRESSION MODEL IN STATA

We will now open a Stata data file and estimate an OLS regression model. This allows us to examine Stata's commands and output and provide guidance on how to test the assumptions of the model. A good source of additional instructions is the *Regression with Stata* web book found at http://www.ats.ucla.edu/stat/stata/webbooks/reg. Stata's help menu (e.g., type `help regress` in the command window) is also very useful.

To begin, open the *GSS* data file (*gss.dta*). This is a subset of data from the biennial *General Social Survey* (see www3.norc.org/GSS+Website). You may use Stata's drop-down menu to open the file. Review the content of the Variables window to become familiar with the file and its contents. A convenient command for determining the coding of variables in Stata is called `codebook`. For example, typing and entering `codebook sei` returns the label and some information about this variable, including its mean, standard deviation, and some percentiles. Other frequently used commands for examining data sets and variables include

<hr/>

using Stata for this purpose, see Hoffmann and Shafer (2015). For more information about using Stata to conduct quantitative research, see Long (2009) and Kohler and Kreuter (2012).

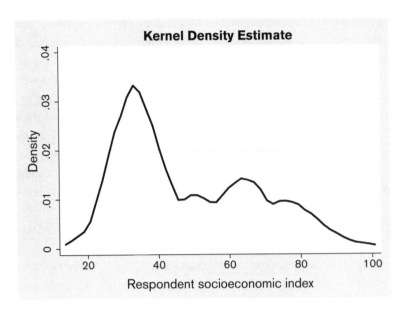

Kernel Density Estimate

FIGURE 1.2

describe, table, tabulate, summarize, graph box (boxplot), graph dot-plot (dot plot), stem (stem-and-leaf plot), hist (histogram), and kdensity (kernel density plot) (see the Chapter Resources at the end of this chapter). Stata's help menu provides detailed descriptions of each. As shown later, several of these come in handy when we wish to examine residuals and predicted values from regression models.

Before estimating an OLS regression model, let's check the distribution of *sei* with a *kernel density graph* (which is also called a *smoothed histogram*). The Stata command that appears below opens a new window that provides the graph in figure 1.2. If *sei* follows a normal distribution, it should look like a bell-shaped curve. Although it appears to be normally distributed until it hits about 50, it has a rather long tail that is suggestive of positive skewness. We investigate some implications of this skewness later.

```
kdensity sei
```

To estimate an OLS regression model in Stata, we may use the regress command.[4] The Stata code in Example 1.1 estimates an OLS regression model that predicts *sei* based on sex (the variable is labeled *female*). The term beta that follows the comma requests that Stata

---

4. If you wish to see detailed information about how this command operates, including its many subcommands and technical details about how it estimates the model, select the **[R] regress** link from the regress help page. Assuming that Stata is found locally on your computer, this brings up the section on the regress command in the Stata reference manual.

furnish *standardized regression coefficients*, or *beta weights*, as part of the output. You may recall that beta weights are based on the following equation:

$$\hat{\beta}_k^S = \hat{\beta}_k \times \frac{\sigma_{x_k}}{\sigma_y}.$$

Whereas unstandardized regression coefficients (the `Coef.` column in Stata) are interpreted in the original units of the explanatory and outcome variables, beta weights are interpreted in terms of *z*-scores. Of course, the *z*-scores of the variables must be interpretable, which is not always the case (think of a categorical variable like *female*).

Example 1.1

```
regress sei female, beta
```

Source	SS	df	MS			
				Number of obs	=	2,780
				F(1, 2778)	=	5.56
Model	2002.36916	1	2002.36916	Prob > F	=	0.0184
Residual	1000216.71	2,778	360.04921	R-squared	=	0.0020
				Adj R-squared	=	0.0016
Total	1002219.08	2,779	360.640186	Root MSE	=	18.975

sei	Coef.	Std. Err.	t	P>|t|	Beta
female	-1.704067	.722596	-2.36	0.018	-.0446983
_cons	48.79006	.5330803	91.52	0.000	.

The results should look familiar. There is an analysis of variance (ANOVA) table in the top-left panel, some model fit statistics in the top-right panel, and a coefficients table in the bottom panel. For instance, the R^2 for this model is 0.002, which could be computed from the ANOVA table using the regression (Model) sum of squares and the total sum of squares (SS(*sei*)): 2,002/1,002,219 = 0.002. Recall that the R^2 is the squared value of the correlation between the predicted values and the observed values of the outcome variable. The *F*-value is computed as *MSReg/MSResid* or 2,002/360 = 5.56, with degrees of freedom equal to *k* and {*n* − *k* − 1}. The adjusted R^2 and the RMSE[5]—two useful fit statistics—are also provided.

The output presents coefficients (including one for the *intercept* or *constant*), standard errors, *t*-values, *p*-values, and, as we requested, beta weights. Recall that the unstandardized regression coefficient for a binary variable like *female* is simply the difference in the expected means of the outcome variable for the two groups. Moreover, the intercept is the predicted mean for the reference group if the binary variable is coded as {0, 1}. Because *female* is coded

---

5. Recall that the RMSE is simply $\sqrt{\text{MSE}}$ or the estimate σ_e. It may also be characterized as the standard deviation of the residuals.

as {0 = male and 1 = female}, the model predicts that mean *sei* among males is 48.79 and mean *sei* among females is 48.79 – 1.70 = 47.09. The p-value of 0.018 indicates that, assuming we were to draw many samples from the target population, we would expect to find a slope of –1.70 or one farther from zero about 18 times out of every 1,000 samples.[6]

The beta weight is not useful in this situation because a one z-score shift in *female* makes little sense. Perhaps it will become more useful as we include other explanatory variables. In the next example, add years of education, race/ethnicity (labeled *nonwhite*, with 0 = white and 1 = nonwhite), and parents' socioeconomic status (*pasei*) to the model.

The results shown in Example 1.2 suggest that one or more of the variables added to the model may explain the association between *female* and socioeconomic status (or does it?— note the sample sizes of the two models). And we now see that education, nonwhite, and parents' status are statistically significant predictors of socioeconomic status. Whether they are important predictors or have a causal impact is another matter, however.

Example 1.2

```
regress sei female educate nonwhite pasei, beta
```

Source	SS	df	MS			
				Number of obs	=	2,231
				F(4, 2226)	=	305.49
Model	293540.757	4	73385.1893	Prob > F	=	0.0000
Residual	534724.324	2,226	240.217576	R-squared	=	0.3544
				Adj R-squared	=	0.3532
Total	828265.081	2,230	371.419319	Root MSE	=	15.499

sei	Coef.	Std. Err.	t	P>|t|	Beta
female	-.2358993	.6584269	-0.36	0.720	-.0061131
educate	3.72279	.1232077	30.22	0.000	.5563569
nonwhite	-2.860583	.9294002	-3.08	0.002	-.0528386
pasei	.0765712	.0191531	4.00	0.000	.0740424
_cons	-4.804311	1.677362	-2.86	0.004	.

The R^2 increased from 0.002 to 0.353, which appears to be quite a jump. Stata's `test` command provides a multiple partial (nested) *F*-test to determine if the addition of these variables leads to a statistically significant increase in the R^2. Simply type `test` and then list the additional explanatory variables that have been added to produce the second model. The result of this test with the three additional variables is an *F*-value of 406.4 (3, 2226 *df*) and

---

6. Consider this statement carefully: we assume what would happen if we were to draw many samples and compute a slope for each. As suggested earlier, this rarely happens in real-world applications. Given this situation, as well as other limitations discussed by statisticians, the p-value and its inferential framework are often criticized (e.g., Gelman, 2015; Hubbard and Lindsay, 2008). Although this is an important issue, we do not explore it here.

a *p*-value of less than 0.0001. Given the different sample sizes, do you recommend using the nested *F*-test approach for comparing the models? How would you estimate the effect of *female* in this model?[7]

Some interpretations from the model in Example 1.2 include the following:

- Adjusting for the effects of sex, race/ethnicity, and parents' *sei*, each 1-year increase in education is associated with a 3.72 unit increase in socioeconomic status.

- Adjusting for the effects of sex, race/ethnicity, and parents' *sei*, each one *z*-score increase in education is associated with a 0.556 *z*-score increase in socioeconomic status.

- Adjusting for the effects of sex, education, and race/ethnicity, each one-unit increase in parents' *sei* score is associated with a 0.077 unit increase in socioeconomic status.

It is useful to graph the results of regression models in some way. This provides a more informed view of the association between explanatory variables and the outcome variable than simply interpreting slope coefficients and considering *p*-values to judge effect sizes. For instance, figure 1.3 provides a visual depiction of the linear association between years of education and *sei* as predicted by the regression model. Stata's `margins` and `marginsplot` post-estimation commands are used to "adjust" the other variables by setting them at particular levels, including placing *pasei* at its mean. The vertical bars are 95% confidence intervals (CIs). What does the graph suggest?

```
qui margins, at(educate=(10 12 14 16 18 20)
    female=0 nonwhite=0) atmeans          * qui suppresses the output
marginsplot
```

Although it should be obvious, note that the graph (by design) shows a linear association. This is because the OLS regression model assumes a linear, or straight line, relationship between education and socioeconomic status (although this assumption can be relaxed). If we know little about their association, then relying on a linear relationship seems

----

7. Consider the different sample sizes. Why did the sample size decrease? What variables affected this decrease? How can you ensure that the same sample is used in both analyses? An essential task in regression and other statistical modeling efforts is to always look closely at the sample size and whether it changes from one model to the next. As a hint about how to approach this issue, look up Stata's `e(sample)` option. Use it to estimate the effect of *female* in the reduced sample. Did we reach the wrong initial conclusion regarding the association between *female* and *sei*? Why or why not? Although we do not cover missing data and their implications in this chapter (but see Appendix C), it is an essential topic for statistical analysts.

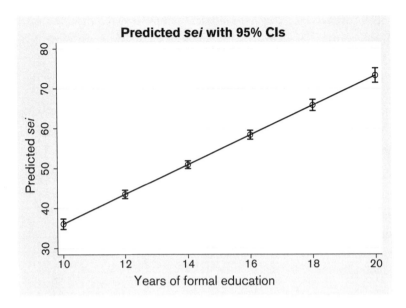

Predicted *sei* with 95% CIs

FIGURE 1.3

reasonable. But it is important to keep in mind that many associations are not linear. Think about what this means given how popular linear regression is in many scientific disciplines.

CHECKING THE ASSUMPTIONS OF THE OLS REGRESSION MODEL

Stata provides many convenient tools for checking the assumptions of OLS regression models. Recall that some of these assumptions are important because, if they are satisfied, we can be confident that the OLS regression coefficients are BLUE. We now briefly examine the assumptions and learn about some ways to examine them using Stata. For a comprehensive review, see Fox (2016).

1. Independence

To be precise, this assumption is that *the errors from one observation are independent of the errors in other observations* ($\text{cov}(\varepsilon_i, \varepsilon_j) = 0$). However, this typically is not the case when the sampling strategy was not randomly driven and there is nesting or clustering among the observations. It is quite common in the social and behavioral sciences given how often surveys are conducted that use clustered or multistage sampling designs. If not taken into account, clustering tends to lead to underestimated OLS standard errors (see Chapter 9). Stata has a host of routines for dealing with complex sampling designs and clustering. In fact, there is a `cluster` subcommand (this means it appears after the main command, usually following a comma)

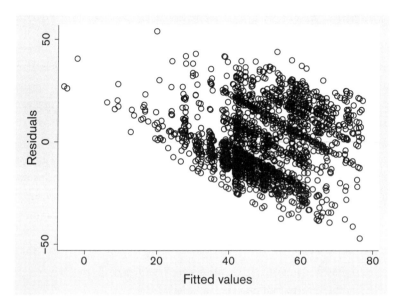

FIGURE 1.4

that may be used to adjust the standard errors for clustering: `regress y x₁ x₂`, cluster(*cluster variable*). We do not pursue this topic any further here, but see Chapters 7 and 8.

2. Homoscedasticity ("Same Scatter")

This involves the assumption that the *variance of the error terms is constant for all combinations of the x*. If this is violated—in other words, if there are *heteroscedastic errors*—then the OLS standard errors are inefficient. Recall that the standard error equation shown earlier included in its denominator the sums of squares for the x variable. This value increases, perhaps substantially, in the presence of heteroscedasticity, thus leading to smaller standard errors, on average. How might it affect one's conclusions about the model's coefficients?

There are several options available to test for heteroscedasticity, some of which examine the residuals from the model. For example, a simple way to build a residual-by-predicted plot is with Stata's `rvfplot` (*residuals vs. fitted plot*) post-estimation command (type `help regress postestimation` for a thorough description of other options). Example 1.3 shows the Stata code to execute this command and figure 1.4 furnishes the graph. Recall that we look for a funnel shape in this graph as evidence of heteroscedasticity (other patterns might also provide evidence). Although no funnel exists, there is a peculiar pattern. Some analysts prefer to graph the *studentized residuals* versus the standardized predicted values. The second part of Example 1.3 provides a set of commands for computing and graphing these predictions. Notice that we use a Stata command called `egen` to create a variable that consists of the standardized values (*z*-scores) of the predicted values.

```
rvfplot                           * after Example 1.2
predict rstudent, rstudent        * studentized residuals
predict pred, xb                  * predicted values
egen zpred = std(pred)            * z-scores of predicted values
twoway scatter rstudent zpred ||
    lowess rstudent zpred
```

The last command in Example 1.3 creates a scatter plot and overlays a *lowess* (locally weighted regression) fit line. The graph (not shown) identifies the same peculiar pattern. Although it might not be clearly indicative of heteroscedasticity, this is an unusual situation that should be examined further.

Stata also offers several numeric tests for assessing heteroscedasticity. These include `hettest` (the Breusch–Pagan test [Breusch and Pagan, 1979]) and White's (1980) test, which is based on the information matrix (`imtest`).

```
hettest                                          * after Example 1.2
```

```
Breusch-Pagan / Cook-Weisberg test for heteroskedasticity
        Ho: Constant variance
        Variables: fitted values of sei

        chi2(1)        =      22.77
        Prob > chi2    =     0.0000
```

```
estat imtest, white
```

Source	chi2	df	p
Heteroskedasticity	60.79	12	0.0000
Skewness	172.28	4	0.0000
Kurtosis	39.23	1	0.0000
Total	272.30	17	0.0000

Both tests shown in Example 1.4 indicate that there is heteroscedasticity in the model (the null hypotheses are that the errors are homoscedastic). These tests are quite sensitive to violations of the normality of the residuals, so we should reexamine this issue before deciding what to do.

We may also estimate Glejser's (1969) test by saving the residuals, taking their absolute values, and reestimating the model but predicting these values rather than *sei*. Example 1.5 furnishes this test.

Example 1.5

* after Example 1.3

```
gen absresid = abs(rstudent)
regress absresid female educate nonwhite pasei
```

Source	SS	df	MS		
Model	14.794001	4	3.69850024	Number of obs = 2,231	
Residual	677.18008	2,226	.304213873	F(4, 2226) = 12.16	
				Prob > F = 0.0000	
				R-squared = 0.0214	
Total	691.974081	2,230	.310302279	Adj R-squared = 0.0196	
				Root MSE = .55156	

absresid	Coef.	Std. Err.	t	P>|t|	[95% Conf. Interval]	
female	-.0024566	.0234312	-0.10	0.917	-.0484059	.0434927
educate	.0241652	.0043845	5.51	0.000	.015567	.0327635
nonwhite	.0088975	.0330742	0.27	0.788	-.055962	.0737571
pasei	.001275	.0006816	1.87	0.062	-.0000616	.0026116
_cons	.4418138	.0596917	7.40	0.000	.3247565	.558871

We can now conclude with some certainty that education is involved in a heteroscedasticity issue since it appears to have a substantial association with this form of the residuals.

But what do we do in this situation? Some experts suggest using weighted least squares (WLS) estimation (Chatterjee and Hadi, 2006, Chapter 7). Since we have evidence that education is implicated in the problem, we may wish to explore further using a WLS model with some transformation of education as a weighting factor. It is much simpler, though, to rely on the Huber–White sandwich estimator to compute robust standard errors (Fox, 2016, Chapter 12). This is a simple maneuver in Stata. By adding the subcommand robust after the regress command, Stata provides the robust standard errors.

Example 1.6

```
regress sei female educate nonwhite pasei, robust
```

Linear regression

				Number of obs = 2,231
				F(4, 2226) = 277.88
				Prob > F = 0.0000
				R-squared = 0.3544
				Root MSE = 15.499

sei	Coef.	Robust Std. Err.	t	P>|t|	[95% Conf. Interval]	
female	-.2358993	.6587073	-0.36	0.720	-1.527644	1.055846
educate	3.72279	.1338505	27.81	0.000	3.460305	3.985274
nonwhite	-2.860583	.9346099	-3.06	0.002	-4.693381	-1.027784
pasei	.0765712	.020297	3.77	0.000	.0367683	.1163742
_cons	-4.804311	1.713414	-2.80	0.005	-8.164367	-1.444255

Notice that the standard errors in the model in Example 1.6 are generally larger than in the model displayed in Example 1.2. Since we are most concerned with education, it is interesting that its coefficient's standard error increased from about 0.123 to 0.134, even though its p-value indicates that it remains statistically significant. Thus, even in the presence of heteroscedasticity, education appears to be a relevant predictor of *sei* in these data (although other evidence is needed to substantiate this conclusion).

Another popular option for attenuating the effects of heteroscedasticity is to estimate bootstrapped standard errors (consider `help bootstrap`). This is a resampling technique that takes repeated samples, with replacement, from the sample data set and calculates the model for each of these samples. It then uses this information to estimate a sampling distribution of the coefficients from the model, including the standard errors (Guan, 2003). This usually results in estimates that are affected less by heteroscedacity or other problematic issues.

However, it is critical to remember that issues such as heteroscedasticity do not necessarily signal that there is something wrong with the data or the model. It is important to think carefully about *why* there is heteroscedasticity or other seemingly nettlesome issues. Considering why these issues occur may lead to a better understanding of the associations that interest us. Moreover, there might be additional issues directly or indirectly involving heteroscedasticity, so we will explore some more topics after reviewing the other assumptions—as well as a few other characteristics—of the OLS regression model.

3. Autocorrelation

This assumption states that there is *no autocorrelation among the errors*. In other words, they are not correlated based on time or space. Autocorrelation is related to the independence assumption: when errors are correlated, they are not independent. Thus, the main consequence of violating this assumption is underestimated standard errors. Graphs of the residuals across time (or space in spatial regression models) are typically used as a diagnostic tool. Stata also offers a Durbin–Watson test that is invoked after a regression model (`estat dwatson`) and a whole range of models designed to adjust time-series and longitudinal models for the presence of autocorrelation. Two useful regression models are Prais–Winsten and Cochran–Orcutt regression (Wooldridge, 2010), which are implemented in Stata with the command `prais`. There are also tools available for spatial autocorrelation: errors that are correlated across space. Chapter 8 reviews regression models that are appropriate for longitudinal data, which often experience autocorrelation.

4. Collinearity

There is no *perfect collinearity* among the predictors. The model cannot be estimated when this occurs. It should be noted, though, that problems can arise with standard errors and regression coefficients even when there is high collinearity. Recall that the standard error formula shown earlier included the tolerance $(1 - R_i^2)$ in the denominator. This, you may remember, is based on regressing each explanatory variable on all the other explanatory variables. If there is perfect collinearity, then at least one of these R^2 values is 1, and the tolerance is 0. Thus, the standard error for this particular coefficient cannot be estimated. However, even very small tolerance values may lead to odd and unstable results.

The Stata post-estimation command `vif` provides *variance inflation factors* (VIFs) for an OLS regression model. For example, if we follow Example 1.2 by typing `vif`, Stata returns the next set of results.

`vif`   ** after example 1.2*

Variable	VIF	1/VIF
pasei	1.18	0.845523
educate	1.17	0.855440
nonwhite	1.02	0.984092
female	1.00	0.996199
Mean VIF	1.09	

The values in the table show no evidence of multicollinearity. But what might demonstrate evidence of a potential problem? Fox (2016) suggests examining the square root of the VIFs and cautions that the estimates are substantially affected when this value is between 2 and 3. However, some researchers point out that other aspects of the model are just as consequential as collinearity for getting stable results (O'Brien, 2007). In any event, there are some downloadable commands in Stata that provide other collinearity diagnostics, such as condition indices, that are called `collin` and `coldiag2`. Typing and entering `findit coldiag2` in the command line searches for locations that have this command and provide instructions about how to download it.

Assumptions 1–4 are the most consequential for the OLS regression model in terms of the earlier discussion of unbiasedness, efficiency, and consistency. According to the Gauss–Markov theorem, when these assumptions are satisfied, the OLS estimates offer the BLUE among the class of linear estimators (see Lindgren, 1993, 510). No other linear estimator has lower bias, or is more precise, on average, than OLS estimates. Nevertheless, there are some additional assumptions that are important for this model.

5. Error Distribution

The errors are normally distributed with a mean of zero and constant variance. This statement has three parts. The first part is also known as the *normality assumption*. If the second part $(E(\varepsilon_i) = 0)$ is violated, the intercept is biased. The last part of this statement is simply another way of stating Assumption 2. Thus, the key issue here is whether the errors follow a normal distribution. If there is non-normality, we may get misleading regression coefficients and standard errors. Nevertheless, some studies point out that even when the errors are not normally distributed, but large samples are utilized, OLS regression models yield unbiased and

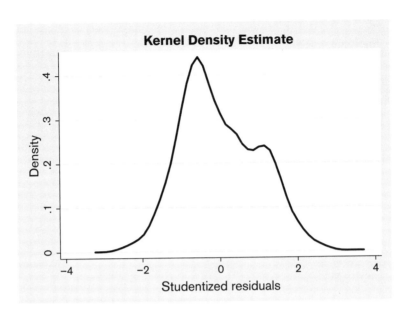

FIGURE 1.5

efficient coefficients (Lumley et al., 2002), with normally distributed intercepts and slopes (Weisberg, 2013). However, the degree of non-normality can make a difference (see Appendix B).

Saving the residuals and checking their distribution with a kernel density plot, a histogram with a normal curve overlaid, or with q–q plots and p–p plots, is the most straightforward method of testing this assumption. Stata has a qnorm and a pnorm command available for these. The pnorm graph is sensitive to non-normality in the middle range of data and the qnorm graph is sensitive to non-normality near the tails. For instance, after estimating the model in Example 1.2, save the studentized residuals and subject them to a kernel density plot, a qnorm plot, and a pnorm plot.

Figures 1.5 and 1.6 provide only a small kernel (pun intended) of evidence that there is non-normality in the tails. But figures 1.5 and 1.7 suggest there is a modest amount of non-normality near the middle of the distribution. Consider the variation from normality in the 0–2 range of the kernel density plot.

```
predict rstudent, rstudent                    * after Example 1.2
kdensity rstudent                        * try it with and without, normal
```

We examine some issues involving distributional problems later in the chapter.

```
qnorm rstudent
```

```
pnorm rstudent
```

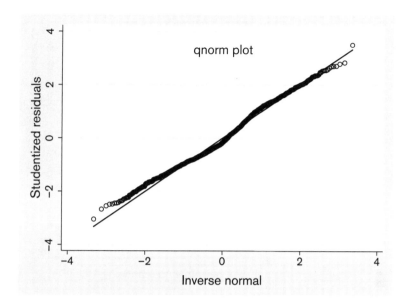

FIGURE 1.6

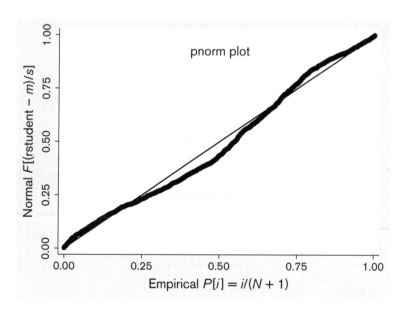

FIGURE 1.7

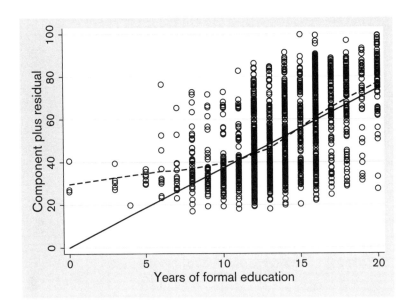

FIGURE 1.8

6. Linearity

The mean value of y for each specific combination of the x is a linear function of the x. A simple way to understand this assumption is to consider that researchers should use linear estimators, like OLS, to estimate linear associations. If the association is not linear, then they should use a nonlinear approach or else there is a risk of obtaining misleading predictions.[8] For example, imagine if an *x* variable and *y* variable have a U-shaped relationship: the slope from an OLS regression model is zero, even though there is certainly an association between them.

In Stata, we may test this assumption by estimating partial residual plots or residual-by-predicted plots and looking for nonlinear patterns. For example, use the post-estimation command below to view a partial residual plot given in figure 1.8. This is also called a *component plus residual plot*. Adding a lowess line makes it easier to detect nonlinearities in the associations. It appears that there are two linear associations between the residuals and education (above and below 10 or 11 years). These might be a concern or suggest some modifications to the model. We consider this issue in a bit more detail later.

```
cprplot educate, lowess
```
* after Example 1.2

---

8. One way to modify the OLS regression model to allow for a particular type of nonlinear association is presented later in the chapter.

7. Specification

For each explanatory variable, the correlation with the error term is zero. If not, then there is specification error in the model. This is also known as *misspecification bias. Omitted variable bias* is one type: an explanatory variable that should be included in the model is not. *Endogeneity bias* is another type. Recall that a property of the residuals is that they are uncorrelated with the predicted values of the outcome variable ($r(\varepsilon, \hat{y}) = 0$), yet the predicted values are a function of the explanatory variables. Therefore, the most consequential omitted variables are those that are correlated with one or more of the explanatory variables. Using the wrong functional form—such as when linear term is used rather than a more appropriate nonlinear term (see the linearity assumption)—is another type. Misspecification bias usually results in incorrect standard errors because it can cause the independence assumption to be violated (recall that the errors are assumed independent). The slopes are also affected when misspecification bias is present. Stata has several approaches for examining specification problems, although none is a substitute for a good theory.

The most straightforward test for OLS regression models is Stata's linktest command. This command is based on the notion that if a regression model is properly specified, we should not be able to find any additional explanatory variables that are statistically significant except by chance (Pregibon, 1980). The test creates two new variables: the predicted values, denoted _hat, and squared predicted values, denoted _hatsq. The model is then reestimated using these two variables as predictors. The first, _hat, should be statistically significant since it is the predicted value. On the other hand, _hatsq should not be significant, because if our model is specified correctly, the squared predictions should not have explanatory power. That is, we would not expect _hatsq to be a significant predictor if our model is specified correctly. So, we should assess the *p*-value for _hatsq.

Example 1.7

* after Example 1.2

linktest

Source	SS	df	MS		
				Number of obs	= 2,231
				F(2, 2228)	= 667.86
Model	310443.029	2	155221.514	Prob > F	= 0.0000
Residual	517822.052	2,228	232.415643	R-squared	= 0.3748
				Adj R-squared	= 0.3742
Total	828265.081	2,230	371.419319	Root MSE	= 15.245

sei	Coef.	Std. Err.	t	P>|t|	[95% Conf. Interval]
_hat	-.2378828	.1478597	-1.61	0.108	-.5278399 .0520744
_hatsq	.0128182	.0015031	8.53	0.000	.0098706 .0157659
_cons	28.19505	3.595472	7.84	0.000	21.14422 35.24587

The results shown in Example 1.7 suggest that there is an omitted variable or some other specification problem. An alternative, but similar, approach is Ramsey's (1969) RESET (*regression specification error test*), which is implemented in Stata using the `ovtest` command.

Example 1.7 (Continued)

ovtest

```
Ramsey RESET test using powers of the fitted values of sei
        Ho:  model has no omitted variables
                 F(3, 2223) =       32.54
                 Prob > F =        0.0000
```

This test has as its null hypothesis that there is no specification error. It is clearly rejected. A third type of test that we do not discuss here, but is available in Stata, is called a Hausman (1978) test. It is more generally applicable than the other two tests. Unfortunately, these tests do not tell us if the misspecification bias is due to omitted variables or because we have the wrong functional form for one or more features of the model. To test for problems with functional forms, we are better off examining partial residual plots and the distribution of the residuals. Theory remains the most important diagnostic tool for understanding this assumption, though.

8. Measurement Error

We assume that y and the x are measured without error. When both are measured with error— a common situation in the social sciences—the regression coefficients are often underestimated and standard errors are incorrect. For example, suppose that only the explanatory variable is measured with error. Here we have x, as we observe it, made up of a true score plus error $\{x_{1i}^* = x_{1i} + v_i\}$. This can be represented with the following:

$$y_i = \alpha + \beta_1 x_{1i}^* + \varepsilon_i = y_i = \alpha + \beta_1 \left(x_{1i} + v_{1i} \right) + \varepsilon_i \, .$$

Distributing by the slope we have the following regression equation:

$$y_i = \alpha + \beta_1 x_{1i} + \beta_1 v_{1i} + \varepsilon_i \, .$$

So the error term now has two components ($\beta_1 v_{1i}$ and ε_i) and x is usually correlated with at least one of them (although, in some cases, it may not be). Thus, measurement error is a form of misspecification bias and violates the independence assumption. But the estimated slope is biased, usually towards zero (*attenuation bias*). The degree of bias is typically unknown, though; it depends on how much error there is in the measurement of x. The standard error of the slope is also incorrect since the sum of squares of x is not accurate.

Unfortunately, variables in the social and behavioral sciences are often measured with error, especially when using survey data. Stata offers several models designed to address measurement error, including two- and three-stage least squares (instrumental variables approaches; see `ivregress`), a command called *eivreg* (errors-in-variables regression) where the analyst may set the reliabilities of the variables measured with error, and several other approaches. Type `search measurement error, all` in the command window (or use the help menu) to examine some of these options. Chapter 10 discusses some commonly used techniques for estimating latent variables, which allow one type of adjustment for measurement error.

9. Influential Observations

Strictly speaking, this is not an assumption of the OLS regression model. However, you may recall that influential observations include outliers and high leverage values that can affect the model in untoward ways. Some leverage points are bad and others are not so bad. Leverage points that fall along the regression line or plane are usually acceptable, even though they are relatively far away from the other values. Bad leverage points are extreme on the joint distribution of the explanatory variables. Their main influence is on the standard errors of the coefficients (consider the standard error equation shown earlier in the chapter; leverage points can affect the denominator). Outliers—also known as *vertical outliers*—are extreme on the outcome variable and do not fall near the regression line or surface. Outliers are especially problematic because they can influence the coefficients to a substantial degree (recall that OLS minimizes the *squared distances* from the observations to the regression line or surface).

Consider figure 1.9, which provides a scatterplot using a small set of data. It shows the effect of an outlier on the slope, with the solid line representing the slope with the outlier and the dashed line representing the slope without the outlier. Of course, the first question we should ask is, why does this outlier exist?

Stata allows us to save studentized residuals, leverage values, and Cook's *D* values (as well as other quantities such as DFFITS and Welch's distances) to check for these types of observations. It also has some automated graphs, such as the `lvr2plot`. This graph examines the leverage values against the squared values of the standardized residuals (which are positive), thus providing a useful way to check for high leverage values and outliers at the same time. Figure 1.10 provides an example from our *sei* regression model.

`lvr2plot`   * *after Example 1.2*

The two reference lines are the means for the leverage values (horizontal) and for the normalized residual squared (vertical). It appears there is a cluster of observations that are relatively high leverage values and one observation in particular that is a substantial outlier (see the right-hand portion of figure 1.10).

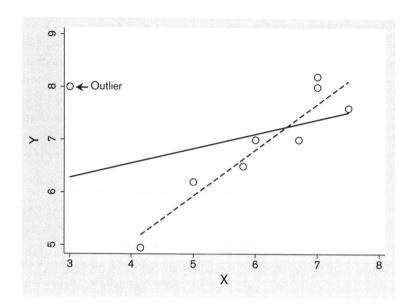

FIGURE 1.9

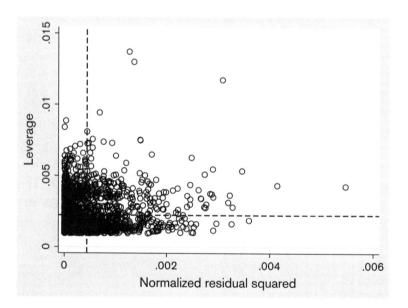

FIGURE 1.10

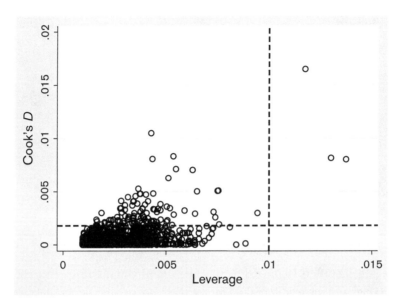

FIGURE 1.11

The Cook's D and leverage values may be saved using the commands given below. It is then a good idea to examine these values with some exploratory methods, such as stem-and-leaf plots and box-and-whisker plots. Another useful visualization is a scatterplot of Cook's D values and leverage values with reference lines at the thresholds of each. A common Cook's D threshold is $4/(n - k - 1)$ or, for our model, $4/(2{,}231 - 4 - 1) = 0.0018$. There are also two frequently used thresholds for determining high leverage values: $2(k + 1)/n$ and $3\hat{h}$ or three times the average leverage value. Given the large sample size, we use the latter threshold, which rounds to 0.01 for our model. Figure 1.11 provides the graph.

```
predict cook, c                                    * after Example 1.2
predict leverage, leverage
twoway scatter cook leverage, yline(0.0018) xline(0.01)
```

The `twoway` subcommand places reference lines in the graph that designate the thresholds. The graph shows a few outliers that are not high leverage values (in the upper left quadrant) and at least three points that are outliers and high leverage points (upper right quadrant). These deserve further exploration (jittering the points using *jitter(5)* as a subcommand may reveal the number of points better).

At this juncture, we may wish to use a robust regression model or a median regression model to minimize the effects of the outliers. As an example, Example 1.8 provides a median regression model in Stata that is invoked with the *quantile regression* (`qreg`) command.

Example 1.8

```
qreg sei female educate nonwhite pasei
```

```
Median regression                                Number of obs =      2,231
  Raw sum of deviations 18357.65 (about 39.700001)
  Min sum of deviations 14061.14                 Pseudo R2     =      0.2340
```

sei	Coef.	Std. Err.	t	P>|t|	[95% Conf. Interval]	
female	-.3982264	.9046558	-0.44	0.660	-2.172284	1.375831
educate	4.482618	.1692831	26.48	0.000	4.150648	4.814587
nonwhite	-2.093388	1.276964	-1.64	0.101	-4.597552	.4107768
pasei	.1090661	.0263158	4.14	0.000	.0574601	.1606721
_cons	-18.73586	2.304638	-8.13	0.000	-23.25532	-14.21639

Median regression differs from OLS regression by minimizing the distances between the predicted values and the median values for the outcome variable. So it is not as affected by outliers (i.e., it is a *robust* technique). Note that the general results of this model are not much different than what we have seen before. The *nonwhite* coefficient has a *p*-value somewhat above the common threshold level, but education and parents' socioeconomic status remain predictive of *sei*. Alternatives to qreg include rreg (robust regression) and mmregress, which has some desirable properties when the model is affected by influential observations. Kneib (2013) provides a good overview of some alternatives to linear regression based on means.

MODIFYING THE OLS REGRESSION MODEL

We have now reviewed how to test the assumptions of the model and discussed a few ideas about what to do when they are violated. However, we focused on simply correcting the model for some of these violations, such as heteroscedasticity and influential observations, with an eye toward getting unbiased or more precise coefficients. But we should also think carefully about model specification in light of what the tests have shown. For example, recall that education and *sei* may be involved in a nonlinear association, that the residuals from the model are not quite normally distributed, and that education is involved in the heteroscedastic errors. These may or may not be related, but are probably worth exploring in more detail. Some detective work also allows us to illustrate alternative regression models.

To begin, remember that the residuals from the model, as well as the original *sei* variable, are not normally distributed. So we should consider a transformation to normality, if we can find one. Recall also that taking the natural logarithm (log$_e$) of a variable with a long tail often normalizes its distribution. In Stata, we may do this with the following command:

```
generate logsei = log(sei)
```

According to a kernel density graph, it appears we have made things worse; there is now a clear bimodal distribution that was induced by the transformation (determine this for yourself; why did it occur?). Nonetheless, for illustrative purposes, reestimate the model using *logsei* as the outcome variable (see Example 1.9).[9]

The directions of the effects are the same, and the standardized effects based on the beta weights are similar. But, of course, the scale has changed so the interpretations are different. For example, we could now interpret the education coefficient in this way:

- Adjusting for the effects of sex, nonwhite, and parents' socioeconomic status, each 1-year increase in education is associated with a 0.074 log-unit increase in socioeconomic status.

Example 1.9

```
regress logsei female educate nonwhite pasei, beta
```

Source	SS	df	MS			
				Number of obs	=	2,231
				F(4, 2226)	=	291.03
Model	120.196619	4	30.0491548	Prob > F	=	0.0000
Residual	229.833462	2,226	.103249534	R-squared	=	0.3434
				Adj R-squared	=	0.3422
Total	350.030081	2,230	.156964162	Root MSE	=	.32132

logsei	Coef.	Std. Err.	t	P>|t|	Beta
female	-.0100858	.0136505	-0.74	0.460	-.0127139
educate	.0743256	.0025543	29.10	0.000	.5403255
nonwhite	-.0800009	.0192683	-4.15	0.000	-.0718826
pasei	.0016843	.0003971	4.24	0.000	.0792274
_cons	2.738645	.0347751	78.75	0.000	.

Another convenient way to interpret regression coefficients when we have a *log-linear model* (since the logarithm of the outcome variable is used, the model is assumed linear on a logarithmic scale) is to use a *percentage change formula* that, in general form, is

$$\% \text{ change} = 100 \times \{\exp(\beta_i) - 1\}.$$

In our example, we can transform the education coefficient using this formula. Taking advantage of Stata's display command (which acts as a high-end calculator), we find

---

9. See http://blog.stata.com/2011/08/22/use-poisson-rather-than-regress-tell-a-friend/, for an alternative approach to this model.

```
display 100 × (exp(0.0743) - 1)
7.713
```

How can we use this number in our interpretation? Here is one approach:

- Adjusting for the effects of sex, nonwhite, and parents' socioeconomic status, each one-year increase in education is associated with a 7.71% increase in the socioeconomic status score.

The nonwhite coefficient may be treated in a similar way, keeping in mind that it is a binary variable.

- Adjusting for the effects of sex, education, and parents' socioeconomic status, socioeconomic status scores among nonwhites are expected to be 7.69% lower than among whites.

It is important to reexamine the assumptions of the model to see what implications logging *sei* has for them. The diagnostics reveal some issues that may need to be addressed. The partial residual plot with education still looks odd. Moreover, there remains a problem with heteroscedasticity according to `imtest`, but not `hettest`. Glejser's test shows no signs of this problem, so perhaps we have solved it. There are still some outliers and high leverage points. These could be addressed using a robust regression technique, such as `qreg` or `mmregress`, or by assessing the variables more carefully.

EXAMINING EFFECT MODIFICATION WITH INTERACTION TERMS

In this section, we briefly review *interaction terms*. These are used when the analyst suspects that there is effect modification in the model. Another way of saying this is that some third variable moderates (or modifies) the association between an explanatory and the outcome variable. For example, if you've worked previously with the *GSS* data set used here, you may recall that education moderates the association between gender and personal income.

Do you have any ideas for moderating effects in our socioeconomic status model? Consider education, parents' socioeconomic status, and socioeconomic status. One hypothesis is that parents' status matters a great deal for one's own status, but this can be overcome by completing more years of formal education. Hence, education may moderate the association between parents' status and one's own status. To test this model, we introduce an interaction term in the model using Stata's factor variables options (`help factor variables`). Since including interaction terms often induces collinearity issues, the regression model is followed by a request for the VIFs (see Example 1.10).

Example 1.10

regress sei female nonwhite c.educate##c.pasei

Source	SS	df	MS				
				Number of obs	=		2,231
				F(5, 2225)	=		245.12
Model	294190.629	5	58838.1259	Prob > F	=		0.0000
Residual	534074.451	2,225	240.033461	R-squared	=		0.3552
				Adj R-squared	=		0.3537
Total	828265.081	2,230	371.419319	Root MSE	=		15.493

sei	Coef.	Std. Err.	t	P>|t|	[95% Conf. Interval]	
female	-.2225939	.6582242	-0.34	0.735	-1.513392	1.068204
nonwhite	-2.928681	.9299653	-3.15	0.002	-4.752371	-1.10499
educate	3.235946	.3204869	10.10	0.000	2.607461	3.86443
pasei	-.077266	.0954341	-0.81	0.418	-.2644152	.1098833
c.educate#c.pasei	.0106026	.0064437	1.65	0.100	-.0020337	.023239
_cons	2.064548	4.498666	0.46	0.646	-6.757475	10.88657

vif

Variable	VIF	1/VIF
female	1.00	0.996049
nonwhite	1.02	0.982143
educate	7.92	0.126332
pasei	29.39	0.034030
c.educate#		
c.pasei	46.35	0.021574
Mean VIF	17.14	

Although we might wish to make some interpretations from this model, we should first pause and consider what the VIFs indicate. These are typically evaluated in models with interaction terms. When we multiply the values of two variables together, the resulting variable is usually highly correlated with its constituent variables. For example, the correlation between *pasei* and the interaction term is 0.923. The VIFs show the implications of the high correlations. Consider the square root of the largest VIF: $\sqrt{46.35} = 6.8$. This is substantial (Fox, 2016). What implication does it have for the model? It is unclear, but we can be confident that the standard errors for these two variables are inflated in the regression model. What is a solution? One approach is to take the z-scores of education and parents' status and reestimate the model using the z-scores of the variables along with the updated interaction term. Example 1.11 provides the steps for carrying this out.

Example 1.11

```
egen zeducate = std(educate)                      * calculate z-scores of educate
egen zpasei = std(pasei)                          * calculate z-scores of pasei
regress sei female nonwhite c.zeducate##c.zpasei
```

Source	SS	df	MS			
				Number of obs	=	2,231
				F(5, 2225)	=	245.12
Model	294190.629	5	58838.1259	Prob > F	=	0.0000
Residual	534074.451	2,225	240.033461	R-squared	=	0.3552
				Adj R-squared	=	0.3537
Total	828265.081	2,230	371.419319	Root MSE	=	15.493

sei	Coef.	Std. Err.	t	P>|t|	[95% Conf. Interval]	
female	-.2225939	.6582242	-0.34	0.735	-1.513392	1.068204
nonwhite	-2.928681	.9299653	-3.15	0.002	-4.752371	-1.10499
zeducate	10.95155	.3619516	30.26	0.000	10.24175	11.66135
zpasei	1.199235	.3820587	3.14	0.002	.4500064	1.948464
c.zeducate#c.zpasei	.5779742	.3512613	1.65	0.100	-.1108601	1.266808
_cons	48.36958	.5122487	94.43	0.000	47.36505	49.37412

```
vif
```

Variable	VIF	1/VIF
female	1.00	0.996049
nonwhite	1.02	0.982143
zeducate	1.18	0.849157
zpasei	1.36	0.735944
c.zeducate#c.zpasei	1.15	0.866698
Mean VIF	1.14	

The main implication is that the coefficient for parents' status is now statistically significant. This might be because, before taking the z-scores, its standard error was unduly inflated by collinearity (although there could be other reasons). Moreover, there are positive coefficients for education, parents' status, and their interaction term. This suggests that the slope of the association between parents' status and one's own status is slightly steeper at higher levels of education, although note that the slope for the interaction term is not statistically significant ($p = 0.1$). This fails to support the hypothesis.

Nonetheless, assume that we did find an interesting association. Instead of relying only on the regression coefficients and their directional effects, it is helpful to graph predicted values for the different groups represented by the moderator. Here is one approach to this that uses the predicted values from the OLS regression model and examines three categories

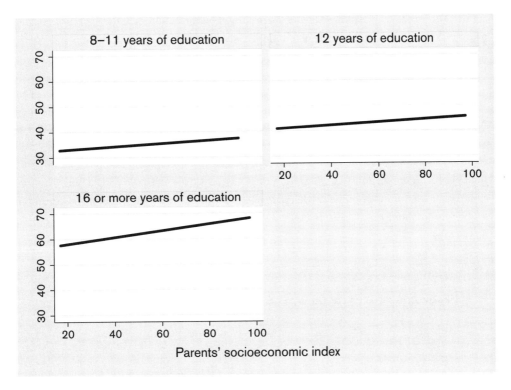

FIGURE 1.12

of education. (The `margins` and `marginsplot` commands could also be used to construct a similar graph.)

```
predict seihat, xb
recode educate (0/7=.)(8/11=1)(12=2)
   (13/15=.)(16/max=3), generate(cateducate)
```

after Example 1.10

We have divided the education variable into three parts: those who did not graduate from high school, those who graduated from high school but did not attend college, and those who graduated from college, some of whom may have attended graduate school. We then request graphs that include linear fit lines for each parent socioeconomic status–*sei* association (see figure 1.12). This provides a simple way to compare the slopes for particular education groups to see if they appear distinct.

```
twoway lfit seihat pasei, by(cateducate)
```

There are only slight differences in these three slopes. Recall that, even after taking care of the collinearity problem, the coefficient for the interaction term was not statistically significant, so the graphs should not be surprising. The combination of these graphs and the unremarkable interaction term in Example 1.11 should persuade us that there is not much, if any, effect modification in this situation.

ASSESSING ANOTHER OLS REGRESSION MODEL

As a final exercise in this review of OLS regression, use the *USData* data set (*usdata.dta*) and assess a model that examines violent crimes per 100,000 population (*violrate*) as the outcome variable and the following explanatory variables: unemployment rate (*unemprat*), gross state product (*gsprod*), and state migration rate (*mig_rate*) (see Example 1.12). Then, examine some of the assumptions of the model, judge whether there are violations, and consider what we might do about them. The following examples provide some guidance for executing these steps.

Example 1.12

```
regress violrate unemprat gsprod mig_rate
```

Source	SS	df	MS			
				Number of obs	=	50
				F(3, 46)	=	10.07
Model	1408073.67	3	469357.889	Prob > F	=	0.0000
Residual	2143815.79	46	46604.6911	R-squared	=	0.3964
				Adj R-squared	=	0.3571
Total	3551889.46	49	72487.5399	Root MSE	=	215.88

violrate	Coef.	Std. Err.	t	P>|t|	[95% Conf. Interval]	
unemprat	71.64414	25.02161	2.86	0.006	21.27825	122.01
gsprod	80.38644	19.76907	4.07	0.000	40.59336	120.1795
mig_rate	.0176516	.0084813	2.08	0.043	.0005795	.0347236
_cons	29.22665	132.9562	0.22	0.827	-238.4003	296.8536

```
cprplot mig_rate, lowess
```

What does the graph in figure 1.13 suggest? What might we do about it? Check the other explanatory variables also.

```
rvfplot
```

What does the residuals-by-fitted plot show (figure 1.14)? What other tests that are related to the test that this graph provides would you like to see? Is there a problem and, if yes, what should we do about it?

Check for multicollinearity. Are there any problems? How can you tell?

What does the following test indicate about an assumption of the model? Which assumption? Should we have any additional concerns regarding this assumption after using this test?

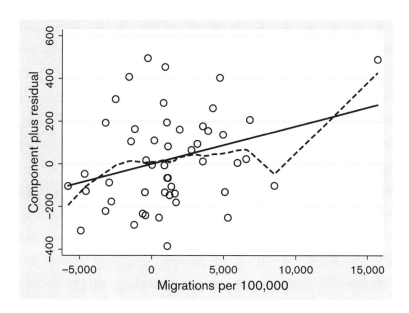

FIGURE 1.13

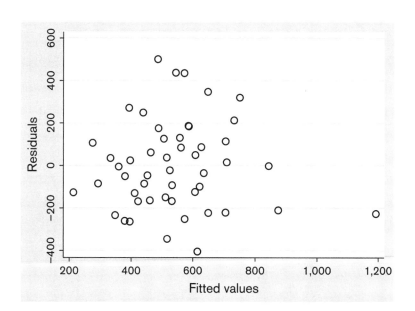

FIGURE 1.14

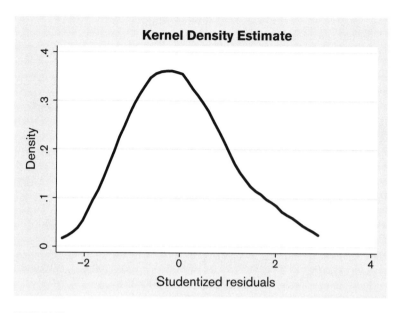

Kernel Density Estimate

FIGURE 1.15

linktest

Source	SS	df	MS		Number of obs	=	50
					F(2, 47)	=	17.59
Model	1520753.83	2	760376.915		Prob > F	=	0.0000
Residual	2031135.63	47	43215.6516		R-squared	=	0.4282
					Adj R-squared	=	0.4038
Total	3551889.46	49	72487.5399		Root MSE	=	207.88

violrate	Coef.	Std. Err.	t	P>|t|	[95% Conf. Interval]	
_hat	2.110521	.7097014	2.97	0.005	.6827859	3.538256
_hatsq	-.0008759	.0005424	-1.61	0.113	-.0019671	.0002153
_cons	-319.5151	221.2434	-1.44	0.155	-764.5994	125.5693

kdensity rstudent   ** rstudent = studentized residuals*

Consider figure 1.15. It shows the distribution of the studentized residuals from our model. Is there any cause for concern? Why or why not?

Finally, after saving two types of values using Stata's `predict` post-command, generate figure 1.16. What does it represent? What is the source of the two reference lines? What do you conclude based on this graph? What should we do in this situation?

```
twoway scatter cook leverage, yline(0.087) xline(0.16)
    mlabel(state)
```

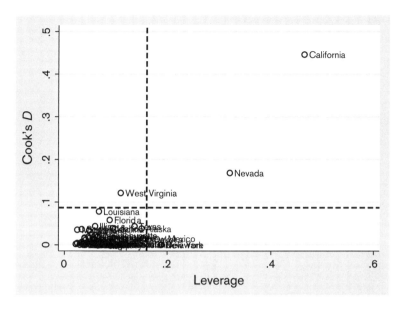

FIGURE 1.16

Do you have any concluding thoughts about this OLS regression model?

Consider the following as a final step in this exercise. Why is California an influential observation in the model? The answer is that it has a very large gross state product value. Examine the distribution of gross state product. Then, as shown in Example 1.13, take its natural logarithm (plus one to avoid negative values) and reestimate the model, check the partial residual plot with the log of gross state product, and reestimate the influential observations graph.

Example 1.13

```
gen loggsprod = log(gsprod + 1)
regress violrate unemprat loggsprod mig_rate
```

Source	SS	df	MS			
				Number of obs	=	50
				F(3, 46)	=	14.13
Model	1703254.4	3	567751.468	Prob > F	=	0.0000
Residual	1848635.05	46	40187.7185	R-squared	=	0.4795
				Adj R-squared	=	0.4456
Total	3551889.46	49	72487.5399	Root MSE	=	200.47

violrate	Coef.	Std. Err.	t	P>|t|	[95% Conf. Interval]	
unemprat	79.36568	22.95864	3.46	0.001	33.15234	125.579
loggsprod	300.4235	58.33766	5.15	0.000	182.9959	417.8511
mig_rate	.0169867	.0077277	2.20	0.033	.0014316	.0325418
_cons	-117.0315	128.1596	-0.91	0.366	-375.0033	140.9403

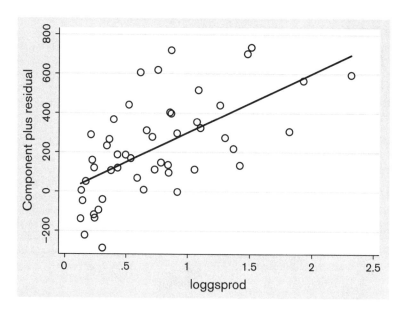

FIGURE 1.17

```
cprplot loggsprod                                    * see figure 1.17

predict cook, c
predict leverage, leverage
twoway scatter cook leverage, yline(0.087) xline(0.16)
  mlabel(state)
```

Nevada remains a high leverage point and is now the most extreme outlier (see figure 1.18). It is not clear why, but an in-depth exploration shows that it has a relatively high migration rate compared to other states. This may lead to the high leverage value. Dropping Nevada from the model leads to a better fit and fewer influential observations, but migration rate is no longer a statistically significant predictor (see Example 1.14 and figure 1.19).

However, before we close the case on migration and violent crimes, we need to think carefully about what we have done. Do we have sufficient justification for dropping Nevada from the analysis? Are there other steps we should take first before doing this? Do not forget that dropping legitimate observations is a dubious practice. It is much better to consider the source; for example, why does Nevada have a migration rate that appears to be substantially higher than in other states? Does the association between violent crimes and the migration rate have any particular implications for a state such as Nevada? There are numerous paths one might take to understand the variables and the model better.

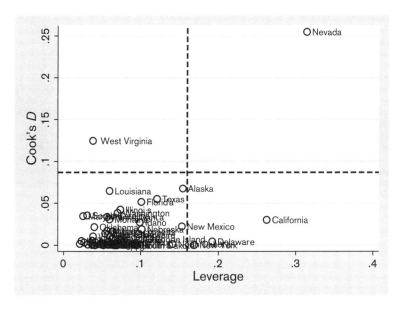

FIGURE 1.18

Example 1.14

```
regress violrate unemprat loggsprod mig_rate if state
  ~= "Nevada"
```

Source	SS	df	MS			
				Number of obs	=	49
				F(3, 45)	=	13.85
Model	1624737.77	3	541579.256	Prob > F	=	0.0000
Residual	1759165.58	45	39092.5684	R-squared	=	0.4801
				Adj R-squared	=	0.4455
Total	3383903.35	48	70497.9864	Root MSE	=	197.72

violrate	Coef.	Std. Err.	t	P>|t|	[95% Conf. Interval]	
unemprat	77.8264	22.66651	3.43	0.001	32.17372	123.4791
loggsprod	295.7535	57.62004	5.13	0.000	179.7007	411.8062
mig_rate	.0095692	.0090626	1.06	0.297	-.0086839	.0278222
_cons	-103.7971	126.7036	-0.82	0.417	-358.9913	151.3971

```
predict cook, c
predict leverage, leverage
twoway scatter cook leverage if state ~= "Nevada",
  yline(0.091) xline(0.204) mlabel(state)
```

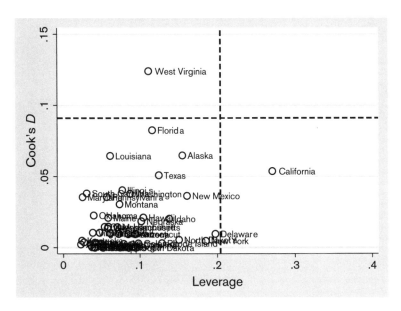

FIGURE 1.19

Finally, why West Virginia (see figure 1.19)? Sort by the Cook's D values and list the variables, predicted values, Cook's D values, and leverage values. This shows that West Virginia had fewer violent crimes than expected, but it is also relatively high on unemployment. Perhaps something about this is causing it to be an outlier. Given all this effort, should we rely on an OLS regression model to analyze state-level violent crimes or is an alternative model preferable? We do not investigate further here, but you may wish to.

FINAL WORDS

The OLS regression model is a powerful and frequently used statistical tool for assessing associations among variables. If the assumptions discussed in this chapter are satisfied, it can provide good estimates of the coefficients with which to judge these associations. However, meeting these assumptions can be challenging, even though some analysts note that only the first four are critical for achieving the BLUE property. Moreover, even when the errors are not distributed normally, the OLS regression model provides pretty accurate and efficient results. However, there are alternative models that can also provide good results but do not make some of these assumptions. Some of these are also designed for the common situation in the social and behavioral sciences when the outcome variable is not continuous and the relationship between variables is not, strictly speaking, linear. In this situation, linear models may provide questionable results, so, as discussed in subsequent chapters, these alternative models can be valuable.

EXERCISES FOR CHAPTER 1

1. The GPA data set (*gpa.dta*, available on the book's website) includes 20 observations from a sample of college students. Examine the variables to make sure you understand what they are measuring. Then, complete the following exercises.
   a. Construct a scatterplot with *gpa* on the *y*-axis and *sat_quan* on the *x*-axis. What does the scatterplot suggest about their association? Are there any unusual patterns shown in the scatterplot?
   b. Estimate a simple OLS regression model that uses *gpa* as the outcome variable and *sat_quan* as the explanatory variable. Interpret the intercept and the unstandardized coefficient for *sat_quan*.

2. Estimate the following three OLS regression models that each uses *gpa* as the outcome variable:
   a. Use only *hs_engl* as the explanatory variable.
   b. Use *hs_engl* and *sat_verb* as explanatory variables.
   c. Use *hs_engl*, *sat_verb*, and *sat_quan* as explanatory variables.
   d. Interpret the unstandardized coefficient for *hs_engl* from all three models.
   e Interpret the R^2 from the model in 2c.

3. Please describe what happened to the association between *hs_engl* and *gpa* as we moved from the first to the second to the third model. Speculate, in a conceptual way, why this change occurred.

4. Using the model in 2c, check the following OLS regression assumptions:
   a. Normality of the residuals.
   b. Homoscedasticity.

5. Check the model for influential observations.

CHAPTER RESOURCES: SOME USEFUL STATA COMMANDS

Rather than describing each, here is brief list of several of the most useful commands for data management, exploratory analysis, and regression modeling. Type `help command name` in Stata's command window for more information on each. The Stata User Manuals provide even more information.

```
findit Stata command or statistical procedure
help Stata command or statistical procedure
set memory xx
use "filename.dta"
insheet using "filename.txt"
save "filename.dta"
clear (be careful, this removes data from memory!)
describe
edit
browse
qui command (omits the output)
list varnames
codebook varnames
```

```
reshape data type, variables
recode varname
replace varname
generate newvar = f(oldvarname)
egen newvar = f(oldvarname)
drop varname
summarize varnames, detail
table varname or varname1 varname2
tabulate varname or varname1 varname2
histogram varname, normal
graph bar varname, over(group variable)
graph dot varname, over(group variable)
graph box varname, over(group variable)
kdensity varname
scatter varname1 varname2
lowess varname1 varname2
correlate varlist
spearman varlist
mean varname, over(group variable)
ci varname
ttest varname1 = varname2 or varname, by(group variable)
prtest varname1 = varname2
ranksum varname, by(group variable)
median varname, by(group variable)
regress y-varname x-varnames
predict newvar, type (e.g., residual, predicted value)
test varnames
glm y-varname x-varnames, link(link function) family(distribution)
```

## TWO

Categorical Data and Generalized Linear Models

In Chapter 1, we considered the ordinary least squares (OLS) regression model. This model is highly appropriate for relationships among the explanatory and outcome variables that are linear, when the outcome variable is continuous, and when the errors are normally distributed with constant variance. But the OLS regression model does not lend itself well to the way that social science data are usually collected and variables are constructed. Occasionally, we may satisfy or avoid some of the assumptions of the model by considering logged outcome variables, using robust standard errors, or some other approach, but we still make assumptions about continuity in the outcome variable.

As an example, consider a variable that is concerned with whether or not sample respondents have been divorced. Or a variable that is based on a question that asks people their attitudes toward some social or political topic: "Do you strongly agree, agree, disagree, or strongly disagree that the United States should give humanitarian aid to Cuba?" Other types of variables are coded as nominal variables, count variables, feeling thermometers, and in other ways that do not lend themselves to continuity or normality. In general, these often constitute *noncontinuous outcome variables*. Moreover, in a regression context the relationship between predictors and these types of outcome variables is typically not linear, so the OLS regression model does not usually work well.

This chapter introduces several types of noncontinuous variables and discusses some particular probability distributions that they may follow. It also introduces a class of models—of which the linear regression model is merely one type—that are designed for a variety of forms of outcome variables. First, though, we take a brief tour of categorical variables and contingency tables.

A BRIEF REVIEW OF CATEGORICAL VARIABLES

Recall that the types of data and variables used in statistical models are often classified as continuous or discrete. Continuous variables can, theoretically, take on any numeric value within a specific interval. Discrete variables, on the other hand, take on only integer values. Additional distinctions are often made, primarily when describing different types of discrete variables. For example, *nominal variables* are those that consist of categories that cannot be ordered in a logical and consistently acceptable manner. A common example in the social sciences is ethnic background or status: a variable may classify respondents into (a) Caucasian, (b) Latino, (c) African-American, and (d) Asian-American. When unordered discrete variables are used as explanatory variables in a regression model, they are typically converted into *dummy variables* (also known as *indicator variables*). The way dummy variables are set up is crucial: there are as many dummy variables as there are categories of a discrete variable. Continuing the ethnic status example, we should convert the four categories into four dummy variables, each coded as 0 if a respondent is not in that particular group and 1 if that person is in that particular group. Some of the examples provided later use dummy variables and show the importance of specifying a *reference category*.

Ordinal variables are discrete variables that can be ordered in a logical fashion, such as when one uses a Likert scale to measure sample members' attitudes toward some political issue. As discussed later, there are also discrete variables that count some process or outcome. This presentation makes a general distinction between continuous variables and categorical variables, the latter of which fall into ordered or unordered categories. Some statisticians also define categorical data or variables as those that identify qualitatively distinct groups (e.g., females and males), but with categories that are assigned numeric labels (e.g., 0 = male, 1 = female).

As discussed in Chapter 1, OLS regression models assume that the outcome variable is measured on a continuous scale. Explanatory variables may be either continuous or categorical, although, as mentioned earlier, categorical variables are usually converted into dummy variables before they are used in a regression model.

Categorical variables are generally examined using frequency and contingency tables. For example, the following provides a Stata-generated frequency table of a variable from the GSS data set (*gss.dta*) called *polviews*. The variable, which is based on self-reports, places respondents into liberal, moderate, and conservative groups.

Example 2.1

tabulate polviews

think of self as liberal or conservative	Freq.	Percent	Cum.
extreme liberal	59	2.15	2.15
liberal	303	11.05	13.20
slight liberal	334	12.18	25.38
middle of the road	1,044	38.07	63.46
slight conservative	451	16.45	79.91
conservative	458	16.70	96.61
extreme conservative	93	3.39	100.00
Total	2,742	100.00	

We might also examine the distribution of this variable using a bar graph, categorical plots, and various other graphical approaches (Cox, 2004).

It is interesting that, if we examine the frequency distribution of this variable, it is close to a normal distribution, with a peak in the middle (*middle of the road*) and a diminishing number of observations on either side of this peak. However, it is still a categorical variable, with numeric, integer values from 1 to 7 (consider `codebook polviews`).

The frequency column allows us to estimate the percentage or the proportion in each category. For instance, about 11% (a proportion of 0.11) report liberal when asked about their political views. If treated as a multinomial outcome (see the discussion later in the chapter), the *polviews* variable has seven expected values (probabilities) and seven variances. For example, the probability (expected value) of conservative is 0.167, with a variance of $0.167 \times (1 - 0.167) = 0.14$. Moreover, if we assume a sample and wish to generalize to the population, there are inferential statistics available for categorical data.

Suppose we also wish to predict what type of person falls into a particular category. Although we will learn much more about regression models for categorical outcome variables in subsequent chapters, the first step is to examine a *contingency table*. Contingency tables are useful for examining whether the presence of one characteristic is contingent (more or less likely to appear) on the presence of another characteristic. For example, are sex and political views associated or are they statistically independent? The contingency table in Example 2.2 helps determine this.

The row frequencies and chi-square (χ^2) test provide information with which to judge whether females and males have different reporting patterns on political views. For example, note that, overall, about 55% of the sample members are female. Yet about 63% of those reporting liberal are females. Similarly, only about 45% of the sample is male, yet almost 52% of the extreme conservatives are male. Combined with the χ^2, there is tentative evidence that political views are contingent, in some way, on the sex of the respondent. There are various measures of association and significance tests available for contingency tables that may be used to add to this evidence base (Agresti, 2013). As discussed in later chapters, there are also regression models designed for different types of categorical outcome variables. However, we require additional background information before introducing these models.

Example 2.2

`tabulate polviews female, row chi2`

think of self as liberal or conservative	sex of respondent male	female	Total
extreme liberal	33	26	59
	55.93	44.07	100.00
liberal	111	192	303
	36.63	63.37	100.00
slight liberal	143	191	334
	42.81	57.19	100.00
middle of the road	447	597	1,044
	42.82	57.18	100.00
slight conservative	217	234	451
	48.12	51.88	100.00
conservative	222	236	458
	48.47	51.53	100.00
extreme conservative	48	45	93
	51.61	48.39	100.00
Total	1,221	1,521	2,742
	44.53	55.47	100.00

Pearson chi2(6) = 19.5102   Pr = 0.003

GENERALIZED LINEAR MODELS

As mentioned earlier, the OLS regression model is designed best for outcome variables that are continuous. However, OLS is not the only estimation technique used in linear regression modeling. There are also versions of it that are subtypes of a more general class of regression models. In particular, generalized linear models (GLMs) allow us to generalize the linear regression approach to accommodate many types of outcome variables (McCullagh and Nelder, 1989). We no longer have to assume that the outcome variable is continuous or that the errors are normally distributed. But how does this work? Think about the functional form of the linear regression model presented at the outset of Chapter 1. We can write it compactly as

$$E(y) = \mu = \sum_{k=0}^{K} B_k x_k \ .$$

The Greek letter μ (*mu*) represents the expected value, or the conditional mean of y. We can make this equation more general by using the Greek letter η (*eta*) to represent the linear predictor. So, in the standard linear regression context, we may substitute η for μ in the equation:

$$\eta = \sum_{k=0}^{K} B_k x_k \ .$$

This approach may be generalized so that a set of explanatory variables is used to linearly produce the η, which are designed to predict the outcome variable, y. Recall that regression models allow us to estimate predicted values, and this is what we are concerned with here.

LINK FUNCTIONS

The key for a GLM is to somehow link the two outcomes represented as η and μ. We must find a function that does this, $g(\mu)$, so that $\eta = g(\mu)$. There are numerous ways to specify this so-called *link function*. The link function in GLMs specifies a nonlinear transformation of the predicted values so that their distribution is one of several special members of the exponential family of distributions (e.g., normal [Gaussian], lognormal, gamma, Poisson, binomial, etc.) (Hardin and Hilbe, 2012). The link function is therefore used to model responses when an outcome variable is assumed to be nonlinearly related to the explanatory variables.

One possibility is to simply specify that $\eta = \mu$. This is known as the *identity function* and allows us to estimate a linear regression model. Assuming the errors are homoscedastic and several of the other assumptions are met (see Chapter 1), we may use OLS to come up with the best linear unbiased estimators (BLUE), although, as mentioned earlier and discussed later, other estimation techniques are also available. But most forms of μ are not amenable to this approach because they are not derived from outcome variables that are continuous. We must therefore find other link functions so that the model may linearly produce η.

The most common link function, besides the identity link, is known as the *logit link*. It is specified as

$$\eta = \ln\left[\frac{\mu}{1-\mu}\right] = logit(\mu) \ .$$

The expression ln is the natural or Naperian logarithm. The logit link function is used when the outcome variable is distributed as binomial. In practical terms, we use it when the outcome variable is binary or dichotomous, typically coded as {0, 1} to represent no or yes, false or true, employed or not employed, alive or dead, or some other measure with two outcomes. The principal GLM that uses the logit link is called *logistic regression*.

The *inverse normal link* leads to the *probit regression model*, which is also appropriate for binary outcome variables:

$$\eta = \Phi^{-1}(\mu).$$

The Greek letter Φ (*phi*) represents the standard normal distribution, so it should be clear how to transform values from this distribution into something tangible. Since it is measured using z-scores, we may find probabilities based on the predicted values from this function using a table of z-scores or Stata's `normprob` function, which transforms z-scores into probabilities (e.g., `display normprob(1.96)` = 0.975).

The *natural logarithm link* is typically used with regression models that have outcome variables that count some process. Two of these models are called *Poisson regression* and *negative binomial regression*. The link function is characterized by

$$\eta = \ln(\mu).$$

Finally, we may also generalize the logit link function as

$$\eta_j = \ln\left(\frac{\mu_j}{\mu_J}\right).$$

In this function there are J categories (typically more than two). So we are concerned with the expected value of some category relative to the overall number of categories. This link function is used with the *multinomial logistic model* and is appropriate when the outcome variable consists of an unordered (nominal) set of responses (e.g., religious groups; marital status categories; car models).

Before going too far into this examination of GLMs, it is important to recall that, when we estimate regression models, we assume that the models pull out the systematic part of the outcome variable—that is, the part that is related to the explanatory variables—and then whatever is left over—the random component—follows a particular distribution:

$$y_i \;=\; \underset{\text{systematic}\atop\text{component}}{\nearrow}\quad \alpha + \beta_1 x_1 \quad \underset{\text{random}\atop\text{component}}{\nwarrow}\; +\varepsilon_i \; .$$

This is why we test the distribution of the residuals in a linear regression model: to determine whether they follow a normal distribution. And we will continue to assess residuals throughout this book.

PROBABILITY DISTRIBUTIONS AS FAMILY MEMBERS

A second element of GLMs is often called the *family*. When we assume a particular link function, we identify a particular member of a family of distributions. One way to think about this is that the random component of the outcome variable presumably follows a distribution that is specified as a particular family member. Although random variables may follow numerous probability distributions, we address only a few of the most common. Since you are familiar with the normal (Gaussian) distribution from your work with OLS regression models, it is not covered in this overview.

The Binomial Distribution

The binomial distribution concerns replications of a Bernoulli sequence of trials. Bernoulli trials have only two outcomes, which for convenience are labeled 0 for an "unsuccessful" outcome and 1 for a "successful" outcome. So the binomial distribution addresses how many successes occur in n trials. For example, we may flip a coin 10 times and wish to determine how many times heads come up (how many successes). Or we might be interested in how many adolescents have used marijuana in a sample of 100.

The *probability mass function* (PMF) of the binomial distribution is

$$p(i) = \binom{n}{i} p^i (1-p)^{n-i}.$$

The PMF of a distribution is a function that assigns a probability to each value of a discrete random variable within its range of application (e.g., $p(5) = 0.3$) (Ross, 2014).[1] The i in the PMF for the binomial distribution represents the number of successes. Recall that the variables stacked in brackets indicate a combinatorial or the number of ways we can order n things in i ways. If we are interested in five coin flips and wish to know how many ways three heads can occur, we can use the combinatorial

$$\binom{5}{3} = \frac{5!}{3!(5-3)!} = \frac{5 \times 4 \times 3 \times 2 \times 1}{3 \times 2 \times 1 \times (2 \times 1)} = 10.$$

Thus, there are 10 ways in which five coin flips can result in three heads (e.g., HHHTT, HHTHT, etc.). (Note: in Stata, `display comb(5,3)` returns 10.)

Suppose we flip a coin four times and the probability of heads is 0.5. The PMF provides the following probabilities of particular numbers of heads.

i	0	1	2	3	4
$p(i)$	0.0625	0.250	0.375	0.250	0.0625

---

1. The analogue for continuous random variables is called the *probability density function* (PDF). The PDF assigns a probability that y falls into an interval of the data space:

$$p(a \le y \le b) = \int_a^b f_y \, dy.$$

The reason we use an interval is because the probability that y is any specific value in this function is zero.

Note that the distribution is symmetric when the probability is 0.5. Hence, we can claim that the probability of finding two heads in four flips of a coin is 0.375 (try to compute this in Stata using the `display` command). This suggests that, if we were to repeatedly flip a coin four times, we expect to find exactly two heads in about 37.5% of those sequences of coin flips. Of course, this is a theoretical exercise and has led to widespread criticisms about the reality of so-called frequentist statistics.

Other examples of binomial distributions are the number of hits by softball players out of a finite number of at bats, the number of free throws basketball players make out of the number attempted, or whether or not a lottery ticket yields a winning number. The *expected value*, or mean, of the binomial distribution is $n \times p$ and the variance is $n \times p \times (1 - p)$. Thus, when the probability of heads is 0.5 and we have 60 trials, we expect to see, on average, about 30 heads, with a variance of 15. The binomial distribution is used in logistic regression models (see Chapter 3).

The Multinomial Distribution

The multinomial distribution generalizes the binomial distribution. Rather than only two possible outcomes, suppose we have three or more. Examples include the color of newborn babies' eyes (blue, brown, or green); whether a basketball player makes a 2-point shot, a 3-point shot, or misses; or whether a 40-year-old person has never married, is currently married, or is currently separated or divorced.

The PMF of the multinomial distribution is

$$p\{x_1 = n_1, x_2 = n_2, \ldots, x_r = n_r\} = \frac{n!}{n_1! n_2! \ldots n_r} p_1^{n_1} p_2^{n_2} \ldots p_r^{n_r} .$$

In this function, r is the number of possible outcomes or groups (e.g., 2-pointer, 3-pointer, miss), the p are the probability of each outcome, and the n are the frequency with which each outcome is observed. For example, in competitive games Jocelyn makes 2-pointers 40% of the time, 3-pointers 10% of the time, and misses 50% of the time. What is the probability that, in the next game, if she shoots six times she will make two 2-point shots, one 3-point shot, and miss three shots?

$$\frac{6!}{2!1!3!}(0.4)^2(0.1)^1(0.5)^3 = 0.12 .$$

Thus, we expect Jocelyn to have games in which she takes six shots to make them in this combination of ways about 12% of the time. If we figured out all the different possibilities for six shots and added them up, they would sum to 1.

Rather than one expected value and one variance, there are now as many as there are categories. For example, the mean for the number of 2-point shots is $\{6 \times 0.4\} = 2.4$ and the variance is $\{6 \times 0.4 \times (1 - 0.4)\} = 1.44$.

A practical use of the multinomial distribution is when we are faced with categorical outcome variables that cannot be ordered logically. For example, a voting survey may ask people

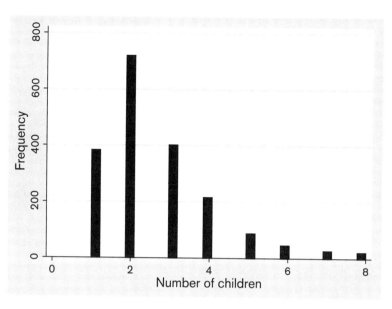

FIGURE 2.1

if they plan to vote for Hopkins, Sanders, or Jacobson; some studies examine whether people prefer to commute to work by bus, car, or bicycle; and marketing surveys may wonder whether people prefer chocolate, vanilla, or butterscotch pudding. These types of variables may be analyzed using a *multinomial logistic regression model*. This is where the multinomial distribution comes into play. We can also generalize the inverse normal function to develop a *multinomial probit regression model* (see Chapter 5).

The Poisson Distribution

The Poisson distribution is concerned with counting events. We may wish to determine how many earthquakes a county experiences in a year; how many times a stretch of road has accidents over several months; or how many times a person has committed a criminal act in the past 10 years. We often find, if we compute a histogram or a kernel density plot of a count variable, that it has what is called a *rapidly descending tail*. For example, if we look at a sample of adults and ask how many children they have, we would likely find that most had one or two, but relatively few had four or more. Using the GSS data set (*gss.dta*) and restricting the sample to those who are currently married and have any children, figure 2.1 shows the distribution of the number of children.

```
hist childs if marital ~= 5 & childs>0, frequency
```

This is not a normal distribution. Count variables rarely are. In fact, theoretically they are not continuous because they cannot be less than zero and they consist of whole numbers.

The Poisson distribution is often used to predict counts of rare events. It was originally developed to study jury decisions in France but became famous because it was used to examine Prussian soldiers killed by mule kicks in the late 1800s. The PMF of the Poisson distribution is

$$p(i) = e^{-\lambda} \frac{\lambda^i}{i!}.$$

The Greek symbol λ (*lambda*) is known as the *rate* of the distribution. It must be a positive number. One way to think about it is that it represents the average number of events over a specified period of time (e.g., the average number of cases of chicken pox in a city over a series of 1-year periods).

To see how this distribution works, imagine that the rate of earthquakes in a California city is two per year. In other words, $\lambda = 2$. Then, ask the following: what is the probability of three earthquakes in the next year? Plugging the various possibilities in the Poisson PMF, we find

I	0	1	2	3	4	5
$p(i)$	0.1353	0.2707	0.2707	**0.1804**	0.0902	0.0361

An interesting, yet unsurprising, result is that the expected probability peaks at the value of λ. This is the expected value, so we should have anticipated as much. But another interesting thing about the Poisson distribution is that the variance is also λ. Assuming that the mean equals the variance is often a restrictive proposition, but we must live with it if we wish to rely on the Poisson distribution.

The Poisson distribution is used with *Poisson regression* (see Chapter 6). When an outcome variable is measured as a count, especially when it gauges a relatively rare event, Poisson regression is often used. Count variables are quite common in the social sciences: the number of stressful events that occur in a year; the number of delinquent acts youth commit per month; the number of extracurricular activities students engage in during their senior years; the number of appointments to the US Supreme Court presidents make; the number of species that die off per year; and so forth.

The Negative Binomial Distribution

The negative binomial distribution is similar to the Poisson distribution since it is used with count variables. However, it assumes that the variance exceeds the mean, which is reasonable in most situations. Many count variables follow this pattern, which is called *overdispersion*. Studies of drug use and criminal behavior, for example, often reveal overdispersed count variables (e.g., Osgood, 2000).

Because of this, many analysts prefer the *negative binomial (NB) regression model* to the Poisson regression model. The NB model uses the NB distribution. This distribution is concerned with how many trials we must go through to get a certain number of successes. For example, we may need 100 signatures to get our candidate on the ballot. How many doors must we knock on? If we know that the probability that any random door yields a signature is 0.6, then we can calculate the number of doors we need to go to using the NB PMF:

$$p(i) = \binom{i-1}{r-1}(1-p)^{i-r} p^r.$$

The terms include p for the probability, i for the number of trials expected, and r for the number of successes we need to have. Thus, if we needed 10 signatures and the probability of getting a signature from a random house is 0.6, the probability we would need to go to exactly 20 houses is

$$p(20) = \binom{20-1}{10-1}(1-0.6)^{10} 0.6^{10} = 0.059.$$

Repeating this process over and over again, we expect that we would need to knock on 20 doors to get 10 signatures about 5.9% of the time. The following table shows some additional values for i.

i	10	15	20	25	30
$p(i)$	0.006	0.124	0.059	0.009	0.0007

Similar to the Poisson and other distributions, NB probabilities peak around a mean, but they tail off rapidly. The mean of the NB distribution is r/p and the variance is $r(1 - p)/p^2$. Thus, for our exercise the mean is $10/0.6 = 16.67$.

Seeing the connection between the NB distribution and outcome variables is not as clear as it was for the other distributions. One way to think about it is to imagine how many people we must observe or survey to come up with a certain number of outcomes, or how many times an average person (trials) engages in some event, such as marijuana use.

For rare events the NB model is often preferred to the Poisson model because the former allows for overdispersion. For example, when $r = 5$ and $p = 0.10$ the mean of the NB distribution is 50 and its variance is 450. Even when the mean equals the variance—a situation known as *equidispersion*—the NB regression model and the Poisson regression model tend to reach the same conclusions.

There are many other distributions that are also used in GLMs, such as the gamma, the exponential, and the Weibull. But we will address mainly those that have been described in this section; see, however, Chapter 7.

HOW ARE GLMS ESTIMATED?

As discussed in Chapter 1, analysts usually rely on OLS to estimate a linear regression model. OLS is relatively simple to compute and its properties have been understood for many decades. However, it is limited in the types of models it can be used to estimate. A much more flexible and general estimation technique is *maximum likelihood* (ML). There are many good overviews of ML estimation (MLE) (see Fox, 2010), and it requires a good grounding in calculus to understand fully, so we cover just the basics.

To begin with, think about probabilities. A common problem in elementary statistics courses is to find the probability of getting some particular sequence of heads when flipping a coin. Using the binomial PMF we may estimate, for instance, the probability of getting three heads in five flips of a coin. If the coin is balanced, the probability of heads is 0.5. So the probability calculation is straightforward. Similarly, as we have seen, if we know the mean number of earthquakes per year in a California city, determining the probability of a certain number in the next year, assuming environmental conditions have not changed too much, is easy to do with the Poisson PMF.

A more realistic situation in statistics is that we observe the number of trials, such as when we survey sample members, and the number of "successes" (we define, a priori, what a "success" is), such as the number of people who have attended college, and then we need to estimate a probability or some similar statistic (e.g., λ, β, μ). For example, if we get three heads in five flips of a coin, what is the likelihood that the probability of heads for this coin is 0.5?

MLE is used to find the value of the parameter—such as p, λ, or β—that makes the *observed data most likely*. With probability distributions, we typically allow the i to vary and then check the probabilities. This is what we did earlier. With likelihood functions we hold the i and the n constant and allow the parameters to vary. For instance, in a binomial model the value of p is allowed to vary in the *likelihood function* (actually a closed form solution is used):

$$L\left(p \mid i = 3, n = 5\right) = \binom{5}{3} p^{3} \left(1 - p\right)^{5-3}.$$

Suppose we plug 0.5 for p in this equation. The result of our calculation is then 0.313 (confirm this). Now imagine plugging in several possible p (0.2, 0.3, . . ., 0.9) and going through the calculations. The ML estimate is the estimated value of p that maximizes the likelihood of observing a specific set of outcomes, in particular the sample data we observed. The maximum value is typically derived with calculus by setting the partial derivative of the likelihood function with respect to p to zero and solving (see below), but computers use an iterative approximation approach to arrive at ML estimates. It is often easier to use the log of the likelihood function because addition is simpler than multiplication (e.g., $3 \times 5 = 15$; $\ln(3) + \ln(5) = \ln(15)$), so we usually see values of the log-likelihood reported in statistical models that rely on ML.

For our coin-flipping exercise we could also rely on the closed form solution mentioned earlier to approximate p:

$$\frac{\partial \ln\left(L(\hat{p})\right)}{\partial \hat{p}} = 0 \quad Ex.\ \frac{3}{5} : \frac{5}{p} - \frac{3}{1-p} = 0 = \hat{p} = \frac{5}{8} = 0.625 \,.$$

Or we might try out some values for p in the likelihood function and make some educated guesses, as in the following table. In other situations, we could test out values of β or other parameter estimates.

$n = 5$				
i	1	2	3	4
$p = 0.4$	0.259	0.346	0.230	0.077
$p = 0.5$	0.156	0.313	0.313	0.156
$p = 0.6$	0.077	0.230	0.346	0.259
$p = 0.7$	0.028	0.132	0.309	0.360

Since we obtained three heads in five flips we are most interested in the column with 3 as the header. Work down the column and find the largest likelihood value. The values peak in the row with $p = 0.6$, so this is the ML estimate of p for our coin. Of course, we would want to have many more trials before reaching any conclusions about the probability of heads with this coin.

In fact, this demonstrates a common criticism of MLE: it requires relatively large samples to yield unbiased estimators. For example, the ML variance (and hence standard error) estimate in a regression model is biased. However, it also has the following desirable statistical properties:

- *Consistency*: as the sample size grows larger, the bias diminishes.
- *Asymptotically efficient*: the ML estimators have the smallest variance among consistent estimators (*asymptotically* refers to what happens when it approaches the limit).
- *Asymptotically normally distributed*: this allows for powerful hypothesis testing. If the estimators follow a normal distribution, then comparing estimates is straightforward.

There is no strict rule of thumb regarding how large a sample should be before we can be confident that these properties hold.[2] This depends on a number of factors, however.

---

2. Some have claimed, it seems, that a regression model should include about 20–30 observations per explanatory variable in order to yield sufficiently low bias and high efficiency. However, whether this holds true depends on other characteristics of the model (e.g., sample size, distribution of the random component of the outcome).

Generally, though, as the number of explanatory variables increases, the sample size should also increase to gain the benefits of MLE.

MLE may be used to estimate several types of regression models, including, as shown later in this chapter, the linear regression model. Technical details about likelihood functions and parameter estimation for many of the regression models presented in subsequent chapters may be found in books on GLMs (e.g., Hardin and Hilbe, 2012; McCullagh and Nelder, 1989), as well as online resources. However, just to give an initial flavor of regression models estimated with MLE, consider the PDF of the normal (Gaussian) distribution:

$$p(y_i) = \frac{\exp^{-(y_i - \mu)^2 / (2\sigma^2)}}{\sigma\sqrt{2\pi}}.$$

You may recall that μ represents the mean and σ represents the variance of the distribution. Suppose we have a data set and we wish to predict y based on a set of explanatory variables (x_i). Therefore, we substitute $\mathbf{x\beta}$ for μ, with σ_ε^2 representing the residual variance. Assuming a single explanatory variable, though, the likelihood function—the function we wish to maximize—is

$$L(\beta, \sigma_\varepsilon^2 | y) = \prod_{i=1}^{N} \frac{\exp^{-(y - \mathbf{x\beta})^2 / (2\sigma_\varepsilon^2)}}{\sigma_\varepsilon \sqrt{2\pi}}$$

Rather than showing the intercept as α, we assume it is represented as a column vector of ones that is part of $\mathbf{\beta}$. As mentioned earlier, it is easier computationally to work with logarithms, so it is simpler to maximize the log-likelihood:

$$\ln L(\beta, \sigma_\varepsilon^2 | y) = -\frac{n}{2}\ln(2\pi) - \frac{n}{2}\ln(\sigma_\varepsilon^2) - \frac{1}{2\sigma_\varepsilon^2}(\mathbf{y} - \mathbf{x\beta})'(\mathbf{y} - \mathbf{x\beta}).$$

Maximizing the log-likelihood is still a difficult task and specific methods are needed (e.g., the Newton–Raphson method), so we let software such as Stata estimate the results. However, it turns out that solving part of log-likelihood function using derivatives yields the OLS equation for the coefficients $(\mathbf{\beta})$; thus, MLE and OLS result in the same intercept and slope estimates. The variance component—which is used to estimate the standard errors—is based on

$$\sigma_\varepsilon^2(\beta) = \frac{1}{n}(\mathbf{y} - \mathbf{x\beta})'(\mathbf{y} - \mathbf{x\beta}) = \frac{\varepsilon'\varepsilon}{n}.$$

Compare this with the residual variance estimated with OLS (see Chapter 1). As suggested earlier, the MLE-derived variance is a biased estimator (note the n in the denominator; the unbiased estimator uses $\{n - k - 1\}$), but, given the quality of consistency that MLE enjoys, the bias diminishes with larger sample sizes. Thus, under MLE, parameter estimates have the desirable properties described earlier.

HYPOTHESIS TESTS WITH ML REGRESSION COEFFICIENTS

This section focuses on how an ML regression model is used for hypothesis tests for coefficients. Recall that the regression coefficient divided by its standard error in an OLS regression model is assumed distributed as a t random variable. The fact that ML estimators are asymptotically normal suggests that, given a large enough sample (and assuming random sampling), the regression coefficients divided by their standard errors follow a standard normal distribution, or a z-distribution. Hence, we may use a z-test to check whether the null hypothesis of $\beta_k = 0$ is supported or not. Stata, as well as other statistical programs, provides z-values associated with ML regression coefficients. The programs also provide p-values for the z-values.

Although it has become less common in recent years, some programs provide a Wald (1939) statistic in addition to z-values. The Wald statistic looks very similar to a t-value or a z-value, but the quantity is squared so that it is distributed χ^2:

$$W = \left(\frac{\beta_k - \beta^*}{\text{se}(\beta_k)} \right)^2 .$$

Notice how similar this looks to the computation for a standard t-value in OLS regression models. The $\beta*$ is typically zero, although other values could be used.

A nice property of Wald statistics is that they may be used rather easily to compare two regression coefficients for statistical equivalence. The following shows how this is formulated:

$$W = \frac{(\beta_1 - \beta_2)^2}{\{\text{se}(\beta_1)\}^2 + \{\text{se}(\beta_2)\}^2 - 2\text{cov}(\beta_1, \beta_2)} .$$

As long as the covariance among the regression coefficients is available, the Wald statistic may be computed. However, it is much simpler in Stata to use the `test` command. In Chapter 1 we discussed using this command after an OLS regression model for conducting a partial F-test. It may also be used to compute Wald statistics for comparing two ML regression coefficients. After estimating an ML regression model in Stata, type

`test varname1 = varname2`

This sets up a test that examines whether the difference between two coefficients is zero (the null hypothesis) or something other than zero. So, if we presume there is a difference, then we look for a statistically significant χ^2 value. Be careful, though; the coding of the explanatory variables affects what we can conclude from this test. If the coding of the variables differs, it is usually much easier to find statistically significant differences. In other words, the variables should have the same measurement scale.

TESTING THE OVERALL FIT OF ML REGRESSION MODELS

With an OLS regression model, we use F-tests, R^2 values, adjusted R^2 values, RMSEs, and a few other statistics to determine how well the model fits the data. Another way of thinking about this is that we examined whether the set of explanatory variables predicts the variability in the outcome any better than the mean.

In ML regression models explained variance is not as clear-cut because we typically do not have continuous outcome variables to work with. Moreover, even with ML linear regression models we do not partition the variance like we do with OLS regression models. Instead, we maximize a likelihood value. Tests of model fit typically focus on comparing log-likelihood values. If we add statistically significant information to the model, we should see the log-likelihood value change also.

Before seeing how to compute model fit statistics, we need to define three types of models.

1. The *null model* contains no explanatory variables ($y = \alpha$). Hence, it is an *intercept-only* or a *mean-only* model since, in this case, the mean is considered the best linear predictor of the outcome variable. This model is labeled as M_N.

2. The *hypothesized model* includes the explanatory variables. However, it could include a subset of these, which makes it the *constrained* model, or it could include all of the explanatory variables, which makes it the *unconstrained* model. The term constrained is used because leaving explanatory variables out of the model is identical to claiming that their slope is zero—in other words, we have constrained them to be zero. On the other hand, placing variables in a model suggests that we have unconstrained them by allowing their effects to be estimated. The null model is a fully constrained model since it assumes that all the slopes are zero. Moreover, it is important to consider that when we use this terminology constrained models are always *nested* within unconstrained models. The constrained model is labeled as M_C and the unconstrained model as M_U.

3. The *full or saturated model* has one parameter for each observation. It is as if we had perfect information about all characteristics of the model; we can see the full data matrix. Yet it violates a key tenet of statistics: we wish to have summary information—a limited number of features—with which to predict outcomes, not every bit of information available. So the full model is not helpful in our quest to summarize patterns in the data. Yet its perfection gives us a yardstick with which to compare models and see how close we can get to perfect prediction. The full model is labeled as M_F.

An interesting characteristic of the full model is that its log-likelihood value is zero (since, with perfect prediction, the likelihood is one). Therefore, we may compare the log-likelihood of our model, which is typically the hypothesized or unconstrained model, against a yardstick of zero. This is how a measure called the *Deviance* operates. It is derived using the following formula:

$$\text{Deviance} = 2 \times \ln L\left(M_F\right) - 2 \times \ln L\left(M_U\right) = 0 - 2 \times \ln L\left(M_U\right) = -2 \times \ln L\left(M_U\right).$$

Hence, it reduces to $-2 \times$ log-likelihood for M_U. The Deviance is distributed χ^2, so it is subject to significance testing. However, like other ML estimators, it does not operate well in small samples.

The log-likelihood values from constrained and unconstrained models (recall that the former is nested within the latter) can also be compared to determine if one model fits the data better than the other. The likelihood ratio (LR) χ^2 test is computed using the following formula:

$$\text{LR test} = \left\{2 \times \ln L\left(M_U\right)\right\} - \left\{2 \times \ln L\left(M_C\right)\right\}.$$

The resulting quantity is distributed χ^2 with degrees of freedom (*df*) equal to the difference in the number of explanatory variables from the two models. Therefore, if we add two explanatory variables to a model and conduct this test, the *df* for the test is 2. Some programs report the LR test statistic as G^2. Type `help lrtest` in Stata to see how this test is executed.

The Wald test discussed earlier may also be used as a model comparison procedure. In this context, it is similar to the likelihood ratio χ^2 test. To invoke the Wald test in Stata, use the test command that follows the regression model. As mentioned earlier, we also use this command to conduct a partial *F*-test after Stata's OLS regression routine. However, it also works for ML regression models by directly computing the Wald test. It is a simple matter of typing `test` *varname1 varname2* (or whatever variables you added to the model) after estimating the unconstrained model; Stata returns a χ^2 value, the *df*, and the *p*-value. Like the partial *F*-test, the null hypothesis is that the two models are equally predictive of the outcome variable, so the addition of the extra explanatory variables does not increase the ability of the model to fit the data. Another way of thinking of this is that we are testing whether the additional explanatory variables have jointly zero slopes (e.g., $\beta_3 = \beta_4 = 0$).

There are also a number of measures of model fit that have been proposed for ML regression models. These are designed, rather crudely, to mimic OLS R^2 values. In fact, they are often called *pseudo-R^2 values*. The logic behind these measures is that we can examine the proportional improvement of a hypothesized model relative to the null model by comparing their deviance values. Others have been proposed comparing proportional changes in log-likelihood values.

For example, a fit measure that is called simply *pseudo-R^2* is calculated as

$$\text{Pseudo-}R^2 = \frac{\text{Deviance}\left(M_N\right) - \text{Deviance}\left(M_U\right)}{\text{Deviance}\left(M_N\right)}.$$

Thus, we are comparing the null model and the unconstrained model. Another popular measure is called the *McFadden R²* (1973). It is similar to an adjusted R^2 since it takes into account the number of explanatory variables in the model (designated k):

$$\text{McFadden } R^2 = 1 - \frac{\ln L\left(M_U\right) - \left(k+1\right)}{\ln L\left(M_N\right)}.$$

However, we need to be careful using these measures. Presenting them can be risky because we may give the illusion that they assess the proportion of explained variance. Yet they do not. Instead, they are simply rough measures of model improvement over the null model.

The final way to assess model fit provided in this chapter is with information measures. The two most commonly used are Akaike's information criterion (AIC) and the Bayesian information criterion (BIC) (Kuha, 2004). Both are based on the deviance values of the models. And both have a simple rule of thumb for judging whether one model fits the data better than another model: the smaller the value of the AIC or BIC, the better the model fits the data. A claimed benefit is that AICs and BICs may be used to compare non-nested models. This may be valid, but the analyst must be careful not to take this too far and compare models with different sample sizes or that use different data.

There are several formulas for these measures, but here are two common ways to compute them:

$$\text{AIC} = \frac{\text{Deviance}\left(M_U\right) + 2\left(k+1\right)}{n},$$

$$\text{BIC} = \text{Deviance}\left(M_U\right) - \left\{\left(n-k-1\right) \times \ln\left(n\right)\right\}.$$

Stata computes some variation of these automatically in several ML regression procedures or if the `estat ic` command is requested after an ML regression command. However, the formulas used to compute the AIC and BIC differ depending on the particular procedure (this has to do with how the n and k are identified and used; moreover, in some Stata commands the BIC equation adds the second term to the Deviance rather than subtracting it). So they may appear as quite different values. As long as we remember the principle that smaller is better, however, we can compare the fit of ML regression models.

There are a few other ways to see how well an ML estimated model fits the data (Millar, 2011), but several of these are used primarily for models with categorical outcome variables. Thus, they are discussed in later chapters.

EXAMPLE OF AN ML LINEAR REGRESSION MODEL

In Chapter 1, we estimated an OLS regression model with socioeconomic status scores as the outcome variable and female, nonwhite, education, and parents' socioeconomic status as explanatory variables (see Example 1.2). Here we use the same set of outcome and explanatory variables to explore an ML linear regression model. Stata has a command called `glm`, which, not surprisingly, is shorthand for *generalized linear model*. Check Stata's help menu for information about this command. It operates like `regress`, but we must specify two characteristics of the model: the link function and the family or distribution name. For example, the Stata command to reestimate our *sei* model using MLE is shown in Example 2.3. This is utilized as the unconstrained model; in this case it serves as the hypothesized model (M_U). Note that we specify the identity link and the Gaussian family (these are the default options, so if we did not list them, Stata would still return an ML linear regression model).

Example 2.3

```
glm sei female nonwhite educate pasei, link(identity)
    family(Gaussian)
```

```
Generalized linear models                    No. of obs      =       2,231
Optimization     : ML                        Residual df     =       2,226
                                             Scale parameter =    240.2176
Deviance         =   534724.3236             (1/df) Deviance =    240.2176
Pearson          =   534724.3236             (1/df) Pearson  =    240.2176

Variance function: V(u) = 1                  [Gaussian]
Link function    : g(u) = u                  [Identity]

                                             AIC             =    8.321661
Log likelihood   = -9277.812597              BIC             =    517561.4
```

	Coef.	OIM Std. Err.	z	P>|z|	[95% Conf.	Interval]
female	-.2358993	.6584269	-0.36	0.720	-1.526392	1.054594
nonwhite	-2.860583	.9294002	-3.08	0.002	-4.682173	-1.038992
educate	3.72279	.1232077	30.22	0.000	3.481307	3.964272
pasei	.0765712	.0191531	4.00	0.000	.0390318	.1141107
_cons	-4.804311	1.677362	-2.86	0.004	-8.09188	-1.516742

One of the first things to notice is that the output lists *Optimization: ML*, which tells us that the model was estimating using ML. The output also provides the Deviance value (534,724) and the log-likelihood (ln$L$) value (−9,277.8; because of computational differences with the formula shown earlier, it cannot be used to compute the Deviance). It also lists an AIC and a BIC. The standard errors are listed as OIM because they are based on the *observed information matrix* (see Gould et al., 2010, for additional information regarding Stata's ML routines).

The regression coefficients and *p*-values do not differ from what we obtained from `regress` in Example 1.2. As mentioned earlier, we expect the regression coefficients to be identical and, in large samples, other coefficients are also quite close. In fact, when the following assumptions (see Chapter 1) are met, OLS and ML lead to the same results: (a) the observations are independent, (b) the errors are homoscedastic, and (c) the errors follow a normal distribution. However, standard error estimates from OLS and ML linear models may diverge in smaller samples (recall that the variance in ML regression is biased); and there may be differences across estimates when the errors follow a non-normal distribution (MLE is actually preferred in this situation if large samples are available) (Myung, 2003).

Typing `estat ic` after the model shows something interesting. The AIC and BIC values differ (in Stata 13) from those shown in the `glm` output. This indicates that they are based on different formulas, so we must be careful when comparing regression models: did the AICs and BICs come from the same source?

Example 2.3 (Continued)

```
estat ic
```

Akaike's information criterion and Bayesian information criterion

Model	Obs	ll(null)	ll(model)	df	AIC	BIC
.	2,231	.	-9277.813	5	18565.63	18594.18

Note: N=Obs used in calculating BIC; see **[R] BIC note**.

Rather than spend time interpreting the model, it is useful at this point to show the various fit statistics that are used for GLMs. But we also need some models to compare this one to. We therefore estimate a constrained model that omits education and parents' socioeconomic status (so we assume that their slopes are zero) and a null model (see Example 2.4).

Before doing this, though, consider a problem we ran into in Chapter 1. When we compared the models designed to predict the GSS variable *sei*, we found that a model with just one explanatory variable (*female*) yielded a different sample size than a model with this variable and several other explanatory variables. This occurred because the additional variables had missing values. As you know from previous work, it is essential that our models are comparable. Fortunately, Stata provides some simple solutions to this problem. Here use the `e(sample)` option after the unconstrained model to identify only those observations that make up the analytic sample.

`generate sample = e(sample)` *after Example 2.3*

We may then use the new variable `sample` to select for our subsequent regression models only those observations that were used in the unconstrained model.

Example 2.4

Constrained model (M_C)
```
qui glm sei female nonwhite if sample==1,
  link(identity) family(Gaussian)          * qui suppresses the output
```

Null model (M_N)
```
qui glm sei if sample==1, link(identity) family(Gaussian)
```

As mentioned earlier, a simple way in Stata to conduct a Wald test is to use the `test` command. For example, after estimating the unconstrained model (M_U) type

---

Example 2.4 (Continued)

```
test educate pasei                          * after Example 2.3
(1)  [sei]educate = 0
(2)  [sei]pasei = 0
        chi2(2)  = 1184.90
        Prob > chi2 = 0.0000
```

This test compares the unconstrained model with the constrained model. The χ^2 value is 1,184.9 with two degrees of freedom and a *p*-value of less than 0.0001. We thus have good evidence that the unconstrained model fits the data substantially better than the constrained model that assumes zero slopes for education and parents' status.

The table below shows the three models and their respective fit statistics. Note that the AIC and BIC values and the Deviances were taken directly from the Stata printout rather than from the `estat ic` post-estimation command. All of the evidence points to the constrained model as preferred to the null model and the unconstrained model preferred to the others. Notice that the LR test shows noticeable improvement as we go from one model to the next. Moreover, the AICs and BICs decrease substantially, especially when we estimate the unconstrained model.

In most treatments of GLMs with ML estimation, analysts present some combination of these fit statistics. The AIC and BICs have become very popular over the last several years. However, many people still present LR tests and pseudo-R^2 values. Stata also has some specialized programs that provide additional fit statistics and other information about GLMs. For example, a command called `fitstat` provides several fit statistics (it is one option among several available in a user-written suite of commands called `spost9_ado` [Long and Freese, 2006]). It does not work with the `glm` command, but will work with many of the models discussed in subsequent chapters.

	Null model (*sei* = α)	Constrained model (*sei* = α + female + nonwhite)	Unconstrained model (*sei* = α + female + nonwhite + educate + pasei)
Deviance	828,265.08	819,357.59	534,724.32
LR χ² test (G²)[a]		24.12 (*df* = 2)	976.25 (*df* = 4)
Pseudo-R²		0.011	0.354
AIC	8.756	8.747	8.322
BIC	811,071.3	802,179.3	517,561.4
Sample size	2,231	2,231	2,231

a. Based on Stata's `lrtest` command.

FINAL WORDS

The social and behavioral sciences are replete with noncontinuous variables. In fact, they tend to be the rule rather than the exception. Categorical variables present few problems when they are treated as explanatory variables in regression models, but their use as outcome variables requires us to consider alternative models and estimation techniques. Fortunately, by understanding the likely distribution of the random component of an outcome variable and using an appropriate link function, we may use MLE to develop alternative models. The following chapters provide specific information about several of these models. Moreover, there are a number of excellent books on MLE and GLMs that provide substantially more detail than is provided here (Dobson and Arnett, 2011; Gould et al., 2010; Hardin and Hilbe, 2012; Millar, 2011).

EXERCISES FOR CHAPTER 2

1. Recall that we estimated an OLS regression model as part of the exercises at the end of Chapter 1. This model used the GPA data set (*gpa.dta*) and was described in exercise 2c of Chapter 1. Reestimate this model using MLE. Although the results are similar to those found earlier, why might this model not be suited to these particular data?

2. Consider the following variables. Specify the probability distribution that best describes them. Then, assuming we wished to use them as outcome variables in a regression model, select the most likely link function for each.
    a. A measure of the number of avalanches per year in the Wasatch Mountain Range of Utah.
    b. A measure of whether or not members of a nationally representative sample of adults in the United States smoke cigarettes.
    c. A measure of the temperature (in kelvin) inside a sample of volcanoes in Japan.
    d. A measure of whether members of a sample of workers have either quit, been laid-off, been fired, or remained in their jobs in the past year.
    e. In a sample of adult probationers in Florida, a measure of the number of times they have been arrested in the previous 10 years.

3. Compute the expected values and variances for each of the following variables:
    a. A sample of 1,500 adults in which the probability of alcohol use is 0.65.

b. A sample of 200 adults with the following probabilities of involvement in the workforce: 0.55 of being employed full-time, 0.15 of being employed part-time, 0.10 of being unemployed, and 0.20 of not being in the workforce (e.g., students, homemakers).

c. A sample of 850 adolescents with the following probabilities of low and high self-esteem: 0.25 low self-esteem and 0.75 high self-esteem.

d. A sample of traffic accidents per day along a 20-mile stretch of I-95 in Virginia that yielded the following results:

Number of accidents	Frequency
0	121
1	199
2	21
3	12
4	5
5	4
6	2
7	1

4. You've been asked to collect 12 signatures for a petition that asks the state government for more funds to clean up garbage on public land. The probability of getting a signature from a person approached is 0.40. After finding the mean and variance, answer the following question: what is the probability you will have to approach exactly 30 people to get the 12 signatures?

5. Suppose we survey six people and find that two of them say they read national news blogs every day and the other four say they do not. We wish to determine the MLE of p, the probability of daily national news blog reading among this sample. Use the likelihood function for the binomial distribution to fill in the cells of the following table.

	$i = 2$
$p = 0.1$	
$p = 0.2$	
$p = 0.3$	
$p = 0.4$	

From this table, what is the most likely value of p?

6. Recall that the *USData* data set (*usdata.dta*) contains a set of variables from the 50 states in the United States. In this exercise we are interested in using a linear regression model to predict *violrate*, the number of violence crimes per 100,000 population. We shall use the following explanatory variables: *unemprat, density,* and *gsprod*. The variable labels provide additional information on each variable. Estimate two ML linear regression models. The first model is the null model and the second model includes the three explanatory variables. Use the output from these models to compute the following fit statistics: McFadden R^2 and the pseudo-R^2. Compute the likelihood ratio χ^2 test that compares the null model to the unconstrained model.

Chapter Resources: Characteristics of Some Common Generalized Linear Models

Type of variable	Link function	Probability function	Mean and variance	Regression model
Binomial	$\eta = \log\left[\dfrac{\mu}{1-\mu}\right]$	$p(i) = \binom{n}{C_i} \times p^i(1-p)^{n-i}$ binomial	$E[x] = n \times p$ $\mathrm{Var}[x] = n \times p \times (1-p)$	(Binary) logistic
	$\eta = \Phi^{-1}(\mu)$	$f(x) = \left\{\left[\left[\dfrac{1}{\sqrt{2\pi}\,\sigma}\exp\left[\dfrac{-(x-\mu)^2}{2\sigma^2}\right]\right]\right]\right\}^{-1}$ inverse normal	$E[x] = \mu$ $\mathrm{Var}[x] = \sigma^2$	Probit
Multinomial	$\eta = \log\left[\dfrac{\mu_j}{\mu_J}\right]$	$p\{x_1 = n_1, x_2 = n_2, \cdots, x_r = n_r\} = \dfrac{n!}{n_1!n_2!\cdots n_r!}\, p_1^{n_1} p_2^{n_2} \cdots p_r^{n_r}$	$E[x_j] = n_i \times p_i$ $\mathrm{Var}[x_j] = p_i \times n_i \times (1-p_i)$	Multinomial logistic
Poisson	$\eta = \log(\mu)$	$p(i) = e^{-\lambda}\dfrac{\lambda^i}{i!}$	$E[x] = \lambda$ $\mathrm{Var}[x] = \lambda$	Poisson
Negative binomial	$\eta = \log(\mu)$	$p(i) = \binom{i-1}{C_{r-1}} \times (1-p)^{i-r}\, p^r$, where i = number of trials and r = number of "successes"	$E[x] = r/p$ $\mathrm{Var}[x] = r(1-p)/p^2$	Negative binomial

## THREE

Logistic and Probit Regression Models

Suppose we are interested in examining some variable but it has only two possible outcomes. For example, our outcome variable might be whether a person supports the death penalty, whether a person graduated from high school, or whether a company adopts a particular employee assistance program.

Note that this is a binary variable (sometimes called *dichotomous*) (e.g., 0 = no; 1 = yes). It is distributed as a binomial random variable (see Chapter 2). If it is used as an explanatory variable in a regression model, it presents no problems. But if it is an outcome variable, what should we do?

As an example, consider the *depress* data set (*depress.dta*). We are interested in evaluating life satisfaction. According to Stata's `codebook` command, however, the variable *satlife* has only two outcomes:

codebook satlife

```
        type:  numeric (byte)

       range:  [0,1]                         units:  1
unique values:  2                     missing .:  0/117

  tabulation:  Freq.   Value
                 65   0
                 52   1
```

Life satisfaction is coded as low (0) and high (1) only. Now, suppose we wish to predict this outcome with an OLS regression model using *age* as a predictor.

Example 3.1

regress satlife age

Source	SS	df	MS		
Model	.306493718	1	.306493718		
Residual	28.5823952	115	.248542567		
Total	28.8888889	116	.249042146		

Number of obs	=	117
F(1, 115)	=	1.23
Prob > F	=	0.2691
R-squared	=	0.0106
Adj R-squared	=	0.0020
Root MSE	=	.49854

satlife	Coef.	Std. Err.	t	P>|t|	[95% Conf. Interval]	
age	-.0108174	.0097412	-1.11	0.269	-.0301129	.0084781
_cons	.8499592	.3680681	2.31	0.023	.1208871	1.579031

The regression model shown in Example 3.1 is known as a *linear probability model*. It looks reasonable, but we should also test the assumptions of the model. First, plot the residuals by predicted values using Stata's rvfplot. (Alternatively, save the predicted values [predict pred] and look at a scatterplot: scatter satlife age || line pred age.)

rvfplot

Figure 3.1 shows a problematic but, with some thought, unsurprising result. Note that if we were to place a fit line in the figure and connect the points to the line, we would see a systematic pattern indicative, perhaps, of heteroscedasticity (what does the Breusch–Pagan test indicate?).

Now, examine a normal probability plot of the studentized residuals (see figure 3.2).

predict rstudent, rstudent   * *after Example 3.1*
qnorm rstudent

This is not acceptable. Another problem with the linear probability model is that it is easy to get predicted values that fall outside the range of 0–1.

WHAT IS AN ALTERNATIVE? LOGISTIC REGRESSION

Rather than modeling the outcome variable directly (estimating the mean of y for some combination of x variables), we may estimate the probability that $y = 1$. Similar to linear regression, we assume that some set of x variables is useful for predicting y values, but we say that some combination of x predicts the probability that $y = 1$ (assuming we have coded y as {0, 1}). This is the approach underlying *logistic regression analysis*.

The basic formula for estimating $y = 1$ consists of transforming the regression model to

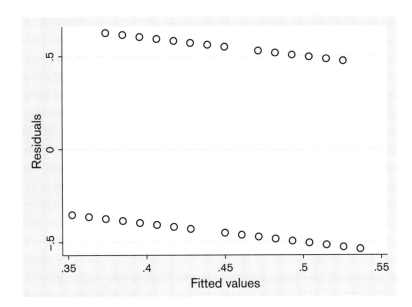

FIGURE 3.1

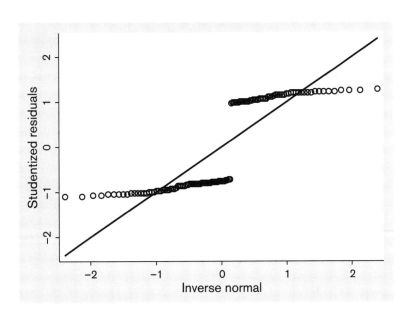

FIGURE 3.2

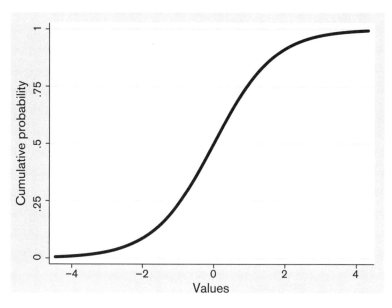

FIGURE 3.3

$$p(y=1) = \frac{1}{1+\exp\left[-\left(\alpha + \beta_x x_1 + \cdots + \beta_k x_k\right)\right]}.$$

This is called the *logistic function* and it is estimated using maximum likelihood (ML). An advantage of this function is that it guarantees that the probabilities range from 0 to 1 as the regression equation predicts values from $-\infty$ to $+\infty$ (to verify this plug in some numbers in Stata using `display 1/[(1 + exp(x)]`).

The other advantage is that if you plot the logistic function, it reveals a shape that looks like that shown in figure 3.3.

Probabilities often behave this way in social research. There are relatively few probabilities that are very rare or very common; rather, most probabilities occur in the middle. However, think about how this compares to a linear model. In a simple linear regression model, for example, we fit a straight line to the association between two variables. And the slope represents a consistent change across values of the explanatory variable. Yet, using the logistic curve (see figure 3.3) the change depends on where along the *x*-axis one wishes to interpret the association. As suggested, larger changes occur in the middle of the distribution. This can be confusing, so we can use other measures to help in interpretation.

The most important of these measures used in logistic regression are *odds* and *odds ratios*. But what do these terms mean? An example is probably the best way to illustrate them. Suppose we determine that the probability of marijuana use among some population of adolescents is 0.25. Another way of saying this is that 25% of adolescents in this population use marijuana. The odds of marijuana use among youths in this population is

$$\text{Odds}_{mj} = \frac{p(\text{marijuana use})}{1 - p(\text{marijuana use})} = \frac{0.25}{1 - 0.25} = \frac{1}{3}.$$

This is equivalent to saying that the odds are "3 to 1" that an adolescent from this population *does not* use marijuana. Or, for every one adolescent who uses marijuana, there are three who do not use marijuana.

An odds ratio is just what it sounds like: the ratio of two odds. The usefulness of this measure is apparent by extending our example. Suppose we wish to compare the odds of marijuana use among male and female adolescents. Then we would be interested in the following:

$$\text{OR}(\text{mj, male v. female}) = \frac{\text{Odds, males}}{\text{Odds, females}} = \frac{p(\text{males})}{1 - p(\text{males})} \Big/ \frac{p(\text{females})}{1 - p(\text{females})}.$$

Suppose that 25% of males but only 10% of females had used marijuana. Then the odds ratio is

$$\text{OR} = \frac{0.25}{0.75} \Big/ \frac{0.10}{0.90} = \frac{1}{3} \Big/ \frac{1}{9} = 3.$$

In other words, the odds of marijuana use among males are three times the odds of marijuana use among females.

What is an odds ratio that suggests that there is no association between one variable and another? This occurs when the odds are equivalent, so the odds ratio equals one. In the example, an odds ratio of one also indicates that there was no difference between females and males in the probability of marijuana use.

If we consider the logistic function that was presented earlier, a useful property is that we may write it as

$$\ln\left[\frac{p(y=1)}{1 - p(y=1)}\right] = \alpha + \beta_1 x_1 + \cdots + \beta_k x_k.$$

Thus,

$$\frac{p(y=1)}{1 - p(y=1)} = \exp(\alpha + \beta_1 x_1 + \cdots + \beta_k x_k).$$

The left-hand term in the first equation is also known as the *log-odds* and should look familiar since it represents the logit link function introduced in Chapter 2. The second equation shows the model in exponential form (recall that the exponential function is the inverse of the natural logarithm).

Since we are interested in distinguishing the systematic and the random components of the data (see Chapter 2), we assume the errors in this regression model (random component) follow a logistic distribution with variance = $(\pi^2/3)$. This latter value is arbitrary, but allows efficient estimation of the model. In a linear regression model, by contrast, the variance of the errors is estimated from the data.

Given these preliminary characteristics of the logistic function and estimation, what is the ML approach to estimating this model? First, modify the earlier equation so that

$$\ln\left[\frac{p(y=1)}{1-p(y=1)}\right] = \alpha + \beta_1 x_1 + \cdots + \beta_k x_k = \text{logit}(\pi_i) = \mathbf{x'\beta}.$$

The term $\mathbf{x}_i$ represents the values of the explanatory variables and $\mathbf{\beta}$ the regression coefficients (including the intercept). This modification allows a concise depiction of the log-likelihood function:

$$\ln L = \sum_{i=1}^{N}\left(y_i \ln(\pi_i) + (n_i - y_i)\left(\log(1-\pi_i)\right)\right).$$

The ML estimates are the set of coefficients that maximize this function (see Chapter 2 for a simple example of what this means). As noted in Chapter 2, maximizing a function such as this one is difficult, so we let software such as Stata provide the results of the maximum-likelihood estimation (MLE).

Returning to the example from the *depress* data set, we wish to predict life satisfaction but it is coded as a binary or dichotomous variable. Suppose we want to examine differences between women and men. A simple way is to examine a cross-tabulation of the *satlife* and *male* variables.

Example 3.2

tabulate male satlife, row

0=female, 1=male	0=low, 1=high 0	1	Total
0	58 60.42	38 39.58	96 100.00
1	7 33.33	14 66.67	21 100.00
Total	65 55.56	52 44.44	117 100.00

Note that we ask for row percentages because this makes it easier to compute the probabilities and odds of high life satisfaction. The probability of high life satisfaction overall is 0.44; the probability of high life satisfaction for males is 0.67; and the probability of high life satisfaction for females is 0.40.

But what is the odds of high life satisfaction? It is 0.44/(1 − 0.44) = 0.79. For males it is 0.67/0.33 = 2; and for females it is 0.40/0.60 = 0.67. Another way is to simply divide the number of highs by the number of lows: $14/7 = 2$ (males) and $38/58 = 0.67$ (females). Thus, the odds ratio that compares males to females is $2/0.67 = 3.0$ (note that some rounding error affects the precision of this estimate). In other words, the odds of high life satisfaction among males are three times the odds of high life satisfaction among females.

A logistic regression model results in log-odds ratios and odds ratios that are simply an extension of this basic 2×2 table. We can reproduce the analysis we just completed using Stata's `logit` command (see Example 3.3).

Example 3.3

```
logit satlife male, or
```

```
Logistic regression                        Number of obs    =        117
                                           LR chi2(1)       =       5.13
                                           Prob > chi2      =     0.0235
Log likelihood = -77.810255                Pseudo R2        =     0.0319
```

satlife	Odds Ratio	Std. Err.	z	P>|z|	[95% Conf. Interval]	
male	3.052632	1.550072	2.20	0.028	1.128364	8.258468
_cons	.6551724	.1367368	-2.03	0.043	.435218	.9862894

The `or` subcommand requests that the point estimates be provided as odds ratios (the Stata command `logistic` provides odds ratios by default). If we omitted this command, Stata returns log-odds ratios, which in this case is 1.12 for *male*. By exponentiating this value ($e^{0.112}$) we find the odds ratio, or 3.05. Regardless of how it is computed, the interpretation of the coefficient is the same. But also note that the difference between women and men is statistically significant ($p = 0.028$).

To convert these results to probabilities (usually a good idea), we may rely on the logistic function to compute them. Rerun the logistic regression equation but omit the `or` subcommand. Then use the coefficients to compute probabilities for females and males using the logistic function shown earlier:

$$p(y_i = 1 | x_i) = \frac{1}{1 + e^{(-x_i\beta)}} \, .$$

For females (coded as *male* = 0), we find

$$1 + \left[\exp(-1 * 0.423) \right]^{-1} = 0.40 \, .$$

For males (coded as *male* = 1), this is

$$1 + \left[\exp(-1 * \{0.423 + 1.116\}) \right]^{-1} = 0.67 \, .$$

Recall from Chapter 1 that Stata features the post-estimation command `margins`. Here it may be used to compute these probabilities, but it is important to specify the level of the explanatory variables (see Example 3.3 (cont.)).

```
margins, at(male=(0 1))
```

	Margin	Delta-method Std. Err.	z	P>|z|	[95% Conf. Interval]	
_at						
1	.3958333	.0499113	7.93	0.000	.298009	.4936577
2	.6666667	.1028689	6.48	0.000	.4650473	.868286

The results of this command show predicted probabilities of 0.396 and 0.667. These agree with our original probabilities for females and males. The `margins` command also provides standard errors, *p*-values, and confidence intervals for the probabilities. The `margins-plot` command—which may be executed directly after the `margins` command—could also be used to graph the predicted probabilities. This comes in useful for presenting the results of these models. Requesting the predicted probabilities (e.g., `predict pred_sat`) and graphing the results with a bar chart (`graph bar pred_sat male`) or a dot plot (`graph dot (mean) pred, over(male)`) can also be informative. We will see some examples of these later.

THE MULTIPLE LOGISTIC REGRESSION MODEL

It is a simple matter to add explanatory variables to the model. The key difference is that we now have *adjusted odds ratios* as a result. Example 3.4 extends the life satisfaction model.

```
logit satlife male age iq, or
```

Logistic regression

Log likelihood = -71.079782

Number of obs	=	109
LR chi2(3)	=	7.39
Prob > chi2	=	0.0604
Pseudo R2	=	0.0494

satlife	Odds Ratio	Std. Err.	z	P>|z|	[95% Conf. Interval]	
male	3.252229	1.683819	2.28	0.023	1.178902	8.971906
age	.9516269	.0453433	-1.04	0.298	.8667792	1.04478
iq	.9375624	.0471589	-1.28	0.200	.8495426	1.034702
_cons	1488.766	8399.956	1.29	0.195	.0234506	9.45e+07

How do we interpret the odds ratio for *male*? Adjusting for the effects of age and IQ, the odds of high life satisfaction among males are expected to be 3.25 times the odds of high life satisfaction among females. Another useful way to interpret the odds ratio is to use the *percentage change approach* discussed in Chapter 1. This approach works well with any model that uses a log or logit link function (see Chapter 2). Recall that the general formula is % change = {exp(β) − 1} × 100. Since the odds ratio is already exponentiated, we simply subtract 1 from it and multiply by 100. For the *male* coefficient, this leads to the following interpretation:

- After adjusting for the effects of age and IQ, the odds of high life satisfaction among males are expected to be 225% higher than the odds of high life satisfaction among females.

There is, however, an important issue to be aware of when attempting to compare constrained and unconstrained logistic regression models (it also applies to probit models). That is, when we add explanatory variables to a logistic regression model, we also change the explained and total variance of the model (recall that the variance of the errors is regarded as fixed). A consequence of this is that we cannot compare the effects of explanatory variables across nested models in the ordinary way. In the previous model, for example, we should not claim that the effect of the variable *male* changed even though its odds ratio went from 3.05 to 3.25. Williams (2011) recommends that *y*-standardization—which may be estimated with the user-written command called `listcoef` (part of a suite of commands called `spost9`)—or a user-written command called `khb` may be useful for comparing coefficients across logistic regression models (type `findit listcoef` or `findit khb`).

Returning to the multiple logistic regression model, how do we interpret odds ratios for continuous variables? Consider age:

- Adjusting for the effects of gender and IQ, for each one-year increase in age the odds of high life satisfaction are *expected to change 0.952 times* (or decrease 4.8% if we use the percentage change formula).

As discussed earlier, it is also useful to compute predicted probabilities. We can either use the logistic function approach or use Stata's `margins` command. In either case, we need to specify the particular values of the explanatory variables. It is often convenient to use the means, which for IQ is 91 and for age is 37. Here is how to compute the predicted probability for females using the logistic function:

$$p\left(\text{satlife} = 1 \middle| \text{male} = 0, \text{age} = 37, \text{IQ} = 91\right) = \frac{1}{1 + e^{\left(-1 \times \left[7.31 + (1.18 \times 0) + (-.0496 \times 37) + (-.064 \times 91)\right]\right)}}.$$

As shown in Example 3.5, using the `margins` command is much simpler and more precise.

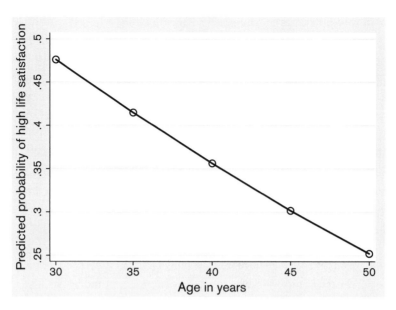

FIGURE 3.4

```
logit satlife i.male age iq, or          * note that male is identified as an
                                           indicator variable in this model

margins male, atmeans
```

	Margin	Delta-method Std. Err.	z	P>|z|	[95% Conf. Interval]	
female	.3840897	.0523627	7.34	0.000	.2814606	.4867188
male	.6697635	.1034043	6.48	0.000	.4670948	.8724322

Thus, we can see that the probability of high life satisfaction for females is about 0.38 and for males about 0.67 for those who are at the mean of age and IQ.

It might also be useful to compare groups defined by continuous variables. For example, we can compare the predicted probabilities for those between the ages of 30 and 50 to see the presumed impact of age on life satisfaction. Once again, ask Stata for predicted probabilities using the margins command. Note that we set *male* = 0 and IQ at its mean (approximately 91.7). We may then use the marginsplot command to graph these results. Figure 3.4 provides this example.

```
qui margins, at(age=(30(5)50) male=0) atmeans
marginsplot, noci                          * noci omits the confidence intervals
```

Based on figure 3.4, what is the expected difference in probabilities for the different age groups? Keeping in mind that probabilities can conceivably range from 0 and 1, how meaningful are these differences? What other factors would you like to know about before reaching any conclusions regarding the association between age and life satisfaction?

MODEL FIT AND TESTING ASSUMPTIONS OF THE MODEL

Stata makes it simple to test model fit with its `estat` command that follows the regression model. We may wish to find the Akaike's information criterion (AIC) and the Bayesian information criterion (BIC) (recall that these are scaled differently in Stata than in some other representations, such as the formulae found in Hoffmann [2004]). Typing `estat ic` following the logistic regression model in Example 3.5 results in the following printout:

Example 3.5 (Continued)

```
Akaike's information criterion and Bayesian information criterion
```

Model	Obs	ll(null)	ll(model)	df	AIC	BIC
.	109	-74.77596	-71.07978	4	150.1596	160.925

```
Note: N=Obs used in calculating BIC; see [R] BIC note.
```

Note that the AIC and BIC are provided, as well as the log-likelihood (`ll`) values for the null and unconstrained models. We may then compare the fit statistics to those of other models to determine the best fitting model. Recall that lower values of the AIC or BIC indicate a better fitting model. As noted earlier, Stata also makes it easy to conduct tests of nested models. Using the `test` command, we may compare a constrained and an unconstrained model using a Wald test. For example, if we wish to test whether a model with just *male* is superior to a model with *male*, age, and IQ, we may use this command:

Example 3.6

```
test age iq                          * after Example 3.5 (Wald test)

( 1)  [satlife]age = 0
( 2)  [satlife]iq = 0

       chi2(  2) =    1.91
     Prob > chi2 =    0.3842
```

We may now see that the addition of age and IQ did not improve the fit of the model (note the relatively large *p*-value). In other words, we do no better at predicting life satisfaction with *male*, age, and IQ than we do with just the *male* variable.

There are some other common fit measures used when we have classified an outcome, such as life satisfaction. One measure is known as *proportional reduction in error* (PRE) (Liebetrau, 2006). It compares the errors from the null model to the errors from the unconstrained model in the following manner:

$$PRE = \frac{\text{error}\left(M_{N}\right) - \text{error}\left(M_{U}\right)}{\text{error}\left(M_{U}\right)}.$$

A similar measure is known simply as *accuracy*. It is defined as the number of accurate predictions divided by the number of total predictions. The *error* or *misclassification rate* is simply {1 – accuracy}. An important decision for these measures is to choose a cutoff point for the predictions. For example, since the logistic (and probit) model allows the predicted probabilities to be estimated, we must decide the probability level at which to classify the outcome. The most common probability cutoff used is 0.50. Stata's post-estimation command *estat classification* uses this as the default:

Example 3.6 (Continued)

```
estat classification
```
                                                           * after Example 3.5

```
Logistic model for satlife
```

Classified	True D	~D	Total
+	19	6	25
−	29	55	84
Total	48	61	109

```
Classified + if predicted Pr(D) >= .5
True D defined as satlife != 0
```

From this table we may compute the accuracy as (19 + 55)/(19 + 55 + 29 + 6) = 0.68, which is an improvement over 0.5 (the result of a random estimate). Stata provides this as a percentage at the end of the printout that results from the classification command. The PRE requires information from the null model. Fortunately, rather than making these computations ourselves, the user-written program PRE will do it for us (type findit PRE). Once you have downloaded this program to Stata, type pre after the model in Example 3.5. The results indicate that the unconstrained model reduces the errors in prediction by 27%.[1] Finally, a relatively new measure of model fit that is designed specifically for binary logistic regression models is called *Tjur's coefficient of discrimination* (Tjur, 2009). It is easy to compute because it involves only the difference between the predicted probabilities of the outcome. Following the model, it may be estimated using the Stata's ttest command:[2]

      predict prob if e(sample)   * after Example 3.5
      ttest prob, by(satlife)

---

1. Stata's pre command also works with linear regression models.
2. This approach is shown and discussed in http://statisticalhorizons.com/r2logistic.

The difference is 0.068, which is slightly higher than the McFadden R^2 of 0.049. But neither measure is particularly impressive.

What should we do about testing some of the assumptions of the model? Stata has a host of post-estimation statistics that may be computed from logistic regression models and used to test several of these. Some most commonly used statistics computed with the predict command include those shown in Example 3.7.

Example 3.7

```
predict dev, deviance          * deviance residuals
predict rstand, rstandard      * standardized Pearson residuals
predict hat, hat               * influence values from hat matrix
                               or Pregibon leverage values; threshold =
                               {3 × average hat value}
predict pred, p                * predicted probabilities
predict deltad, ddeviance      * delta-deviance values: similar to
                               Cook's D values; threshold = 4
```

A complete list is found by searching the Stata help menu for *logit postestimation*. Several of these are discussed in detail in Hosmer et al. (2013). These statistics may then be used to generate several diagnostic graphs, such as those requested in Example 3.7.

Example 3.7 (Continued)

```
kdensity dev                   * distribution of residuals should be
                               approximately normal in large samples
twoway scatter dev pred        * look for heteroscedastic error patterns
twoway scatter dev [x's] ||    * partial residual plot—look
lowess dev [x's]               for nonlinear patterns
twoway scatter rstand hat, xline
  (3*average hat value)        * look for influential observations
twoway deltad pred, yline(4)   * look for influential observations
twoway scatter deltad hat, xline(3*average
  hat value) yline(4)          * look for outliers & extreme leverage points
```

Figure 3.5 furnishes one of these graphs. After saving the delta-deviance values (*deltad*) and the hat values from our life satisfaction model in Example 3.5, we find that the threshold for hat values is $3 \times 0.0428 = 0.1284$. Then, request a scatterplot in Stata with lines to identify the thresholds.

```
twoway scatter deltad hat, xline(0.1284)
  yline(4) jitter(5)           * jitter the points to make them easier to see
```

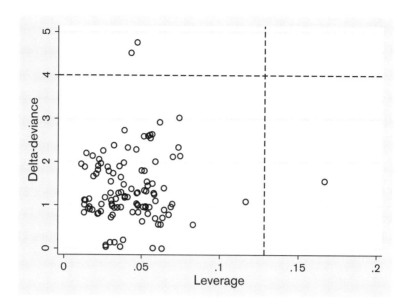

FIGURE 3.5

It appears that there are two outliers with delta-deviance values in excess of four and one high leverage point that is not an outlier (a Pregibon leverage value of approximately 0.175).

Once we have determined if any assumptions have been violated, it is useful to explore the observations that emerge as influential observations or use robust standard errors to obviate heteroscedasticity. For example, considering the results shown in figure 3.5, we should explore what type of observation in the sample emerges as an outlier:

Example 3.8

```
list satlife male age iq if deltad > 4
```

                    * note the 42-year-old females

```
logit satlife male age iq, robust
```

            * this is an option if heteroscedasticity is apparent; or
                      bootstrapped standard errors might be helpful

It is a good idea at this point to try some detective work to practice using logistic regression diagnostics. Consider the following steps in Stata with the GSS data (*gss.dta*):

1.   Regress *volrelig* on *female, age,* and *educate* using logistic regression.

2.   Save the following statistics: deviance residuals, predicted probabilities, hat values, and delta-deviance values.

3.   Look at the deviance by age scatterplot (look for nonlinear pattern).

4.   Look at delta-deviance by predicted probabilities (note the high ΔD values).

5. Look at delta-deviance by hat values (all outliers!).

6. Reestimate the model with *zage* and *zage²* (still plenty of outliers!).

7. Type `twoway lowess pred age` (note the nonlinear association with age).

PROBIT REGRESSION

An alternative to logistic regression that is popular in econometrics is probit regression. There is actually little practical difference between these two approaches; it seems to be a matter of tradition why some disciplines use logistic regression, whereas others use probit regression. In any event, rather than being based on the logistic function, probit regression is based on the standard normal distribution. The cumulative density functions (CDFs)[3] are very similar, however. The standard normal CDF is

$$p(y_i) = F(Z_i) = \frac{1}{\sqrt{2\pi}} \int_{-\infty}^{Z_i} e^{-s^2/2} ds.$$

In this function, s is a normally distributed variable with a mean of zero and a standard deviation of one. The probit model assumes that Z is an unobserved continuous scale that underlies y, its empirical manifestation. In the probit regression model, we assume that the errors follow a normal distribution with variance equal to one (var(ε) = 1). Given the CDF, it should come as no surprise that the log-likelihood required for MLE is a complex function that can be simplified as

$$\ln L = \sum_{i=1}^{N} \left(y_i \ln \Phi(\mathbf{x}'\boldsymbol{\beta}) + (1 - y_i) \ln \left[1 - \Phi(\mathbf{x}'\boldsymbol{\beta}) \right] \right).$$

Recall that the term $\Phi(\cdot)$, which includes the Greek letter *phi*, is used to represent the PDF of the normal distribution. As with the logistic model, the ML estimates are the set of coefficients that maximize this function.

One advantage of the probit model relative to the logistic model is that the coefficients may be transformed directly into probabilities at some level of the x simply by using values from a standard normal distribution table, or a table of z-scores. In fact, whereas the logistic regression model provides log-odds ratios and differences in log-odds for each unit shift in the explanatory variables, the probit model provides differences in terms of z-scores. Example 3.9 presents a simple probit regression model of life satisfaction predicted by the variable *male*.

---

3. The probability density function (PDF), which was discussed in Chapter 2, is the derivative of the CDF.

probit satlife male

```
Probit regression                          Number of obs    =        117
                                           LR chi2(1)       =       5.13
                                           Prob > chi2      =     0.0235
Log likelihood = -77.810255                Pseudo R2        =     0.0319
```

satlife	Coef.	Std. Err.	z	P>|z|	[95% Conf. Interval]	
male	.6948743	.3111687	2.23	0.026	.0849948	1.304754
_cons	-.264147	.1295508	-2.04	0.041	-.5180618	-.0102321

We may compute the probabilities in a number of ways. For example, using a table of z-scores or Stata's normprob function, it is easy to find the probability of high life satisfaction among females and males:

display normprob(−0.264)   * *females*
　.39583
display normprob(−0.264 + 0.6949)   * *males*
　.66667

Similarly, we can use Stata's margins command to compute the probabilities.

margins, at(male=(0 1))   * *output is edited*

	Margin	Delta-method Std. Err.	z	P>|z|	[95% Conf. Interval]	
_at						
1	.3958333	.0499113	7.93	0.000	.298009	.4936577
2	.6666667	.1028689	6.48	0.000	.4650473	.868286

We can work out the probabilities in the same way in more sophisticated models. For example, next we add sleep, age, and IQ to the probit model (see Example 3.10).

Example 3.10

```
probit satlife male sleep age iq
```

Example 3.10

```
Probit regression                          Number of obs    =        104
                                           LR chi2(4)       =      19.20
                                           Prob > chi2      =     0.0007
Log likelihood = -62.181176               Pseudo R2        =     0.1337
```

satlife	Coef.	Std. Err.	z	P>|z|	[95% Conf. Interval]	
male	.6525728	.3592392	1.82	0.069	-.051523	1.356669
sleep	-1.451607	.5272656	-2.75	0.006	-2.485028	-.4181852
age	-.0366609	.0307415	-1.19	0.233	-.0969132	.0235914
iq	-.0262439	.0314311	-0.83	0.404	-.0878478	.03536
_cons	6.330296	3.598143	1.76	0.079	-.7219353	13.38253

We now compute predicted probabilities for various groups. For instance, compare sex and sleeping patterns (note that we set age and IQ at their respective means).

Example 3.10 (Continued)

```
margins, at(male=0 sleep=0) atmeans        * female, poor sleep
   0.359
margins, at(male=1 sleep=0) atmeans        * male, poor sleep
   0.614
margins, at(male=0 sleep=1) atmeans        * female, good sleep
   0.862
margins, at(male=1 sleep=1) atmeans        * male, good sleep
   0.959
```

If we judge the relative merits of one's sex versus getting a good night's sleep, it appears that sleeping is strongly associated with life satisfaction.

As shown earlier, it is also helpful to consider the changes in probability for a particular unit change in a continuous variable. For example, suppose we are interested in age differences. What difference does a 10-year increase in age make? Use the margins command to determine this.

Example 3.10 (Continued)

```
margins, at(male=0 sleep=0 age=(30 40)) atmeans
   30 = 0.469
   40 = 0.329
```

Thus, the shift in probabilities for a 10-year increase in age is {0.469 − 0.329} = 0.14. Note that this is a decrease of 0.14 since age is negatively associated with life satisfaction (yet not statistically significant).

Another way we can estimate a probit model in Stata is to utilize the dprobit command. Rather than differences in z-scores, this model provides the change in probability for a unit change in the explanatory variables at the mean of the explanatory variables. The results of this model are shown in Example 3.11.

The Stata output notes that it is "reporting marginal effects." This is because the model uses the same process as the margins command to estimate probabilities (or changes in probabilities). In addition, the dF/dx label for the coefficient column implies that Stata is providing derivatives, which, you may recall, indicate the rate of change in F relative to the change in x ($\Delta F/\Delta x$). Thus, it is the approximate slope at a particular value of x.

Example 3.11

```
dprobit satlife male sleep age iq
```

Probit regression, reporting marginal effects

Number of obs = 104
LR chi2(4)   =   19.20
Prob > chi2 = 0.0007

Log likelihood = -62.181176

Pseudo R2   = 0.1337

satlife	dF/dx	Std. Err.	z	P>|z|	x-bar	[95% C.I.]
male*	.2531689	.1305738	1.82	0.069	.182692	-.002751 .509089
sleep	-.5780346	.2110085	-2.75	0.006	1.875	-.991604 -.164466
age	-.0145985	.01224	-1.19	0.233	37.7596	-.038588 .009391
iq	-.0104504	.0125141	-0.83	0.404	91.6154	-.034978 .014077
obs. P	.4615385					
pred. P	.4757212	(at x-bar)				

(*) dF/dx is for discrete change of dummy variable from 0 to 1
    z and P>\|z\| correspond to the test of the underlying coefficient being 0

Moreover, pay attention to the note at the bottom of the Stata output. It describes how to interpret the coefficients for the dummy variables *male* and *sleep*. Thus, the difference in probabilities of high life satisfaction for males and females is 0.25, whereas the difference in probabilities between those who report poor sleep and good sleep is 0.48. These differences occur at the means of the other explanatory variables.

The coefficients for continuous variables represent the expected change in the probability for an approximate one-unit increase in the relevant explanatory variable. For instance, the coefficient for age is −0.0146, which may be interpreted as the expected change in the probability of high life satisfaction for each one-year increase in age at the mean of *age*, *male*, *sleep*, and *iq*. However, there is a problem with interpretation: it makes little sense to talk about the mean of *male* and *sleep*, so caution is needed when interpreting these types of effects.

COMPARISON OF MARGINAL EFFECTS: DPROBIT
AND MARGINS USING DY/DX

Although `dprobit` is useful for examining changes in probabilities, it has been superseded by the `probit` command followed by `margins` as a post-estimation command. For example, consider the model in Example 3.12 that uses the Stata-provided data set `nlsw88.dta`.

Example 3.12

```
sysuse nlsw88.dta
dprobit union wage
```

```
Probit regression, reporting marginal effects          Number of obs =     1878
                                                        LR chi2(1)    =    42.20
                                                        Prob > chi2   =   0.0000
Log likelihood = -1025.5265                             Pseudo R2     =   0.0202
```

union	dF/dx	Std. Err.	z	P>|z|	x-bar	[95% C.I.]
wage	.0150651	.0023066	6.51	0.000	7.56542	.010544 .019586
obs. P	.2454739					
pred. P	.2411845	(at x-bar)				

    z and P>|z| correspond to the test of the underlying coefficient being 0

An interpretation of the coefficient is as follows. Each $1 increase in wages per hour is associated with a 0.015 increase in the probability of union membership *at the mean of wages*.

Example 3.12 (Continued)

```
qui probit union wage
margins, dydx(wage) atmeans        * note that this is at the mean of wages
```

```
Conditional marginal effects                 Number of obs     =      1,878
Model VCE    : OIM

Expression   : Pr(union), predict()
dy/dx w.r.t. : wage
at           : wage          =     7.565423 (mean)
```

	Delta-method					
	dy/dx	Std. Err.	z	P>|z|	[95% Conf. Interval]	
wage	.0150651	.0023067	6.53	0.000	.0105442	.0195861

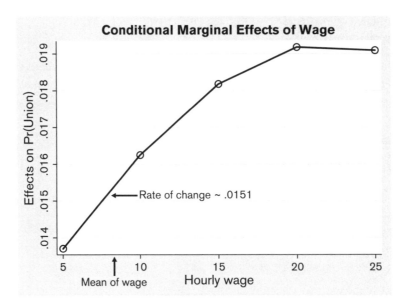

FIGURE 3.6

This coefficient agrees with the dprobit result, so the interpretation is the same. Note, though, that the changes in the probabilities may differ depending on where in the wage distribution it is estimated. For example, consider the graph shown in figure 3.6.

```
qui margins, dydx(wage) at(wage=(5(5)25))
marginsplot, noci
```

The change in the probability of union membership initially increases as hourly wages increase, but flattens out at $20–$25 per hour. See the description of the margins command in the Stata manual for more information and for additional possibilities.

MODEL FIT AND DIAGNOSTICS WITH THE PROBIT MODEL

Examining model fit is now a straightforward exercise with the probit model: use the same approach that was used with the logit model. Since both are estimated with MLE, we may compute AICs, BICs, likelihood ratio tests, PREs, and the various (pseudo) R^2 values.

It is best to check Stata's help menu to determine what diagnostic tools are available. As recommended earlier, look for post-estimation options. Typing help probit postestimation in Stata reveals that we may predict probabilities, linear predictors (z-scores), standard errors of these predictors, deviance residuals, and score statistics. Since there is only one type of residual, we have to rely on it for testing heteroscedasticity and influential

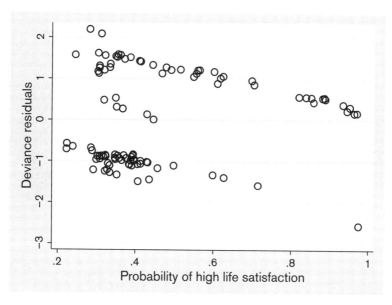

Probability of high life satisfaction

FIGURE 3.7

observations. There is not an option for leverage values. This suggests that using a logistic model is better than relying on a probit model in Stata if one wishes to examine a variety of diagnostic measures.

Returning to the probit model with life satisfaction as the outcome variable, save the predicted probabilities (call them *prob*) and the deviance residuals (call them *dev*) and ask Stata for a scatterplot to examine whether heteroscedasticity is a problem.

```
twoway scatter dev prob, jitter(5)
```

The graph in figure 3.7 reveals a rather peculiar pattern that suggests some problems, perhaps with heteroscedasticity (see the fanning out pattern toward the right-hand portion of the graph; or maybe it is simply an outlier of approximately –2.6). Whatever the problem, it is probably a good idea to reestimate the model with robust standard errors and see whether the results differ. It may also be a good idea to consider the nonlinear association between age and life satisfaction in order to minimize model misspecification. *[But take note: this doesn't help!]*

Finally, Stata has an option called `hetprob` that estimates a probit model corrected for heteroscedasticity. However, the analyst must specify which variables are involved in the heteroscedasticity problem to estimate this model. This may be difficult to determine without substantial information about why the error terms are not constant across some subset of explanatory variables. Moreover, since, technically speaking, the errors from the probit model are unobserved, there are other issues that may have to be addressed.

LIMITATIONS AND MODIFICATIONS

Logistic regression is recommended most often when the probability (and hence the odds) of an outcome is relatively low. The same is not necessarily true for probit regression, however. Yet some have argued that it is possible to have probabilities that are too low. A potential problem when probabilities are very low (or very high), especially when the sample size is small to modest, is *separation* and *quasi-separation*. These occur when the explanatory variables either perfectly predict the outcome or predict most of one category of the outcome. For example, suppose in the life satisfaction model that almost all the males reported high life satisfaction. In this situation, we may have quasi-separation, especially if the sample size is small.

When this sort of problem occurs, the logistic and probit models may not be estimated well by ML (King and Zeng, 2001). In fact, MLE may not be able to find a solution at all. A principal problem, moreover, is that, with very low probabilities, some cell sizes are quite small and this may lead to the small sample bias of MLE (Allison, 2012). For example, suppose a sample of 1,000 (a large sample by most measures) includes an outcome variable that measures, say, deaths from spider bites, that occurred only twice ($p(y)$ = 0.002). Estimating this outcome with ML presents a challenge. Two alternative models are *exact logistic regression* and a modification to MLE known as *penalized maximum-likelihood estimation* (PMLE). Stata has an `exlogistic` command that estimates exact logistic regression models, but it can take a long time to run and should not be used with samples greater than 200 (Leitgöb, 2013). PMLE is implemented with a user-written program called *firthlogit* (type `findit firthlogit`). Example 3.13 shows this command.

Example 3.13

`firthlogit satlife male, or`

The output provides an odds ratio of 2.94, which is similar to the odds ratio given in a normal logistic regression model (3.05). This is because high life satisfaction is not a rare event.

In cases where the event is not rare, some statisticians recommend using a *log-binomial regression model* and estimating relative risks or the ratio of the prevalence for each group (Hilbe, 2009; Selvin, 2004). For example, recall that the prevalence of high life satisfaction is about 0.40 among females and about 0.67 among males.[4] The prevalence ratio is thus 0.67/0.40 = 1.68. In Stata, the log-binomial regression model may be estimated using the `glm` command. For example, since high life satisfaction is not a rare event, and thus might not be the best candidate for logistic regression, we request the log-binomial model with the code shown in Example 3.14.

---

4. To be precise, this is a *point prevalence* since it measures respondents' life satisfaction at only one point in time.

Example 3.14

```
glm satlife male, family(binomial) link(log) nolog eform
```

The output provides a risk ratio of 1.68, which, as already noted, is the ratio of the prevalence of high life satisfaction for males and females. We could also adjust the risk ratio for the effects of sleep, age, and IQ.

The log-binomial model can be difficult to estimate, however, especially with multiple explanatory variables, so some analysts suggest an alternative approach called *robust Poisson regression* (Zou, 2004); it can also be estimated using the glm command (see Example 3.15).

Example 3.15

```
glm satlife male, family(poisson) link(log) nolog eform
    vce(robust)
```

The results are similar to those from the log-binomial model. We address other uses of Poisson regression in Chapter 6.

In addition, when all the variables are measured discretely, many analysts use log-linear models (Dobson and Barnett, 2011). These are computationally similar to the logistic regression model (and may also be used with Poisson variables) and are often preferred when many interactions are hypothesized. They are also useful when the distinction between the explanatory and outcome variables is not well defined. Nonetheless, logistic and probit regression are well suited to a large number of modeling tasks.

FINAL WORDS

Many variables in the social and behavioral sciences are binary, with numerous studies attempting to predict and explain why people make certain choices (went to graduate school: no, yes), hold particular attitudes (favor the death penalty for murder: no, yes), or experience certain events (parents divorced: no, yes). Logistic and probit models offer a reasonably good way to predict these types of outcomes. However, they should not be used—or reported—without understanding important issues such as the difference between odds and probabilities and what it means for variables to be related in a nonlinear manner. Moreover, many researchers who use these models fail to examine the important assumptions that underlie them. Examining assumptions is not just a critical task for linear regression models, but applies to all regression models. Finally, it is important to be aware of alternative models that may be more suitable for outcomes that are relatively common or particularly rare.

EXERCISES FOR CHAPTER 3

1. Consider the following 2 × 2 table. Determine the odds and probabilities of marijuana use among males and females. Then, compute the odds ratio of marijuana use that compares males to females.

	Sex	
Marijuana use	*Male*	*Female*
Yes	10	6
No	30	34

2. The *religion* data file (*religion.dta*) contains a variable called *relschol*, which indicates whether or not survey respondents attend a secondary school affiliated with a religious institution. Check the coding of the values of this variable. Conduct the following analyses:
   a. Compute the overall odds and probability of attending a religious school.
   b. Construct a contingency table of *relschol* by *race*. What are the probabilities that nonwhite students and white students attend religious schools? What are the odds that these two groups of students attend religious schools? What is the odds ratio that compares white to nonwhite students?

3. Estimate two logistic regression models designed to predict *relschol*. In the first model include as an explanatory variable only the variable race. In the second model, add the following explanatory variables: *attend* and *income*. Complete the following:
   a. Based on the first model, what is the odds ratio that compares white to nonwhite students? Compare this to the odds ratio you computed in 2b.
   b. What are the AICs and BICs from the two models? What do they suggest about which model offers a more appropriate fit to the data? What does the classification table show?
   c. Interpret the odds ratio for *race* from the second model.
   d. Predict the probabilities of nonwhite and white students for those who attend religious services five days per month and report a family income of $20,000–$30,000.
   e. Check the distribution of the deviance residuals from the second model. What do they suggest about the model?

4. Reestimate the models in exercise 3 with probit regression. Compute predicted probabilities from the first model, followed by predicted probabilities from the second model for nonwhite and white students who share the following characteristics: attend religious services five days per month and report a family income of $20,000–$30,000.

## FOUR

Ordered Logistic and Probit Regression Models

Although there are many examples of binary or dichotomous outcome variables used throughout the research community, it is also common to have variables with multiple categories. Thus, we need to have some flexibility in our generalized linear models to accommodate them. Consider the following two types of categorical variables: those that can be ordered in some reasonable or logical way and those that cannot. If there is reasonable ordering, a variable is typically designated as *ordinal*. This is the topic of this chapter. In the next chapter, we address categorical variables that are unordered, which, as mentioned in Chapter 2, are frequently designated as *nominal*.

As an example of an ordinal variable, consider educational achievement. In the United States we often think about educational milestones, which are categorical in nature: high school graduate, college graduate, postgraduate degree. Opinions about social and political issues are also frequently addressed using ordinal scales: "Do you strongly agree, agree, disagree, or strongly disagree that we should allow prayers in public schools?" Survey researchers often find it more efficient and cost-effective to provide categories as responses to questions about behaviors, such as "How many times have you used marijuana in the past year? Zero times, 1–2 times, 3–4 times, or 5 or more times?"[1]

In each of these examples, we may wish to predict the ordinal outcome. How might we do this with a regression model? One idea is to code the midpoints of categorical responses.

---

1. In some research projects, continuous variables are collapsed into a set of ordinal categories, such as when test scores are converted into quartiles. This is used on occasion for highly skewed variable, for ease of interpretability, or for simplification of presentation. However, it also leads to a substantial loss of statistical power, especially when there are relatively few categories or high skewness (Taylor et al., 2006). There are clearly trade-offs inherent in such a decision.

This is used occasionally when a continuous variable, such as income or years of education, is collapsed into a categorical variable. However, a problem occurs at the high end: How does one, for instance, decide on a reasonable number when an income category is $100,000 or more? Moreover, the analyst is still left with a non-normal and noncontinuous variable, which presents problems for a linear regression model. An alternative approach recommended by some analysts is known as *optimal scaling*: it transforms categorical variables into interval scales (Casacci and Pareto, 2015).

One rule of thumb that is popular in the social and behavioral sciences is to treat an ordinal variable as continuous if it has seven or more categories (some claim five is sufficient). However, this approach depends on the distribution of the categories. If the distribution does not approximate a normal distribution, then biased estimates are likely. Moreover, we must make a strong assumption about the "distance" between categories. When measuring, say, age in years, we know that the distance between 10 and 20 is 10 times the distance between 1 and 2. For ordinal variables, this is rarely, if ever, the case (what is the distance between agree and disagree?). Given these constraints, most analysts prefer to use a regression model that is designed specifically for ordinal outcomes (Qiao, 2015).

As an initial example, in order to evaluate the association between support for corporal punishment and years of education we use the variable *spanking* from the GSS data set (*gss. dta*). Stata's codebook command reveals its coding (see Example 4.1).

Example 4.1

codebook spanking

```
Spanking            favor spanking to discipline child

        type:  numeric (byte)
       label:  spanking

       range:  [1,4]                          units:  1
unique values:  4                         missing .:  980/2,903

   tabulation:  Freq.    Numeric  Label
                  512          1  strongly agree
                  890          2  agree
                  357          3  disagree
                  164          4  strongly disagree
                  980          .
```

Note that we have an ordinal set of responses to this inquiry. In addition, there are a substantial number of missing values that should be considered at some point (but not now).

Think about how to model the association between attitudes toward spanking and years of formal education. Suppose we decide to estimate a linear model. Figure 4.1 provides a scatterplot with a fit line that demonstrates this association.

twoway scatter spanking educate || lfit spanking educate

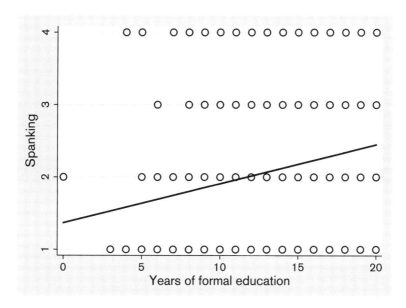

FIGURE 4.1

An ordinary least squares regression model confirms a positive association between education and spanking attitudes (see Example 4.2).

Example 4.2

regress spanking educate

Source	SS	df	MS			
				Number of obs	=	1,918
				F(1, 1916)	=	62.38
Model	47.5292154	1	47.5292154	Prob > F	=	0.0000
Residual	1459.86651	1,916	.761934504	R-squared	=	0.0315
				Adj R-squared	=	0.0310
Total	1507.39572	1,917	.786330581	Root MSE	=	.87289

spanking	Coef.	Std. Err.	t	P>|t|	[95% Conf. Interval]	
educate	.0540175	.0068393	7.90	0.000	.0406042	.0674309
_cons	1.367326	.0936701	14.60	0.000	1.18362	1.551032

But what do the regression diagnostics show? Not surprisingly given the patterns found in figure 4.1, there is a hint of heteroscedasticity and non-normal residuals. Figure 4.2 provides a smoothed histogram (kernel density plot) of the studentized residuals.

```
predict rstudent, rstudent
kdensity rstudent
```

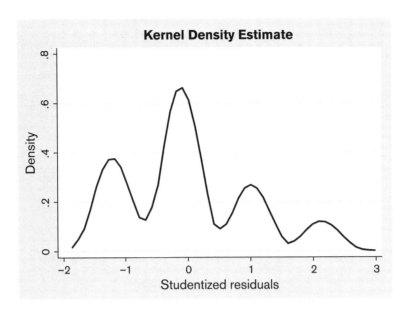

Kernel Density Estimate

FIGURE 4.2

If we are still skeptical about the seriousness of the problems, figure 4.3 shows a normal probability plot (qnorm) of the studentized residuals. These residuals are clearly not normally distributed. In fact, it is easy to see the effect of having a four-category variable rather than a continuous variable in these graphs.

qnorm rstudent   * see figure 4.3

THE ORDERED LOGISTIC REGRESSION MODEL

When confronted with an outcome variable such as spanking attitudes, we may use an *ordered logistic* or an *ordered probit* model to examine its predictors. These are also known as *ordinal logistic* or *ordinal probit* models. As with logistic or probit regression, we must make an assumption about the errors (the part that is left unexplained by the model—the random component; see Chapter 2). For the ordered logistic, we assume a logistic distribution, whereas for the ordered probit we assume a normal distribution for the errors, which we can examine using residuals from the estimated model. Similar to the models described in Chapter 3, we also assume that the variance of the errors is fixed to $\pi^2/3$ in the logistic model and to one in the probit model.

The next important issue is to get some sense of the "cut-points" of the distribution. These cut-points are simply the points at which the categories are separated. An easy way to think about this is in terms of probabilities. What is the probability that someone falls in one group relative to their probability of falling into another group or into any other group? For

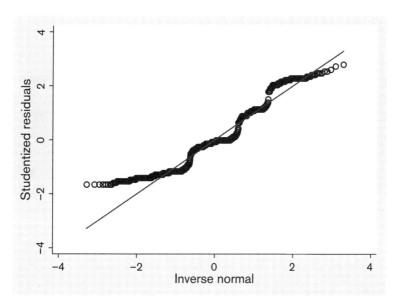

FIGURE 4.3

instance, what is the probability that a person with 12 years of education falls into the "agree" category versus her probability of falling into the other categories?

We will simplify this line of inquiry somewhat by examining the probabilities that males and females support these different attitudes about spanking. The contingency table in Example 4.3 is used to examine the joint distribution.

Example 4.3

```
tabulate female spanking, row
```

sex of respondent	favor spanking to discipline child strongly	agree	disagree	strongly	Total
male	243	388	156	56	843
	28.83	46.03	18.51	6.64	100.00
female	269	502	201	108	1,080
	24.91	46.48	18.61	10.00	100.00
Total	512	890	357	164	1,923
	26.63	46.28	18.56	8.53	100.00

By considering the row percentages, it is simple to compute the probabilities for the different attitudes. The probability that males strongly agree is 0.288; the probability that females strongly agree is 0.249 (or about 25% of females strongly agree that spanking is appropriate). The probabilities of strongly disagreeing are 0.066 for males and 0.100 for females. If we compute the probabilities across the rows they sum to one.

Estimating odds is trickier, however. In order to derive odds from a categorical outcome, there should be a clear reference category. In the dichotomous case, this was simple: The zero (or, typically, "no") category served as a convenient reference category. But what do we do in the case of an ordinal variable? There are actually several different types of ordered logistic regression models (Agresti, 2010), but the most popular is known as the *proportional odds* or *cumulative logit* model. The assumption underlying this model is that the odds of one category relative to the others are roughly equal and cumulative. An example is the best way to show what this means. Using the raw numbers in the frequency table, the odds that a female strongly disagrees versus the other three outcomes is (consider using `tabulate spanking female, column`)

Odds(females, SD vs. SA, A, D) = 108/(269 + 502 + 201) = 0.11.

Similarly, the odds that a male strongly disagrees versus the other three outcomes is

Odds(males, SD vs. SA, A, D) = 56/(243 + 388 + 156) = 0.07.

What then is the odds ratio for females to males?

Odds ratio for females vs. males = 0.11/0.07 = 1.57.

Interpreting this odds ratio can be difficult, but keep in mind that it only differs from what we have seen before in the reference category. Thus, a straightforward interpretation is the odds that females strongly disagree that spanking is acceptable are about 1.57 times the odds that males strongly disagree that spanking is appropriate as a way to discipline a child.

The assumption of proportional odds means that we are presuming that the odds ratios are constant across sex regardless of which comparison we make. For example, slide the scale over, as it were, and compute another set of odds and odds ratios:

Odds(females, SD, D vs. A, SA) = (108 + 201)/(502 + 269) = 0.40.

Similarly, the odds that a male disagrees versus the other two outcomes is

Odds(males, SD, D vs. A, SA) = (56 + 156)/(388 + 243) = 0.34.

What then is the odds ratio for females to males?

Odds ratio for females vs. males = 0.40/0.34 = 1.18.

The next logical step is to slide it over again and compute a final set of odds and odds ratios:

Odds(females, SD, D, A vs. SA) = (108 + 201 + 502)/269 = 3.01.

Similarly, the odds that a male strongly disagrees versus the other three outcomes is

Odds(males, SD, D, A vs. SA) = (56 + 156 + 388)/243 = 2.47,

Odds ratio for females vs. males = 3.01/2.47 = 1.22.

Thus, we now have three odds ratios: 1.57, 1.18, and 1.22. It might seem that an ordered logistic model would provide an odds ratio that was the simple average of these three values (OR = 1.32). Actually, the ordered logistic model provides a weighted average of these three values. Relatively less weight is given to the odds ratios that are based on larger denominators.

Another way to understand what the cumulative aspect of the model implies is to consider a cumulative probability C_{ij}, where the ith person is in the jth category of the outcome variable. This may be represented as

$$C_{ij} = p\left(y_i \le j\right) = \sum_{k=1}^{j} p\left(y_i = k\right).$$

This can then be transformed into a cumulative logit as follows:

$$\text{logit}\left(C_{ij}\right) = \ln\left(C_{ij}/\left(1 - C_{ij}\right)\right).$$

The ordered logistic model then estimates the cumulative logit as a linear function of the explanatory variables, as represented by

$$\text{logit}\left(C_{ij}\right) = \alpha_j - \boldsymbol{\beta x}$$

Note that the intercept (α) is subscripted with a j, whereas the slopes are not. This is because there is a different intercept—or cut-point—for each cumulative category of the ordinal outcome. In addition, the slopes are subtracted from the intercept, or the point at which we pass from one category to another. This affects the interpretations only when we predict probabilities from the ordered logistic equation.

Example 4.4 provides an ordered logistic regression model with *spanking* as the outcome variable and *female* as the explanatory variable.

Example 4.4

ologit spanking female, or

Ordered logistic regression

Number of obs	=	1,923
LR chi2(1)	=	5.87
Prob > chi2	=	0.0154
Pseudo R2	=	0.0012

Log likelihood = -2365.1653

spanking	Odds Ratio	Std. Err.	z	P>|z|	[95% Conf. Interval]	
female	1.229004	.1047268	2.42	0.016	1.039968	1.452401
/cut1	-.9004503	.0693994			-1.036471	-.7644299
/cut2	1.107179	.0709705			.9680792	1.246278
/cut3	2.491789	.0957675			2.304088	2.67949

As with the logit command, placing or at the end of the ologit command provides odds ratios instead of log-odds ratios. It appears that the first odds ratio we calculated (1.57) contributes the least to the computation of the overall odds ratio in ologit. This is not unexpected since it is based on relatively large denominators and small numerators.

The interpretation of the odds ratio from this model is as follows:

· The odds of strongly disagreeing that spanking is appropriate among females are about 1.2 times the odds of strongly disagreeing that spanking is appropriate among males.

But we could also say that the odds of strongly disagreeing or disagreeing among females are 1.2 times the odds of strongly disagreeing or disagreeing among males. Alternatively, one could say that the odds are expected to increase by about 20%. The assumption of proportional odds makes it difficult to offer a specific interpretation of the odds ratio in the ordered model, but it might be helpful to realize that one group has a higher odds of falling into a higher (numerically speaking) category. For our model, females have a higher odds of falling into the disagreement end of the scale. Chapter 5, which addresses models that directly distinguish the categories of the outcome, provides a more intuitive view of the odds ratio.

The cut values shown in the output are separate intercepts for each category, but are not used the same way as the intercept in the logit command output. If you imagine that the probabilities range from 0 (always agreeing) to 1 (always disagreeing), then someone who crosses a threshold (or cut-point) from, say, strongly agreeing to one of the other outcomes can be seen as moving along the probability scale. The values of the cut-points in ologit represent these points of movement from one category to another in terms of log-odds.

Exponentiating them shows the point in the distribution of the odds at which females (the zero category) move from one category to another.

The assumption of proportional odds may be too strong and is often violated in ordered logistic regression models. Unfortunately, Stata does not provide a direct way to test the assumption. SPSS does under what it terms a *test of parallel lines*. It is a χ^2 test with the null hypothesis that the odds ratios are the same across comparisons. There is a downloadable program in Stata called `omodel` that provides the same test (although its label is more precise: *likelihood ratio test*). After downloading this program (`findit omodel`), we simply reestimate the model (see Example 4.5).

Example 4.5

`omodel logit spanking female`   * *output is edited*

```
Approximate likelihood-ratio test of proportionality of odds
across response categories:
        chi2(2) =      3.35
     Prob > chi2 =    0.1874
```

The test compares our ordered model in which the slopes are constrained to be the same and a model that allows them to be different. Thus, the null hypothesis is that constraining them to be the same is appropriate (in a statistical sense). The `omodel` test indicates that we have failed to reject the null hypothesis, so we are relatively safe in assuming that the odds are proportional across the underlying comparisons of the *spanking* variable. Other alternatives include a Brant test that is included in Long and Freese's (2006) `spost9` set of Stata commands and a Stata user-written program called `gologit2` (*generalized logit*—an example is shown later in the chapter—which may also be used to relax the proportional odds assumption; Williams, 2006). For example, after downloading the `spost9` commands and typing `brant` after using `ologit` we find support for the assumption.

Example 4.5 (Continued)

`brant`   * *after Example 4.4*

```
Brant Test of Parallel Regression Assumption
```

Variable	chi2	p>chi2	df
All	3.16	0.206	2
female	3.16	0.206	2

```
A significant test statistic provides evidence that the parallel
regression assumption has been violated.
```

An advantage of this test is that it shows whether each explanatory variable violates the proportional odds assumption using a χ^2 test. If these tests are not available, then it may be useful to simply look at the underlying odds ratios by recoding the outcome variable and using binary logistic regression models.

THE ORDERED PROBIT MODEL

Before moving on to multiple ordered regression models, we first consider an *ordered probit model*. As mentioned in Chapter 3, there is little practical difference between logistic and probit regression models, but it may still be useful to consider this model.

The ordered probit model[2] follows a similar estimation approach as the ordered logistic model, but uses the cumulative distribution function (CDF) from the standard normal rather than the logistic distribution:

$$\Phi\left(C_{ij}\right) = \Phi\left(C_{ij}/\left(1-C_{ij}\right)\right)$$

The ordered probit model is thus estimated as

$$\Phi\left(C_{ij}\right) = \alpha_j - \boldsymbol{\beta x}.$$

As shown earlier in the chapter, the intercept (α) is subscripted, but the slopes are not. Example 4.6 demonstrates an ordered probit model.

Example 4.6

oprobit spanking female

Ordered probit regression

				Number of obs	=	1,923
				LR chi2(1)	=	6.79
				Prob > chi2	=	0.0091
Log likelihood = -2364.7011 | | | | Pseudo R2 | = | 0.0014 |

spanking	Coef.	Std. Err.	z	P>|z|	[95% Conf. Interval]	
female	.1296667	.049761	2.61	0.009	.0321369	.2271965
/cut1	-.5525651	.0412055			-.6333264	-.4718039
/cut2	.6837038	.04174			.601895	.7655127
/cut3	1.446059	.0502016			1.347666	1.544453

---

2. To be precise, this is, like the logistic model discussed earlier in the chapter, a cumulative model, in contrast to other models for ordinal variables (Agresti, 2010).

To understand what `oprobit` is doing, recall our original frequency table with *spanking* and *female*.

sex of respondent	favor spanking to discipline child strongly	agree	disagree	strongly	Total
male	243	388	156	56	843
	28.83	46.03	18.51	6.64	100.00
female	269	502	201	108	1,080
	24.91	46.48	18.61	10.00	100.00
Total	512	890	357	164	1,923
	26.63	46.28	18.56	8.53	100.00

Based on these frequencies, create a table with cumulative probabilities instead of percentages:

favor spanking to discipline child	male	female
strongly agree	0.29	0.25
agree	0.75	0.71
disagree	0.94	0.90
strongly disagree	1.00	1.00

The coefficients from the model along with the cut-points may be used to predict the cumulative probabilities. Recall that untransformed probit coefficients are z-scores, so the cut values in Example 4.6 show where along the z-score distribution the attitude "shifts." Using the Stata `normprob` function (which, you should recall, transforms z-scores into probabilities) with the coefficient and cut values from the `oprobit` output, we find

Example 4.6 (Continued)

```
display normprob(-0.5526)
 .2903
display normprob(-0.5526 - 0.1297)
 .2475
display normprob(0.6837)
 .7529
display normprob(0.6837 - 0.1297)
 .7102
display normprob(1.4461)
 .926
display normprob(1.4461 - 0.1297)
 .906
```

Note that these agree with the cumulative probabilities in the table in above. The reason we subtract the *female* coefficient is because we are backtracking from one (remember the equation shown earlier?).

The `prvalue` command (part of the `spost9` set of commands), if available, is also useful for computing predicted probabilities. Note, however, that there are now four predicted probabilities for males and females because the outcome variable has four possible outcomes.

Example 4.6 (Continued)

prvalue, x(female=0)   ** males*

```
oprobit: Predictions for spanning

Confidence intervals by delta method

                              95% Conf. Interval
    Pr(y=strongly|x): 0.2903  [ 0.2626,    0.3179]
    Pr(y=agree|x):    0.4626  [ 0.4413,    0.4840]
    Pr(y=disagree|x): 0.1730  [ 0.1629,    0.1831]
    Pr(y=strongly|x): 0.0741  [ 0.0603,    0.0879]

       female
x=          0
```

prvalue, x(female=1)   ** females*

```
oprobit: Predictions for spanning

Confidence intervals by delta method

                              95% Conf. Interval
    Pr(y=strongly|x): 0.2475  [ 0.2240,    0.2711]
    Pr(y=agree|x):    0.4627  [ 0.4395,    0.4858]
    Pr(y=disagree|x): 0.1958  [ 0.1836,    0.2079]
    Pr(y=strongly|x): 0.0940  [ 0.0790,    0.1091]

       female
x=          1
```

The `predict` command after `ologit` or `oprobit` may also be used to find predicted probabilities, but we must specify as many predicted probability variables as there are outcome categories.

Example 4.6 (Continued)

predict SA A D SD, pr

It is then helpful to compare summary statistics from these predicted probabilities to the original probabilities to determine how well the model fits the data.

```
summarize SA A D SD if female==0                                    * males
```

Variable	Obs	Mean	Std. Dev.	Min	Max
SA	1,285	.2902806	0	.2902806	.2902806
A	1,285	.4626383	0	.4626383	.4626383
D	1,285	.1730008	0	.1730008	.1730008
SD	1,285	.0740803	0	.0740803	.0740803

```
summarize SA A D SD if female==1                                    * females
```

Variable	Obs	Mean	Std. Dev.	Min	Max
SA	1,618	.2475462	0	.2475462	.2475462
A	1,618	.4626771	0	.4626771	.4626771
D	1,618	.1957555	0	.1957555	.1957555
SD	1,618	.0940212	0	.0940212	.0940212

Finally, we may use Stata's margins command and specify the particular outcome we wish to examine (or all the outcomes, if preferred). For example, the following provides the probabilities of strongly agreeing for males and females.

Example 4.6 (Continued)

```
margins, at(female=(0 1)) predict(outcome(1))
```

```
Adjusted predictions                    Number of obs     =       1,923
Model VCE     : OIM

Expression    : Pr(spanking==1), predict(outcome(1))

1._at         : female          =             0

2._at         : female          =             1
```

	Margin	Delta-method Std. Err.	z	P>|z|	[95% Conf. Interval]	
_at						
1	.2902806	.0141112	20.57	0.000	.2626232	.3179381
2	.2475462	.0120001	20.63	0.000	.2240264	.271066

Or try margins, dydx(female) predict(outcome(1)). This provides the difference in probabilities of strongly agreeing between males and females (−0.042).

There is a useful user-written command, mfx2 (consider findit mfx2), which computes marginal probabilities for all category comparisons. It offers an alternative to the margins command in this situation.

MULTIPLE ORDERED REGRESSION MODELS

Extending the ordered regression models to multiple explanatory variables is straightforward. The interpretations are still general and the proportional odds or parallel regression assumption still apply, however (see Example 4.7).

Example 4.7

ologit spanking female educate polviews *or is omitted*

Ordered logistic regression

Number of obs	=	1,821
LR chi2(3)	=	111.18
Prob > chi2	=	0.0000
Pseudo R2	=	0.0247

Log likelihood = -2198.2519

spanking	Coef.	Std. Err.	z	P>|z|	[95% Conf. Interval]	
female	.2532114	.0882542	2.87	0.004	.0802364	.4261863
educate	.1152978	.0156366	7.37	0.000	.0846506	.1459451
polviews	-.2214535	.0324833	-6.82	0.000	-.2851195	-.1577875
/cut1	-.2976692	.2670656			-.821108	.2257697
/cut2	1.784511	.2706421			1.254063	2.31496
/cut3	3.19266	.2792901			2.645261	3.740058

The odds ratio for female is exp(0.253) = 1.29, the odds ratio for educate is exp(0.1153) = 1.12, and the odds ratio for political views (coded from liberal to conservative) is exp(−0.221) = 0.80. We may thus claim the following:

- Adjusting for the effects of education and political views the odds that females disagree that spanking is appropriate are 1.29 times (or 29% higher than) the odds that males disagree that spanking is appropriate.[3]

- Each 1-year increase in education is associated with a 12% increase in the odds of disagreeing that spanking is appropriate, adjusting for the effect of sex and political views.

- Finally, each one-unit increase in political views is associated with a 20% decrease in the odds of disagreeing that spanking is appropriate, adjusting for the effects of sex and education.

---

3. Recall that, as explained in Chapter 3, we should not directly compare the coefficients for the *female* variable across the two models. Moreover, since the sample sizes are slightly different in the two models, there is an additional reason why it is risky to compare the coefficients.

The oprobit model is shown in Example 4.8. One interpretation from this output is that, adjusting for the effects of sex and political views, each 1-year increase in education is associated with a 0.065 z-score increase in the probability that a person disagrees that spanking is appropriate.

Example 4.8

oprobit spanking female educate polviews

```
Ordered probit regression                    Number of obs   =      1,821
                                             LR chi2(3)      =     108.21
                                             Prob > chi2     =     0.0000
Log likelihood = -2199.7348                  Pseudo R2       =     0.0240
```

spanking	Coef.	Std. Err.	z	P>|z|	[95% Conf. Interval]	
female	.1519324	.0515127	2.95	0.003	.0509695	.2528954
educate	.0650949	.009079	7.17	0.000	.0473004	.0828895
polviews	-.1255041	.0187131	-6.71	0.000	-.1621811	-.088827
/cut1	-.211023	.1557104			-.5162099	.0941639
/cut2	1.056828	.1570686			.7489796	1.364677
/cut3	1.83651	.1597796			1.523348	2.149673

Either model will produce predicted probabilities using Stata's predict command (remember to predict four variables), the prvalue command, or the margins command. Here is an example of prvalue for the variable *female*:

Example 4.8 (Continued)

prvalue, x(female=0 educate=mean polviews=mean)

```
oprobit: Predictions for spanking

Confidence intervals by delta method

                                95% Conf. Interval
    Pr(y=strongly|x): 0.2854 [ 0.2570,    0.3139]
    Pr(y=agree|x):    0.4729 [ 0.4505,    0.4953]
    Pr(y=disagree|x): 0.1723 [ 0.1622,    0.1824]
    Pr(y=strongly|x): 0.0693 [ 0.0557,    0.0830]

        female    educate    polviews
x=           0   13.50961   4.1724327
```

```
prvalue, x(female=1 educate=mean polviews=mean)

oprobit: Predictions for spanking

Confidence intervals by delta method

                                 95% Conf. Interval
      Pr(y=strongly|x): 0.2362  [ 0.2122,    0.2601]
      Pr(y=agree|x):    0.4724  [ 0.4478,    0.4970]
      Pr(y=disagree|x): 0.1995  [ 0.1870,    0.2120]
      Pr(y=strongly|x): 0.0920  [ 0.0766,    0.1073]

           female    educate    polviews
    x=          1   13.50961   4.1724327
```

Note, for example, that the probability that males and females differ in their views on corporal punishment is slightly higher in the strongly agree category. However, the sex differences, even though the coefficient for *female* has a small *p*-value (0.003), are not dramatic. This demonstrates the importance of evaluating predicted values—rather than simply claiming that a coefficient is statistically significant—to understand the association between explanatory and outcome variables.

It is also useful to use values or ranges from the other variables to compute predicted probabilities. For instance, it appears that education is consequential for attitudes toward spanking. Use the `prvalue` command to see how attitudes shift from 10 years of education to 16 years of education. Compute these first for males with average political views.

Example 4.8 (Continued)

```
prvalue, x(female=0 educate=10 polviews=mean)

oprobit: Predictions for spanking

Confidence intervals by delta method

                                 95% Conf. Interval
      Pr(y=strongly|x): 0.3676  [ 0.3279,    0.4072]
      Pr(y=agree|x):    0.4561  [ 0.4375,    0.4747]
      Pr(y=disagree|x): 0.1326  [ 0.1256,    0.1396]
      Pr(y=strongly|x): 0.0437  [ 0.0321,    0.0553]

           female    educate    polviews
    x=          0         10   4.1724327
```

```
prvalue, x(female=0 educate=16 polviews=mean)

oprobit: Predictions for spanking

Confidence intervals by delta method

                                  95% Conf. Interval
        Pr(y=strongly|x): 0.2330   [ 0.2043,    0.2618]
        Pr(y=agree|x):    0.4720   [ 0.4472,    0.4968]
        Pr(y=disagree|x): 0.2013   [ 0.1887,    0.2139]
        Pr(y=strongly|x): 0.0936   [ 0.0757,    0.1115]

           female    educate   polviews
   x=            0        16  4.1724327
```

There is a substantial difference in the strongly agree category. Note that the following `mar-gins` commands give the same results. However, as shown earlier, we may wish to examine each outcome separately, as shown next.

Example 4.8 (Continued)

```
margins, at(educate=(10 16) female=(0 1) ) atmeans
   predict(outcome(1))
margins, at(educate=(10 16) female=(0 1) ) atmeans
   predict(outcome(2)). . .
```

Furthermore, using the `marginsplot` command after each of these lines of code provides useful information. For example, consider outcome 1 (strongly agree). Figure 4.4 provides a graph of the predicted probabilities of strongly agreeing that spanking is appropriate. This helps us visualize the changes in the predicted probabilities of strongly supporting spanking for males and females with different levels of formal education.

It is also interesting to compare probabilities in the extreme categories. For example, the predicted probability of strongly agreeing that spanking is acceptable shifts from 0.42 among those with 8 years of education to 0.16 among those with 20 years of education, a difference of 0.26 or 62%. Thus, we might interpret this result in the following manner:

- Adjusting for the effects of sex and political views, those with a postgraduate degree (perhaps a doctorate or MD) are predicted to be 62% less likely than those with 8 years of education to strongly agree that spanking children is an appropriate disciplinary measure.

A visual way of examining these differences is by creating bar graphs or dot plots with predicted probabilities represented as bars or points along an axis (see figure 4.5).

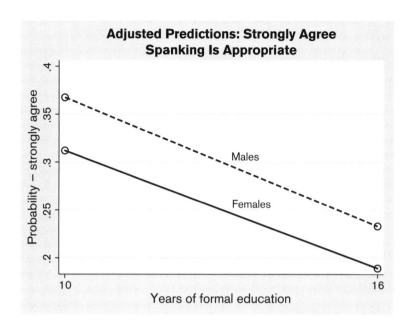

FIGURE 4.4

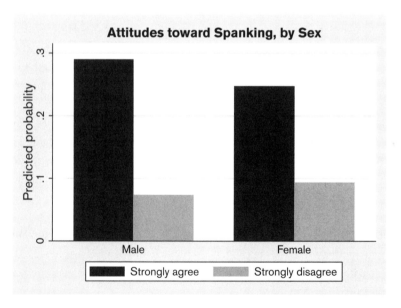

FIGURE 4.5

```
    graph dot (mean) SA SD, over(female)          * not shown
    graph bar SA SD, over(female)                 * see figure 4.5
```

This graph provides more information than the graph shown in figure 4.4 since it includes not only the predicted probabilities for strongly agreeing, but also those for strongly disagreeing.

Recall that it is important to examine the proportional odds assumption directly when using these types of ordered models. Using the brant command after ologit, for example, we find little evidence that the assumption is violated in this model.

Example 4.8 (Continued)

brant   * after Example 4.7

```
Brant Test of Parallel Regression Assumption

    Variable  |   chi2   p>chi2    df
   -----------+------------------------
         All  |   8.80   0.185     6
   -----------+------------------------
      female  |   2.64   0.268     2
     educate  |   1.17   0.558     2
    polviews  |   4.85   0.089     2


A significant test statistic provides evidence that the parallel
regression assumption has been violated.
```

MODEL FIT AND DIAGNOSTICS WITH ORDERED MODELS

Looking at the post-estimation options for Stata's ologit and oprobit commands indicates that, as expected, we may compute Akaike's information criterion (AIC), Bayesian information criterion (BIC), and likelihood ratio tests. For example, compare the two models that we estimated. The estat ic command following ologit for these models shows the following (use e(sample) to select the same sample for both models):

Example 4.8 (Continued)

female only

```
Akaike's information criterion and Bayesian information criterion

  Model  |     Obs  ll(null)  ll(model)    df      AIC         BIC
 --------+------------------------------------------------------------
      .  |   1,923 -2368.098  -2365.165     4   4738.331    4760.577

         Note: N=Obs used in calculating BIC; see [R] BIC note.
```

`female, educate, and polviews`

Akaike's information criterion and Bayesian information criterion

Model	Obs	ll(null)	ll(model)	df	AIC	BIC
.	1,821	-2253.84	-2198.252	6	4408.504	4441.547

Note: N=Obs used in calculating BIC; see **[R] BIC note**.

According to the AICs and BICs, the second model (unconstrained in this context) fits the data better than the first model. A likelihood ratio test confirms this (χ^2 = 102.79, df = 2, $p <$ 0.001). However, if we estimate the proportional reduction in errors from each model, we find that neither of them provides a good fit to these data (confirm this).

Unfortunately, there is a paucity of diagnostic tools directly available for the `ologit` or `oprobit` commands. The `predict` post-command provides only predicted probabilities, the linear predictor, and the standard error for the linear predictor (see `help ologit postestimation` or `help oprobit postestimation`). Thus, we cannot directly examine heteroscedasticity or influential observations in Stata. One recommendation, however, is to look for extreme probabilities, such as those close to zero or close to one. These may present problems for ordered logistic models.

If we simply cannot survive without examining residuals and such, an indirect approach is to test each underlying logistic regression model and use the techniques discussed in Chapter 3. For example, we might estimate a logistic regression model with a recoded variable that compares strongly agree to the other three outcomes, save the residuals, the predicted probabilities, and the influence statistics, and use them to examine assumptions of the model. Example 4.9 demonstrates one possible approach.

Example 4.9

```
recode spanking (4=1)(1/3=0), generate(newspank)
logistic newspank female educate polviews
predict dev, deviance
predict hat, hat
predict pred, p
predict deltad, ddeviance
twoway scatter deltad hat, xline(0.154) yline(4)
```
                                              * are there influential observations?

Figure 4.6 provides one scatterplot that may be useful. Can we detect heteroscedasticity? Or does it confirm the assumption of homoscedasticity? Examine the delta-deviance and hat values (graphing them is useful). Are there influential observations?

```
twoway scatter dev pred
```
                                              * check for heteroscedasticity

We would then go through this exercise for the other two comparisons (SD, D vs. A, SA; SD, D, A vs. SA) and, perhaps, come up with different conclusions for each. This is time-consuming, but the careful researcher would wish to put in the effort.

There is one more important point about these models. Suppose the proportional odds assumption is violated and the shift in odds is not statistically similar across categories. Most analysts recommend choosing an alternative model, such as a generalized ordered logit model (using Stata's `gologit2` command) or a stereotype model (using Stata's `logit` command). Using `gologit2` with the `autofit` option is especially convenient because it tests the proportional odds assumption for each explanatory variable and it allows the coefficients to vary if the assumption is violated (Williams, 2006). Example 4.10 shows how to execute this for the unconstrained model.

Example 4.10

```
gologit2 spanking female educate polviews, autofit
```

spanking	Coef.	Std. Err.	z	P>|z|	[95% Conf.	Interval]
strongly_agree						
female	.2532114	.0882542	2.87	0.004	.0802364	.4261864
educate	.1152978	.0156366	7.37	0.000	.0846506	.1459451
polviews	-.2214535	.0324833	-6.82	0.000	-.2851195	-.1577875
_cons	.2976691	.2670656	1.11	0.265	-.2257698	.821108
agree						
female	.2532114	.0882542	2.87	0.004	.0802364	.4261864
educate	.1152978	.0156366	7.37	0.000	.0846506	.1459451
polviews	-.2214535	.0324833	-6.82	0.000	-.2851195	-.1577875
_cons	-1.784512	.2706421	-6.59	0.000	-2.314961	-1.254063
disagree						
female	.2532114	.0882542	2.87	0.004	.0802364	.4261864
educate	.1152978	.0156366	7.37	0.000	.0846506	.1459451
polviews	-.2214535	.0324833	-6.82	0.000	-.2851195	-.1577875
_cons	-3.192661	.2792901	-11.43	0.000	-3.740059	-2.645262

(output edited)

Notice that the coefficients for each of the explanatory variables are identical in each panel. This is because, under the proportional odds assumption (which is satisfied in this analysis), we assume that the coefficients distinguishing the groups defined by the outcome are the same regardless of which comparison we make. However, if the assumption is not met for some comparison, `gologit2` provides different coefficients to capture this.

In the next chapter we examine perhaps the most common alternative to ordered regression models when the proportional odds or parallel regression assumption is violated:

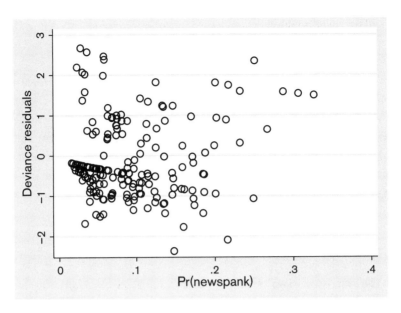

FIGURE 4.6

multinomial logistic regression. Agresti (2010, 2013) provides much more information about ordinal variables and ordered regression models. For example, he discusses several alternatives to the cumulative category models that we examine in this chapter, including an adjacent categories model and a continuation ratio model (type `findit ocratio` in Stata). An advantage of these two models is that neither assumes proportional odds.

FINAL WORDS

As mentioned at the outset of this chapter, ordered regression models are appropriate for outcome variables that are measured as reasonably or logically ordered categories. These are quite common in the social and behavioral sciences, with many studies of opinions and attitudes regarding social and political issues, categorized rankings on tests into quartiles or deciles, qualitative judgments of health risk (e.g., low, medium, and high risk of heart disease), and educational benchmarks such as graduating from high school, college, and graduate school. Although it is helpful to have models designed specifically for these types of outcomes, some researchers continue to use linear regression models. Whether this is appropriate or not depends on a number of factors. For instance, as mentioned early in the chapter, it depends on the distribution of the categories (Casacci and Pareto, 2015); it also depends on the sample size.

As an example, consider the two models using the *GSS* data shown in Example 4.11.

Example 4.11

regress polviews female age divorce

polviews	Coef.	Std. Err.	t	P>|t|	[95% Conf. Interval]
female	-.1588855	.0521874	-3.04	0.002	-.2612162	-.0565547
age	.0065033	.0015765	4.13	0.000	.0034121	.0095946
divorce	.0865906	.0577175	1.50	0.134	-.0265837	.1997649
_cons	3.842612	.1355222	28.35	0.000	3.576876	4.108348

ologit polviews female age divorce, or

polviews	Odds Ratio	Std. Err.	z	P>|z|	[95% Conf. Interval]
female	.7990789	.0552616	-3.24	0.001	.6977879	.9150734
age	1.008316	.0021001	3.98	0.000	1.004208	1.01244
divorce	1.110355	.0846058	1.37	0.170	.9563194	1.289202

The outcome variable is comprised of seven categories from extreme liberal (coded 1) to extreme conservative (coded 7). Even though the coefficients are, of course, quite different, if we are looking simply for general patterns in the data, then we might reach the same conclusions regardless of the model used. Females appear to be less likely to report that they identify with the conservative end of the spectrum, whereas older people seem less likely to report being on the liberal end.

Whether one is divorced makes little difference to one's political views, at least as it is measured here. Moreover, the linear regression model shows little evidence of heteroscedastic errors and the distribution of the residuals, though a bit bumpy, are not too far from normal (determine this for yourself). So perhaps a linear regression model is appropriate in this situation. Nevertheless, it is important to consider two things before we adopt this perspective too easily. First, look at the distribution of *polviews*: it appears to be quite symmetrical. Second, the data set is relatively large, which works to the advantage of a linear regression model (Lumley et al., 2002). Nevertheless, one should exercise care when considering which model to use. If a researcher prefers to use a linear regression model, then the ordinal outcome variable should be evaluated carefully and the model's assumptions examined accordingly.

1. The *environ* data set (*environ.dta*) contains survey data from a study of environmental attitudes among a sample of Presbyterian Church members in the United States. Use the variable *env_ch* as the outcome variable in the following exercises. Consider its label, which provides information about the question upon which it is based. Consider also the response categories, which are based on a Likert-type scale.

   a. Construct a contingency table that cross-classifies *env_ch* by *gender*. Based on this table, compute the probabilities that males and females disagree completely (DC) *or* disagree somewhat (DS) with the question about environmental issues.

   b. Using the table, compute odds ratios for females relative to males using the following comparisons: {AC vs. AS, NDA, DS, and DC}; {AC and AS vs. NDA, DS, and DC}; {AC, AS, and NDA vs. DS and DC}; and {AC, AS, NDA and DS vs. DC}.

   c. Estimate an ordered logistic regression model with *env_ch* as the outcome variable and *gender* as the explanatory variable. Compare the odds ratio from this model to the odds ratios computed in Exercise 1b. Comment on their similarities and differences.

2. Estimate ordered logistic and ordered probit regression models that use *env_ch* as the outcome variable and the following explanatory variables: *gender, literal, age,* and *educate*.

   a. Interpret the odds ratios for *gender, literal,* and *age* from the ordered logistic model.

   b. Compute the predicted probabilities (using either model) that males and females fall into each of the categories of *env_ch* at the following levels of the other explanatory variables: *age* = 52, *educate* = graduated from college, and *literal* = no.

   c. Create a graph (your choice of the type) that compares the predicted probabilities of "disagree somewhat" and "agree somewhat" for males and females who did and did not graduate from college (four groups).

   d. Test the proportional odds (parallel regression) assumption for the ordered logistic model. What does it show?

## FIVE

Multinomial Logistic and Probit Regression Models

As mentioned at the beginning of the last chapter, another type of categorical variable is often labeled *nominal*. When considering a nominal variable, the natural "ordering" of the categories is not clear or cannot be derived logically. In addition, when the proportional odds or parallel regression assumption of the ordered regression model (logistic or probit) is violated, one option is to treat the ordinal outcome variable as unordered and consider a regression model that is appropriate for this type of categorical variable. One frequently used regression model for unordered outcome variables—or what some researchers call polytomous outcome variables—is *multinomial logistic regression*. Less widely used is the *multinomial probit model*. These are the focus of this chapter.

What are some examples of unordered or nominal variables? Suppose we wish to predict what occupation people with different education levels find themselves in. Or we want to predict what set of personal or family characteristics predict the likelihood of cohabitation, being married, or remaining single among adults. In a study I completed several years ago, I tried to predict the odds that various groups of workers would be exposed to different types of drug testing programs, such as random, preemployment, or no testing. Since ranking these outcomes is not a straightforward or logical exercise, I opted to treat them as unordered and use a multinomial logistic regression model to predict each one based on explanatory variables such as occupation, education, and marijuana use.

In order to get a better sense of the multinomial logistic regression model, it may be helpful to revisit the multinomial distribution. Recall from Chapter 2 that the multinomial distribution is a more general form of the binomial distribution. Rather than computing the probability or the odds of falling into one category (e.g., drug user) versus another (nondrug user), we compute the probability or odds of falling into one of three or more categories. Recall the basketball shooting example.

The probability mass function of the multinomial distribution is given by

$$p\left(y_1 = n_1, y_2 = n_2, \ldots, y_r = n_r\right) = \frac{n!}{n_1! n_2! \ldots n_r!} \times p_1^{n_1} p_2^{n_2} \ldots p_r^{n_r}.$$

The letter r in this function is the number of possible outcomes (e.g., random, preemployment, or no drug testing; blue, brown, or green eyes), p indicates the probability of each outcome, and n denotes the frequency with which we observe the particular outcome.

Suppose, for example, that we have a sample of 10 workers and we know from previous research that there are the following probabilities of drug testing: random = 0.05, preemployment = 0.15, and no testing = 0.80. We then observe the following distribution in our sample: one worker reports random testing, three report preemployment testing, and six report no drug testing. From the multinomial distribution function, the probability of this particular outcome is

$$\frac{10!}{1! 3! 6!} \times 0.05^1 0.15^3 0.80^6 = 0.037.$$

If the original (or a priori) probabilities were accurate, we would expect to find this specific combination of drug testing results about 4 times out of every 100 samples. Computing all the combinations given a sample of 10 (e.g., {4, 4, 2}, {2, 8, 0}), the probabilities sum to 1 ($\Sigma p_i = 1$). As discussed in Chapter 2, in many statistical modeling exercises we do not know these predetermined probabilities.[1] Rather, we usually have a sample of, say, workers and we hope to estimate how many report different types of drug testing. We can then estimate probabilities and odds, as well as odds ratios if we have explanatory variables. In our example, perhaps we want to determine if construction workers have a higher odds or probability of being exposed to random testing than, say, accountants.

THE MULTINOMIAL LOGISTIC REGRESSION MODEL

A common option used to predict multinomial outcomes is a multinomial logistic regression model. A straightforward way to understand this model is to imagine a set of binary logistic regression models that cover all the categories of a multinomial outcome variable. The models, for instance, might compare, say, random testing to no testing and preemployment testing to no testing. In other words, we have $j - 1$ binary logistic regression models to

---

1. One of the hallmarks of Bayesian analysis, however, is that researchers may use "prior beliefs" (such as presumed a priori probabilities) as part of the estimation process to determine probabilities of outcomes (see Congdon, 2014).

examine *j* outcomes. Why not just do this, then? Imagine you have six outcomes. This would require five models. Other variables with more outcomes would require even more models. It might quickly get difficult to keep them straight.

Another problem with running a bunch of binary models is that sampling variability affects the results. We would have to compare a bunch of models with different sample sizes and a variety of binomial distributions. This approach is quite inefficient. Some category comparisons may have relatively large standard errors.

To get around this problem, the multinomial logistic regression model simultaneously estimates a set of binary logistic regression models. Thus, it is more efficient than a series of separate binary models. Specifically, consider the following equation:

$$\ln\left(\frac{\pi_{ij}}{\pi_{ij^*}}\right) = \alpha_j + \beta_j x \quad j \neq j^*$$

Similar to the equation shown in Chapter 3, the π are a shorthand way of expressing the probabilities that an observation falls into category *j* relative to category *j**. The *j** indicates a reference category against which the other categories are compared. Note that we use the natural logarithm as a link function and that the slopes (β) are subscripted with a *j*. This indicates that there are separate slopes for each category *j*; thus, the multinomial logistic model differs from the ordered logistic model. The log-likelihood for this model is simply an extension of the log-likelihood shown in Chapter 3, but we sum not just over the entire sample $\sum_{i=1}^{N}$, but also over the categories: $\sum_{J}^{J-1}$ (Czepiel, n.d.).

To see how this operates, examine the GSS data set (*gss.dta*) and consider the variable *polview1* (see Example 5.1). Here we have what would probably be considered a nominal variable (even though we used a variation of it as an ordinal outcome variable at the end of Chapter 4!). Although our partisanship may make us wish to place these into a "natural" order, most political scientists would likely blanch at such an effort.

Example 5.1

```
codebook polview1

        type:  numeric (byte)
        label:  polview1

        range:  [1,3]                          units:  1
unique values:  3                           missing .:  161/2,903

    tabulation:  Freq.   Numeric  Label
                   696         1  liberal
                 1,044         2  moderate
                 1,002         3  conservative
                   161         .
```

Next, consider the association between political views and the variable *nonwhite*, which is coded as 0 = white and 1 = nonwhite, using a contingency table.

Example 5.2

```
tabulate nonwhite polview1, row
```

white or nonwhite	political views categorized liberal	moderate	conservat	Total
white	546 24.34	853 38.03	844 37.63	2,243 100.00
non-white	150 30.06	191 38.28	158 31.66	499 100.00
Total	696 25.38	1,044 38.07	1,002 36.54	2,742 100.00

Based on Example 5.2, we may compute a host of probabilities and odds. For instance, the overall probability of being a conservative is 0.365, whereas the overall probability of being a liberal is 0.254. The three probabilities in the *Total* row sum to 1.00.

Since we are on our way to a logistic regression model, though, we need to try to make sense of some of the odds that may be computed with this table. Recall that when we are dealing with odds, we should have a clear comparison group if we wish to use the simple equation: $p/[1 - p]$. If the comparison is between marijuana use and no marijuana use, for instance, calculating the odds is effortless: 0.25/0.75 = 0.33 (see Chapter 3). But how do we do this in the political views table? How should we compute the odds of being a white liberal, for example? Is it simply 0.24/[1 – 0.24] = 0.316? Similarly, what are the odds of being a non-white moderate? Answer: 0.38/[1-0.38] = 0.613?

A problem with this approach is that we are not using all the information available. We are computing odds based on only two groups, rather than three. Odds ratios are similarly limited by this approach. For example, the odds ratio comparing white conservatives to non-white conservatives is 1.30. We may interpret this in the following way:

- The odds of being conservative among whites are expected to be 1.3 times (or 30% higher than) the odds of being conservative among nonwhites.

The problem again is that we are not using all the information available. We are basing odds ratios on only two categories, in this case conservative versus nonconservative. So a more accurate interpretation for this odds ratio is that it reflects the odds of being conservative versus not being conservative.

There are at least two solutions to this problem. First, we may compute all the odds for whites and nonwhites and then compute all of the odds ratios (there are three). We might find differences across the two racial categories in the odds of being liberal or conservative, but not moderate (the odds ratio is almost exactly one). Second, we can take full advantage of the categories by selecting one category as the comparison group. This is also known as choosing a *reference category* or *reference group*. In essence, we have already done this in the binary logistic case; it's just that with only two categories, it is a more straightforward decision.

Try this with the political views table. Use "moderate" as the reference category. This is reasonable since researchers often choose the modal category to serve this purpose. Rather than three odds ratios, we reduce it to only two.

Nonwhite or white	Odds
Nonwhite	
Liberal vs. moderate	150/191 = 0.79
Conservative vs. moderate	158/191 = 0.83
White	
Liberal vs. moderate	546/853 = 0.64
Conservative vs. moderate	844/853 = 0.99
Odds ratios	
0.79/0.64 = 1.23	
0.83/0.99 = 0.84	

It is now clear that nonwhites have the higher odds of reporting liberal versus moderate and lower odds of reporting conservative versus moderate relative to whites. What does a multinomial logistic regression model provide? This model is estimated in Stata using the `mlogit` command (see Example 5.3).

Example 5.3

```
mlogit polview1 nonwhite, rrr
```

```
Multinomial logistic regression          Number of obs   =     2,742
                                          LR chi2(2)      =      9.16
                                          Prob > chi2     =    0.0103
Log likelihood = -2966.5208               Pseudo R2       =    0.0015
```

polview1	RRR	Std. Err.	z	P>|z|	[95% Conf. Interval]	
liberal						
nonwhite	1.226914	.1497947	1.68	0.094	.9658076	1.558612
_cons	.6400938	.0350818	-8.14	0.000	.5748991	.7126817
moderate	(base outcome)					
conservative						
nonwhite	.8360463	.098646	-1.52	0.129	.6634316	1.053573
_cons	.989449	.0480384	-0.22	0.827	.8996364	1.088228

The setup is slightly different than the setup for the `logit` command. Rather than asking for `or` we ask for `rrr`. This is shorthand for *relative risk ratio*. However, we may interpret these as odds ratios. If we omitted this subcommand, we would find log-odds ratios. Notice that Stata somehow knew we wanted moderate as the reference category (or *base outcome*, as Stata labels it). The program does not know how to do this automatically. Rather, Stata chooses the most frequent—or modal—outcome when specifying a reference category. To set the category to something else, use the subcommand `baseoutcome(y)`, where y could be the numeric code for liberal or conservative.

What do the results show? First, notice that they agree with our crude calculations of the odds ratios. This depends on the coding of the nonwhite variable, though. Again, we find that the odds of reporting liberal versus moderate among nonwhites are expected to be 1.23 times (or 23% higher than) the odds of reporting liberal versus moderate among whites. Similarly, the odds of reporting conservative versus moderate among nonwhites are expected to be 0.84 times (or 16% lower than) the odds of reporting conservative versus moderate among whites. The *p*-values suggest that these odds ratios are not statistically significantly different from one, however.

As with the other logistic regression models discussed earlier, we can also compute probabilities and changes in probabilities. Type `help mlogit postestimation` to determine what is available. First, notice that `estat ic` is available, so we may judge model fit. The `test` command is similarly available. Second, the `margins` command is listed. What does it provide? If we type `margins, at(nonwhite=0) predict(outcome(1))`, the program returns a value of 0.243. What is this value? It is the probability that white respondents report liberal. Here, we have told Stata the specific outcome, or the code 1 to designate the liberal category.

We may also use the `prvalue` command (if it has been downloaded) and request the relevant probabilities for all three groups.

Example 5.3 (Continued)

```
prvalue, x(nonwhite=0)

mlogit: Predictions for polview1

Confidence intervals by delta method

                              95% Conf. Interval
    Pr(y=liberal|x):   0.2434   [ 0.2257,    0.2612]
    Pr(y=conserva|x):  0.3763   [ 0.3562,    0.3963]
    Pr(y=moderate|x):  0.3803   [ 0.3602,    0.4004]

        nonwhite
    x=         0
```

```
prvalue, x(nonwhite=1)

mlogit: Predictions for polview1

Confidence intervals by delta method

                               95% Conf. Interval
    Pr(y=liberal|x):    0.3006  [ 0.2604,    0.3408]
    Pr(y=conserva|x):   0.3166  [ 0.2758,    0.3574]
    Pr(y=moderate|x):   0.3828  [ 0.3401,    0.4254]

        nonwhite
x=             1
```

This is promising since we now have predicted probabilities for all six possible groups. Another method is to save the predicted probabilities using the `predict` command. Recall that, since there are multiple groups, we must predict the probability of reporting liberal, moderate, and conservative.

Example 5.3 (Continued)

predict liberal moderate conservative, pr

pr requests probabilities

table nonwhite, contents(mean liberal mean moderate mean conservative)

```
white or
nonwhite     mean(liberal)  mean(moderate)  mean(conser~e)

    white        .243424        .3802943        .3762818
non-white        .3006012       .3827655        .3166333
```

The `table` command in which we ask for the means for liberal, moderate, and conservative is often a helpful way to view and compare the probabilities (although it does not provide confidence intervals). Notice that the names given to the three variables are arbitrary, but it is much easier to keep them straight if we name them after their value labels from the original *polview1* variable. We can see, therefore, that nonwhites, relative to whites, prefer to label themselves liberal, whereas whites appear more comfortable with the label conservative. We may also present the differences in probabilities: {0.301 − 0.243 =} 0.058 for liberal and {0.317 − 0.376 =} 0.06 for conservative. In addition, graphing the probabilities makes it easier to visualize the differences. For example, see figure 5.1.

graph bar liberal conservative, over(nonwhite)

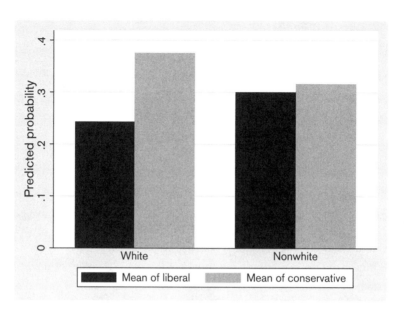

FIGURE 5.1

THE MULTIPLE MULTINOMIAL LOGISTIC REGRESSION MODEL

It is simple to add other explanatory variables to the model to create a multiple logistic regression model. We first drop the crude racial category variable and then add sex, age, and education since each is likely to predict one's propensity to commit to an ideological label.

Example 5.4

```
mlogit polview1 female age educate, rrr
```

Multinomial logistic regression

```
                                        Number of obs   =      2,737
                                        LR chi2(6)      =      75.43
                                        Prob > chi2     =     0.0000
Log likelihood = -2928.4707             Pseudo R2       =     0.0127
```

polview1	RRR	Std. Err.	z	P>|z|	[95% Conf. Interval]	
liberal						
female	1.117224	.1121759	1.10	0.270	.9176452	1.36021
age	.9927973	.0030642	-2.34	0.019	.9868098	.9988212
educate	1.127454	.0203866	6.63	0.000	1.088197	1.168128
_cons	.1698857	.0520324	-5.79	0.000	.0932077	.3096436
moderate	(base outcome)					
conservative						
female	.8080522	.0723264	-2.38	0.017	.6780331	.9630037
age	1.003236	.0026838	1.21	0.227	.9979899	1.00851
educate	1.076356	.0172631	4.59	0.000	1.043048	1.110729
_cons	.3494262	.0953958	-3.85	0.000	.204631	.5966772

There are some interesting results in the model shown in Example 5.4. For instance, the coefficient for *female* is not statistically significant in the first panel of the output, but it is in the second part. The coefficient for *age* has the opposite effect, whereas the coefficient for education has a similar effect in both models. Why does this occur? If we think about this model as two separate (but simultaneously estimated) logistic regression models, it is not that surprising. Age predicts whether a person commits to liberal versus moderate, but not whether a person reports conservative versus moderate. Females have lower odds of reporting conservative versus moderate (since *female* is coded as 0 = male, 1 = female), but roughly the same odds as males of reporting liberal versus moderate. Additional years of education are associated with higher odds of reporting liberal or conservative versus moderate.

It is useful to change the reference category to conservative to understand these results better (see Example 5.5).

Example 5.5

```
mlogit polview1 female age educate, rrr baseoutcome(3)
```

Multinomial logistic regression			Number of obs	=	2,737
			LR chi2(6)	=	75.43
			Prob > chi2	=	0.0000
Log likelihood = -2928.4707			Pseudo R2	=	0.0127

polview1	RRR	Std. Err.	z	P>|z|	[95% Conf. Interval]	
liberal						
female	1.382614	.1385499	3.23	0.001	1.136065	1.682669
age	.9895947	.0030795	-3.36	0.001	.9835773	.9956489
educate	1.047473	.0187144	2.60	0.009	1.011428	1.084802
_cons	.4861848	.1489621	-2.35	0.019	.266687	.8863413
moderate						
female	1.237544	.1107689	2.38	0.017	1.038418	1.474854
age	.9967742	.0026665	-1.21	0.227	.9915616	1.002014
educate	.9290603	.0149007	-4.59	0.000	.9003098	.9587289
_cons	2.861835	.7813011	3.85	0.000	1.675948	4.886845
conservative	(base outcome)					

Conservative serves as the reference group in this model (consider how it's labeled in the third panel). We can now see the association between sex and the outcomes more clearly. The odds of reporting liberal or moderate rather than conservative among females are expected to be higher than the odds of reporting these among males, adjusting for the effects of age and education. Here are some more precise interpretations of these results:

- *female, liberal versus conservative*: Adjusting for the effects of age and education, the odds of reporting liberal rather than conservative are expected to be 38% higher among females than among males.

- *female, moderate versus conservative*: Adjusting for the effects of age and education, the odds of reporting moderate rather than conservative are expected to be 23% higher among females than among males.

- *education, liberal versus conservative*: Adjusting for the effects of sex and age, each 1-year increase in education is associated with a 5% increase in the odds of reporting liberal rather than conservative.

- *education, moderate versus conservative*: Adjusting for the effects of sex and age, each 1-year increase in education is associated with a 7% decrease in the odds of reporting moderate rather than conservative.

It is evident that these models can quickly become cumbersome if there are many groups to compare. But it simplifies things substantially to always remember that there is a set of binary logistic regression models underlying the multinomial regression model.

It is also useful to once again look at predicted probabilities using the `prvalue`, `margins`, or `predict` commands. Example 5.5 continues with an example of using the `prvalue` command after the model.

Example 5.5 (Continued)

```
prvalue, x(educate=12 female=0 age=mean)

mlogit: Predictions for polview1

Confidence intervals by delta method

                                95% Conf. Interval
    Pr(y=liberal|x):    0.2080   [ 0.1836,    0.2323]
    Pr(y=moderate|x):   0.4008   [ 0.3709,    0.4307]
    Pr(y=conserva|x):   0.3912   [ 0.3616,    0.4209]

          female       age    educate
x=             0  44.683961         12
```

```
prvalue, x(educate=16 female=0 age=mean)

mlogit: Predictions for polview1

Confidence intervals by delta method

                                95% Conf. Interval
    Pr(y=liberal|x):    0.2663   [ 0.2374,    0.2951]
    Pr(y=moderate|x):   0.3176   [ 0.2877,    0.3474]
    Pr(y=conserva|x):   0.4161   [ 0.3842,    0.4481]

          female       age    educate
x=             0  44.683961         16
```

Observe, for instance, that the difference between the predicted probabilities of being a liberal for males of 12 and 16 years of education is 0.266 − 0.208 = 0.058, or a difference of about 28%.[2] Thus, the probability of reporting liberal among college graduates is expected to be about 28% higher than among those who only graduated from high school.

Using predicted probabilities (e.g., `predict liberal moderate conservative`) also provides some interesting comparisons.

Example 5.5 (Continued)

```
table educate if (educate == 10 | educate==12 |
  educate==14 | educate==16) & (29 < age < 50) &
  (female==0), contents(mean liberal mean moderate mean
  conservative)
```

years of formal education	mean(liberal)	mean(moderate)	mean(conser~e)
10	.2512209	.3806313	.3681479
12	.2532986	.3807211	.3659804
14	.2538878	.3807465	.3653657
16	.2495965	.3805611	.3698424

These are predicted probabilities for males between the ages of 30 and 49. Education is associated with a substantial increase in the probability of reporting liberal, a substantial decrease in the probability of reporting moderate, and a modest increase in the probability of reporting conservative among this particular group. This shows a bit more clearly the association between education and self-ascribed political views.

These results may also be shown with a graph, although with this many comparisons it may get a little too busy. Nevertheless, figure 5.2 provides a crude method that changes the results in Example 5.5 into a bar graph.

```
recode educate(10=1)(12=2)(14=3)(16=4)(else=.),
  gen(cateduc)
graph bar liberal moderate conservative if (29<age<50)
  & female==0, over(cateduc)
```

Finally, how well does this model fit the data? Stata provides a pseudo-R^2 measure that is quite low (0.03). The user-written proportional reduction in error (PRE) command indicates that we have reduced the errors in prediction in political views by about 8.1%. The PRE

---

2. We may also use the `margins` command to compute the probabilities: `margins, at(female=0 educate=(12 16)) atmeans`.

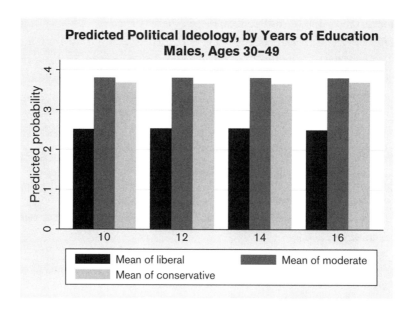

FIGURE 5.2

output also provides an estimate of the accuracy of the model: 43%. However, the accuracy of the multinomial logistic model can be misleading because we now have more than two possible outcomes in a classification table. A better approach is to, first, take the sum of the squared proportions in each predicted category. From the PRE classification table, this is $(164/2,737)^2 + (1,524/2,737)^2 + (1,049/2,737)^2 = 0.460$. Second, a common rule of thumb is that the model accuracy should be at least 25% better than chance. Therefore, multiply 0.460 by 1.25 ($0.460 \times 1.25 = 0.575$). Finally, compare the accuracy of the model to this standard. Although we want it to exceed 0.575, our accuracy falls well below this threshold. Coupled with other fit statistics, it is safe to conclude that the model provides a relatively poor fit to the data.

Example 5.5 (Continued)

pre   * assumes the user-written pre command has been downloaded to Stata

```
Model reduces errors in the prediction of polview1 by    8.14%

                 Prediction of polview1
  polview1 |        1          2          3 |    Total
-----------+---------------------------------+----------
         1 |       74        345        277 |      696
         2 |       44        665        332 |    1,041
         3 |       46        514        440 |    1,000
-----------+---------------------------------+----------
     Total |      164      1,524      1,049 |    2,737

Model predicts polview1 correctly 43% of the time
```

THE MULTINOMIAL PROBIT REGRESSION MODEL

The multinomial probit model was developed many years ago, but it was difficult to estimate. Advances in statistical computing algorithms have reached the point where it is now feasible to estimate, however. Stata's mprobit model can be relatively slow to converge, though, especially with large data sets such as the GSS. The log-likelihood function used in maximum-likelihood estimation, similar to the log-likelihood function for the multinomial logistic model, requires summation not only across the sample space, but also across the j categories.

Example 5.6 shows the mprobit command and output.

Example 5.6

mprobit polview1 female age educate

```
Multinomial probit regression              Number of obs   =      2,737
                                           Wald chi2(6)    =      73.21
Log likelihood = -2928.9532                Prob > chi2     =     0.0000
```

polview1	Coef.	Std. Err.	z	P>|z|	[95% Conf. Interval]	
liberal						
female	.0766549	.0770243	1.00	0.320	-.0743099	.2276196
age	-.005332	.0023484	-2.27	0.023	-.0099348	-.0007292
educate	.0903159	.0135897	6.65	0.000	.0636806	.1169511
_cons	-1.34089	.2325244	-5.77	0.000	-1.796629	-.8851501
moderate	(base outcome)					
conservative						
female	-.1713265	.0724478	-2.36	0.018	-.3133214	-.0293315
age	.0024791	.0021712	1.14	0.254	-.0017764	.0067345
educate	.0602557	.0128755	4.68	0.000	.0350202	.0854912
_cons	-.8559933	.21982	-3.89	0.000	-1.286833	-.4251539

As with the probit results in the previous chapters, these coefficients are measured as differences in z-scores and can be transformed to probabilities rather easily. The prvalue command, the margins command, and the predict command offer considerable flexibility in terms of utilizing predicted probabilities from this model. For example, examine the predicted probabilities for males and females using prvalue:

Example 5.6 (Continued)

prvalue, x(female=0 age=mean educate=mean)

```
mprobit: Predictions for polview1

   Pr(y=liberal|x):    0.2304
   Pr(y=conserva|x):   0.4008
   Pr(y=moderate|x):   0.3688

        female       age    educate
  x=         0  44.683961  13.492145
```

```
prvalue, x(female=1 age=mean educate=mean)

mprobit: Predictions for polview1

    Pr(y=liberal|x):     0.2703
    Pr(y=conserva|x):    0.3414
    Pr(y=moderate|x):    0.3883

            female         age      educate
    x=           1   44.683961   13.492145
```

The probability of reporting conservative is expected to be about 0.06 points higher among males than among females.

The `margins` command is also useful for identifying the relevant probabilities. For example,

Example 5.6 (Continued)

```
margins, at(female=(0 1)) atmeans predict(outcome(1))
margins, at(female=(0 1)) atmeans predict(outcome(2))
margins, at(female=(0 1)) atmeans predict(outcome(3))
```

Similar to earlier presentations, the `marginsplot` command may then be used to graph the results.

EXAMINING THE ASSUMPTIONS OF MULTINOMIAL REGRESSION MODELS

As can be seen by looking at the help menu for `mlogit` and `mprobit`, there are a limited number of predicted values available for either model. Checking assumptions concerning heteroscedasticity and determining the potential effects of influential observations is therefore not a clear-cut exercise. Some statisticians recommend taking advantage of the fact that a multinomial regression model is similar to estimating a set of $j - 1$ binary regression models. Thus, we can take each implicit model and test whether it meets the assumptions.

Here is an example of how we might do this with the multiple multinomial logistic regression model estimated earlier in the chapter.

Example 5.7

1. Recode the *polview1* variable so it is a binary variable with 0 = moderate and 1 = liberal (conservative is missing).

```
recode polview1 (1=1)(2=0)(3=.), generate(libmod)
```

2. Estimate a binary logistic regression model with *female*, *age*, and *educate* as explanatory variables.

```
logit libmod female age educate, or
```

```
Logistic regression                          Number of obs   =      1,737
                                             LR chi2(3)      =      57.01
                                             Prob > chi2     =     0.0000
Log likelihood = -1141.0028                  Pseudo R2       =     0.0244
```

libmod	Odds Ratio	Std. Err.	z	P>|z|	[95% Conf. Interval]	
female	1.095387	.1105536	0.90	0.367	.8987909	1.334985
age	.9927611	.003086	-2.34	0.019	.986731	.9988281
educate	1.128462	.0206567	6.60	0.000	1.088694	1.169684
_cons	.1701367	.0524444	-5.75	0.000	.0929857	.3113006

3. Save the predicted values, residuals, and influential observation diagnostic values as we did with our binary regression models in Chapter 3.

```
predict dev, deviance
predict hat, hat
predict prob, pr
predict deltad, ddeviance
```

4. Compute the diagnostic tests we used with the binary logistic regression models. For example, to examine whether there are influential observations that should concern us, check the delta-deviance values versus the hat (leverage) values. The average hat value is 0.0067, so its threshold is 0.020.

```
twoway scatter deltad hat, xline(0.02) yline(4) jitter(5)
```

Figure 5.3 shows quite a few outliers in this model; there are also a few high leverage values. This might be a consequence of the large sample size, but, nonetheless, merits further investigation. One idea is to sort the data by the liberal-moderate variable and then look over the predicted values, the explanatory variables, and the influence statistics using commands such as

```
sort libmod
list libmod prob female age educate deltad hat if
    deltad>4 & deltad~=.                        * output not shown
```

The resulting printout (not shown) indicates some things that might be examined in greater detail. For example, most of the outliers are women who report being moderates. They also

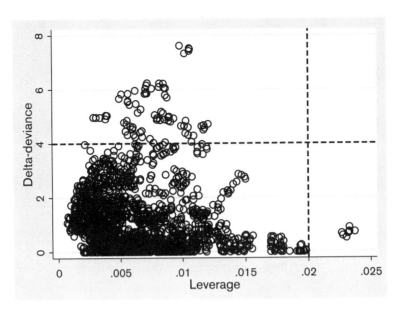

FIGURE 5.3

tend to be in their 30s and 40s and a plurality have 14 years of education. Looking over the predicted probabilities, many of these women have a substantial probability of reporting liberal, yet self-ascribe as moderate. The high leverage values, on the other hand, are all males, aged 21, with 12 years of formal education. (Note: trying sorting by different variables or combinations of variables to look for patterns.)

Another idea is to create a binary variable with 0 = delta-deviance values less than 4 and 1 = delta-deviance values greater than 4. Then, estimate a logistic regression model with this new variable as the outcome variable and *female*, *age*, and *educate* as the explanatory variables (see Example 5.8).

Example 5.8

```
recode deltad (min/3.99999=0)(4/max=1), generate(bidelta)
logit bidelta female age educate
```

```
Logistic regression                          Number of obs   =      1,737
                                             LR chi2(3)      =      11.71
                                             Prob > chi2     =     0.0084
Log likelihood = -303.18199                  Pseudo R2       =     0.0190
```

bidelta	Coef.	Std. Err.	z	P>|z|	[95% Conf. Interval]	
female	.5316479	.2557569	2.08	0.038	.0303737	1.032922
age	-.0183693	.008114	-2.26	0.024	-.0342725	-.0024662
educate	.0445054	.0434285	1.02	0.305	-.0406129	.1296236
_cons	-3.271287	.7354671	-4.45	0.000	-4.712776	-1.829798

It then becomes clear that age and female are involved in the influential observation issue. It is worth exploring these types of observations further to determine why they are relatively extreme on the political views variable.

However, if for some good reason (which is rare—this should be the *last option*) we omit the large delta-deviance values, the results do not change much. The education effect still exists: higher levels of education are associated with a higher odds or probability of reporting liberal rather than moderate. If we examine a binary logistic model that compares conservative and moderate, compute the delta-deviance values, and exclude those over 4, we find a similar effect: higher levels of education are associated with a higher odds of reporting conservative rather than moderate. Both of these results agree with the multinomial logistic model estimated earlier. Of course, there are additional assumptions to consider.

A useful user-written command that tests some aspects of the multinomial logit model is called `mlogtest`, which is part of the `spost9` set of commands written by Long and Freese (2006). For instance, after reestimating the original multinomial logistic model with *female*, *age*, and *educate* as explanatory variables consider Example 5.9.

Example 5.9

`mlogtest, all`   * *after Example 5.4*

The results show many types of tests. For our purposes, we might be interested in whether we should combine categories of the outcome variable. This is often the case when we violate an assumption known as *independence of irrelevant alternatives*. This assumption considers that two or more alternatives may not matter to people, whereas others might, so the choice among alternatives is distinct and their probabilities differ. For example, people may have similar preferences when it comes to buying blue or black socks, but quite different preferences when deciding whether to purchase white socks or colored socks. The `mlogtest` command provides several tests of this assumption (Long and Freese, 2006). One is known as the *Hausman–McFadden test*. It is designed to compare the coefficients from the multinomial model with the coefficients from binary models with each possible comparison (e.g., liberal vs. moderate). The null hypothesis is that the comparisons are independent of the others. The results of the test suggest that the comparisons are independent and therefore it appears that the multinomial model is preferable.

```
**** Hausman tests of IIA assumption (N=2737)

Ho: Odds(Outcome-J vs Outcome-K) are independent of other alternatives.

  Omitted |    chi2   df   P>chi2   evidence
----------+------------------------------------
  liberal |   0.654    4    0.957   for Ho
  conserva |  7.412    4    0.116   for Ho
----------+------------------------------------
```

A second test is simply a likelihood ratio χ^2 test that evaluates whether alternative choices can be combined in a probabilistic sense.

```
**** LR tests for combining alternatives (N=2737)

Ho: All coefficients except intercepts associated with a given pair
    of alternatives are 0 (i.e., alternatives can be collapsed).

Alternatives tested|   chi2   df   P>chi2
-------------------+-------------------------
   liberal-conserva|  29.841   3    0.000
   liberal-moderate|  56.995   3    0.000
   conserva-moderate|  28.364   3    0.000
-------------------+-------------------------
```

The null hypothesis is that the categories may be collapsed. For example, one question is whether we should combine conservative and moderates into one category and simply examine the difference between them and liberals based on sex, education, and age. However, the likelihood ratio test suggests this is not a viable alternative ($\chi^2 = 28.36$, $p < 0.001$), so they should remain separate. In general, the results of these tests indicate that combining the categories would not be a good idea in our model. There are large enough differences in the coefficients to suggest different predictive patterns.

FINAL WORDS

Multinomial logistic and probit models offer a good approach for many types of unordered outcome variables. These variables are commonly used in social and behavioral research and also offer a promising alternative when one analyzes an ordinal outcome variable and finds that the proportional odds assumption is not satisfied. Nevertheless, there are some issues that may temper the use of these multinomial regression models. For instance, we have already briefly touched on the issue of when alternative choices may have unequal probabilities (recall the sock selection scenario). Some analysts suggest that *conditional logistic models* be used in this situation.

Moreover, in some surveys respondents are allowed to choose more than one option (e.g., "choose up to three favorite flavors of ice cream from the following list of nine options"). In this type of situation, Agresti (2013) recommends that analysts use a *marginal logistic model*. Another situation involves when a set of choices is contingent upon an earlier choice. For example, when asking youth whether they use birth control during sexual activities, there are some youth who may not be sexually active so the issue of birth control is not, or may not be, relevant. In other words, those who answer questions about their use of particular methods of birth control are a (presumably) sexually active nonrandom subsample of all youth. Regression models should be adjusted for such sequential stages. Thus, *nested* or *sequential logistic models* are likely needed. The Stata command `nlogit` is useful in this situation. There are also user-written commands that may be appropriate.

EXERCISES FOR CHAPTER 5

1. The *drugtest* data set (*drugtest.dta*) contains information from more than 9,000 adult respondents who participated in the National Survey on Drug Use and Health. Use this data set to examine what types of workers are exposed to different types of drug testing programs. As a first step, construct a contingency table of the variables *drugtest* and *mjuser* (marijuana use in the past year). Then, consider the following exercises.

   a. Compute the probabilities of exposure to each type of testing program for those who report marijuana use in the past year and for those who report no marijuana use in the past year.

   b. Use the no testing group as the reference category. What are the odds of exposure to each of the three types of testing program for the two groups distinguished by *mjuser*?

2. Estimate a multinomial logistic regression model with *drugtest* as the outcome variable and *mjuser* as the explanatory variable. Use "no testing" as the reference category. Interpret the odds ratios from this model.

3. Extend the multinomial logistic regression model estimated in Exercise 2 by adding the following explanatory variables: *south, construct,* and *sales.* Based on this model

   a. Interpret the odds ratios associated with *mjuser, construct,* and *sales.*

   b. Compute and interpret predicted probabilities for each type of drug testing outcome for those who reported marijuana use and for those who reported no marijuana use. Use the following levels of the other explanatory variables: *south* = 0; *construct* = 1, *othwork* = 0, and *sales* = 0.

   c. Compute the deviance, Akaike's information criterion, Bayesian information criterion, accuracy, and PRE for the model. What do the accuracy and PRE suggest about the model?

4. Using the model estimated in Exercise 3, save the delta-deviance values and plot them against the predicted probabilities. What do this plot show? Based on what you have learned thus far about the regression model, discuss what additional steps you might take to improve the model.

## SIX

Poisson and Negative Binomial Regression Models

It is now time to leave the world of odds and probabilities and return to something a bit more tangible. There are many examples of variables in the social and behavioral sciences that are not precisely continuous but are also not precisely categorical. In fact, most "continuous" variables in the social and behavioral sciences are not, strictly speaking, continuous. Years of education, for instance, are not continuous because they consist of only a limited range of positive integers (or zero for a few). But we usually treat this sort of variable as continuous in statistical models.

Often these types of "in-between" variables include the count of some event: how many times a person has visited a physician; how many cases of the measles occur in a particular town or city; or how often hurricanes hit Miami. In sociology, criminology, or economics, we might be interested in how many firms succeed in a particular area of the economy; how many delinquent acts an adolescent commits in a year; how many lynchings occurred per county per year in the postbellum South; or how many times a person is arrested in his lifetime.

Fortunately, many of these types of outcomes occur frequently enough so that their distributions mimic closely a normal distribution. For example, figure 6.1 shows a simulated distribution of the count of monthly traffic accidents on I-15 in an area known as "point of the mountain" during the past 10 years (this location is notorious for its strong crosswinds and blowing snow during the winter months). If this is the distribution of a count variable and we have a relatively large sample, then a linear regression model is usually acceptable, assuming other assumptions are met (see Chapter 1).

Unfortunately, many count variables have distributions with means much closer to zero. Think about what happens as the mean of the distribution approaches zero. One consequence is that the distribution gets cut off at zero since—by most definitions—counts

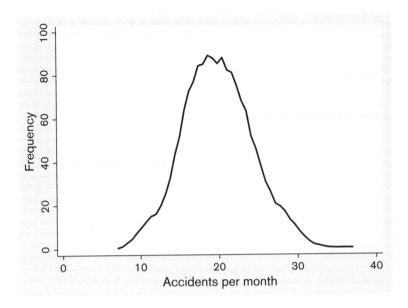

FIGURE 6.1

are nonnegative. For instance, I-15 cannot have −5 accidents in a month. Further, suppose we looked at the average number of accidents per month along a street in Springville, UT (population 31,205 in 2013) over a 10-year period. What would the distribution look like? Figure 6.2 suggests one possibility.

Notice that the normal curve does not fit the distribution in figure 6.2 very well, and not just because this is a bimodal distribution. A large chunk of the mass is located at zero and one accident per month, with relatively little at four or more. This type of distribution is commonly known as the *Poisson distribution*. It was originally developed by Simeon D. Poisson (1837) to study jury decisions in France, but became especially well known when it was used to examine the distribution of Prussian soldiers killed by mule kicks in the late 1800s (von Bortkiewicz, 1898).

The probability mass function (PMF) for the Poisson distribution was shown in Chapter 2, but just as a reminder it is

$$p(i) = e^{-\lambda} \frac{\lambda^i}{i!}.$$

This equation indicates that the probability of observing some count (*i*) is based on a single quantity, λ *(lambda)*, which is also known as the *rate* of the distribution.

Here is a concrete example: suppose the rate of traffic accidents—or the mean number of accidents per month—along the street in Springville is 0.8; then the probability of observing three accidents in the next month is

$$p(3) = e^{-0.8} \frac{0.8^3}{3!} = 0.038.$$

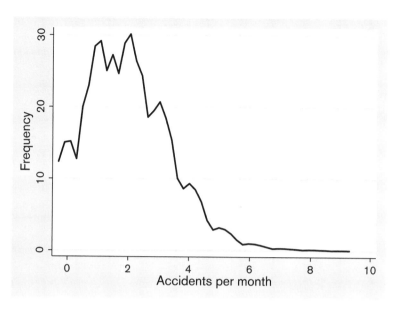

FIGURE 6.2

Hence, if we computed all the probabilities of traffic accidents from zero to infinity (there is no need to go this high!), they would sum to 1. Moreover, we can compute how many months we would expect to see three accidents over a 10-year period: 0.038 × 120 = 4.6 months.

There are two rather unique, yet interrelated, characteristics of the Poisson distribution. First, the Poisson distribution assumes that the mean equals the variance. Recall that in the normal distribution, the mean and variance may have a variety of relationships. The standard normal distribution has a mean of 0 and a standard deviation (and variance) of 1, whereas the binomial distribution has a mean of $\{n \times p\}$ and the variance is $\{n \times p \times (1 - p)\}$, where p is the probability of a "success" and n is the number of trials. Second, the events that make up the Poisson distribution are assumed to be independent. This is a restrictive assumption to make. For instance, the frequency of accidents along I-15 is probably not independent: one accident may set off other accidents. Similarly, a counting of arrests is not an independent matter: the fact that a person is arrested one time may make the likelihood of additional arrests higher (for some poor fellow who cannot control his unlawful habits) or lower (perhaps due to deterrence). One consequence of a lack of independence is that the variance often exceeds the mean for count variables. This situation is known as *overdispersion*. The converse, when the mean exceeds the variance, is known as *underdispersion*. Some researchers lump these two together under the term *extradispersion*. Another important aspect of Poisson variables is that they are concerned with the rate of events over time or over spatial units. In other words, time or space matters.

Suppose we are faced with a count variable. For example, in the *GSS* data set (*gss.dta*) there is a variable called *volteer*, which is a count of the number of volunteer activities respondents claim they were involved in during the past year. Example 6.1 furnishes a frequency distribution of *volteer*. It is apparent that the distribution is not normal. A substantial percentage of respondents reported no volunteer activities in the past year (82%). But does this variable follow a Poisson distribution? A quick way to check this is to look at its summary statistics (see Example 6.2).

Example 6.1

tabulate volteer

number of volunteer activities in past year	Freq.	Percent	Cum.
0	2,376	81.85	81.85
1	286	9.85	91.70
2	133	4.58	96.28
3	64	2.20	98.48
4	19	0.65	99.14
5	11	0.38	99.52
6	7	0.24	99.76
7	6	0.21	99.97
9	1	0.03	100.00
Total	2,903	100.00	

Example 6.2

summarize volteer

Variable	Obs	Mean	Std. Dev.	Min	Max
volteer	2,903	.3334482	.8858374	0	9

The mean is 0.33 and the standard deviation is 0.886; thus, its variance is $0.886^2 = 0.78$. Hence, the variable is overdispersed since its variance exceeds its mean.

Nevertheless, it is often useful to compare a distribution of rare events to a Poisson distribution with the same mean.

hist volteer, normal frequency   * *see figure 6.3*

Comparing figures 6.3 and 6.4, we can see that the *volteer* variable in the GSS data is much more spread out than a true Poisson variable with mean equal to 0.333. The maximum for the *volteer* variable is 9, whereas for a Poisson variable it is 4 (from what we can determine from figure 6.4). There are also proportionally more zeros in the *volteer* variable than predicted by a Poisson distribution.

Although a couple of alternative for overdispersed count variables are examined later in the chapter, we ignore this problem for now and examine a Poisson regression model. Like the

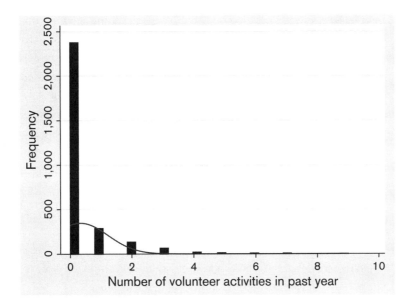

FIGURE 6.3

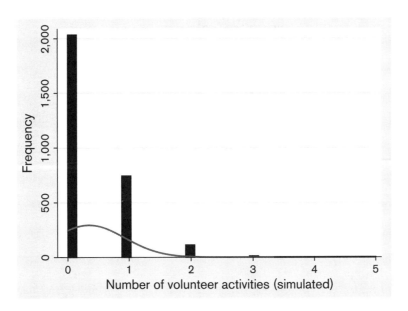

FIGURE 6.4

other regression models, we are interested in determining something about the outcome variable. In logistic models, we are interested in odds and odds ratios, as well as probabilities. In the Poisson model, we are back to something a bit more tangible: predicting the *expected count* for various groups identified by the explanatory variables. But since the Poisson model is a generalized linear model (GLM), we must choose a link function. Fortunately, it is a straightforward choice: the log link. When using a log link, the transformation back to the expected values is simply to estimate the expected log-number of events and then exponentiate (e^x) this quantity to get the predicted counts. The interpretation of coefficients is in terms of differences in expected log-counts or counts, which are typically referred to as rates or incidence rates.[1]

The log-likelihood used in maximum-likelihood (ML) estimation may be represented as

$$\ln L = -\sum_{i=1}^{N} \lambda x_i + \sum_{i=1}^{N} n_i \ln\left(\lambda x_i\right) - \sum_{i=1}^{N} \ln\left(n_i !\right).$$

It should be apparent now that the λ plays a key role in model estimation and, as discussed later, are an important part of the interpretations used for this model.

We begin with a model with one explanatory variable. The first hypothesis to test is that females are more involved than males in volunteer activities (see Oesterle et al., 2004). Before estimating a Poisson regression model, examine *volteer* by *female*. The `bysort` command may be used to request summary statistics for males and females (see Example 6.3).

Example 6.3

```
bysort female: summarize volteer
```

```
-> female = male
```

Variable	Obs	Mean	Std. Dev.	Min	Max
volteer	1,285	.3167315	.8492668	0	7

```
-> female = female
```

Variable	Obs	Mean	Std. Dev.	Min	Max
volteer	1,618	.3467244	.9138824	0	9

---

1. The definition of a rate depends on the discipline, but in epidemiology and demography—where rates are widely used—it refers to the frequency (or incidence) of an event over a specified unit of time, across a clearly defined spatial unit, or, even better, over time within a specific area. Epidemiologists tend to use the term *incidence rate* to clarify this. Thus, we may claim that there are 125 cases of influenza per year in Utah County; another way of saying this is that the incidence rate of influenza is 125 cases per year.

It appears that the mean number of volunteer activities per year among males is lower than among females, but at this point we do not know if this difference is statistically significant. A Poisson regression model helps us answer this question (see Example 6.4).

Example 6.4

poisson volteer female

```
Poisson regression                          Number of obs    =       2,903
                                            LR chi2(1)       =        1.94
                                            Prob > chi2      =      0.1637
Log likelihood = -2460.0677                 Pseudo R2        =      0.0004
```

volteer	Coef.	Std. Err.	z	P>|z|	[95% Conf. Interval]	
female	.0904756	.0651117	1.39	0.165	-.037141	.2180922
_cons	-1.149701	.0495682	-23.19	0.000	-1.246853	-1.052549

It seems that our hypothesis is incorrect. But we'll ignore the p-value for now and consider the model further. Can we reproduce the differences shown in the earlier summary? First, exponentiate the *female* coefficient to see what it gives us: exp(0.090) = 1.09. What does this number mean? Keeping in mind that we used a log-link, one awkward way to interpret this is to say that being a female is associated with an increase in the expected number of volunteer activities per year by a factor of 1.09 or about 9%. We can also use this information to come up with predicted volunteer activities for males and for females. Use the margins command to compute these.

Example 6.4 (Continued)

margins, at(female=(0 1))

```
Adjusted predictions                        Number of obs    =       2,903
Model VCE    : OIM

Expression   : Predicted number of events, predict()

1._at        : female          =            0

2._at        : female          =            1
```

	Margin	Delta-method Std. Err.	z	P>|z|	[95% Conf. Interval]	
_at						
1	.3167315	.0156998	20.17	0.000	.2859605	.3475026
2	.3467244	.0146387	23.69	0.000	.318033	.3754157

The margins output shows that the predicted number of activities for males is 0.317 and for females it is 0.347 (which, of course, are the mean numbers given by the summarize command). What is the ratio of these two rates? It is simply 0.347/0.317 = 1.09. This is what the Poisson regression model gives us. What is the percent difference between these two rates? It is about 9%. And, don't forget, a term used for these rates in the Poisson model is λ. Thus, we are predicting λ for groups represented by the explanatory variables.

THE MULTIPLE POISSON REGRESSION MODEL

As we have seen time and time again, this simple model is not very interesting. Therefore, we will extend the Poisson model to include a larger set of explanatory variables. Suppose we hypothesize that race/ethnicity, education, and personal income predict the number of volunteer activities. But leave the variable *female* in the model. Rather than generating the coefficients in log form, we ask Stata to exponentiate them for us. These are called *incidence rate ratios* (IRR) in Stata and are designated with the `irr` subcommand (see Example 6.5).

Note that the *female* coefficient is now larger and statistically distinct from 1. However, consider that the sample size is much smaller in this model than in the Poisson model estimated earlier (try to determine why this occurred). When the same sample is used in a Poisson model with *female* as the only explanatory variable, its IRR is 1.22 (check this yourself).

Example 6.5

```
poisson volteer female nonwhite educate income, irr
```

```
Poisson regression                              Number of obs   =      1,944
                                                LR chi2(4)      =     101.33
                                                Prob > chi2     =     0.0000
Log likelihood = -1685.4299                     Pseudo R2       =     0.0292
```

volteer	IRR	Std. Err.	z	P>|z|	[95% Conf. Interval]	
female	1.298645	.1010982	3.36	0.001	1.114873	1.51271
nonwhite	.7554988	.0818799	-2.59	0.010	.6109164	.9342987
educate	1.108272	.0159943	7.12	0.000	1.077363	1.140068
income	1.058473	.0165794	3.63	0.000	1.026472	1.091472
_cons	.0424979	.0104031	-12.90	0.000	.0263027	.0686649

The results shown in Example 6.5 suggest that females, whites, those with more years of education, or those with more income tend to volunteer more frequently than others. In particular, we may make the following interpretations:

- Adjusting for the effects of race/ethnicity, education, and income, females are expected to volunteer for about 30% more activities per year than males.

- Adjusting for the effects of sex, education, and income, nonwhites are expected to volunteer for 24% fewer activities per year than whites.

- Each 1-year increase in education is associated with an 11% increase in the number of volunteer activities per year, adjusting for the effects of sex, race/ethnicity, and income.

What are some other characteristics of this model? First, the model fit is suspect, with a pseudo-R^2 measure of about 0.03, a McFadden R^2 of about 0.03, and a PRE of only 2% (try to confirm each of these measures). Nonetheless, we will focus on the association of volunteering with the explanatory variables. In previous chapters, we computed adjusted odds and probabilities. However, we now have returned to an outcome variable with a much clearer scale: it involves counts of events. So it makes sense to consider expected counts for different groups. Another way to phrase this is to evaluate *expected or predicted rates* of events for different groups (or simply different λ). After estimating the model, we may use the `margins` command to see predicted rates for distinct groups represented by the explanatory variables (see Example 6.5 (Continued)).

Thus, the expected number of volunteer activities among females is $[(0.25 - 0.19)/0.19]$ × 100 = 31.5% higher than among males for white, high school graduates with low income (income = 5 is the 10th percentile of the income distribution). (Alternatively, $0.25/0.19$ = 1.315; {1.315 − 1.0} × 100 = 31.5%.)

Example 6.5 (Continued)

```
margins, at(female=(0 1) nonwhite=0 educate=12 income=5)
```

```
Adjusted predictions                      Number of obs    =      1,944
Model VCE    : OIM

Expression   : Predicted number of events, predict()

1._at        : female          =            0
               nonwhite        =            0
               educate         =           12
               income          =            5

2._at        : female          =            1
               nonwhite        =            0
               educate         =           12
               income          =            5
```

	Margin	Delta-method Std. Err.	z	P>|z|	[95% Conf. Interval]	
_at						
1	.1938754	.0211026	9.19	0.000	.1525151	.2352357
2	.2517754	.0239795	10.50	0.000	.2047764	.2987743

What does the `prvalue` command reveal? If we type only the command, the result is

Example 6.5 (Continued)

prvalue

poisson: Predictions for volteer

Confidence intervals by delta method

		95% Conf. Interval	
Rate:	.33491	[.3083,	.36152]
Pr(y=0|x):	0.7154	[0.6964,	0.7344]
Pr(y=1|x):	0.2396	[0.2269,	0.2523]
Pr(y=2|x):	0.0401	[0.0348,	0.0454]
Pr(y=3|x):	0.0045	[0.0035,	0.0054]
Pr(y=4|x):	0.0004	[0.0003,	0.0005]
Pr(y=5|x):	0.0000	[0.0000,	0.0000]
Pr(y=6|x):	0.0000	[0.0000,	0.0000]
Pr(y=7|x):	0.0000	[0.0000,	0.0000]
Pr(y=8|x):	0.0000	[0.0000,	0.0000]
Pr(y=9|x):	0.0000	[0.0000,	0.0000]

	female	nonwhite	educate	income
x=	.5	.1872428	13.865226	9.869856

It has returned the expected rate for those respondents at the mean of *female, nonwhite, educate*, and *income*. It has also provided the probability of each possible outcome for those at the mean of each variable. This makes little sense for *female* and *nonwhite*, however. It is best to specify values rather than let prvalue do it for you.

Example 6.5 (Continued)

prvalue, x(female=0 nonwhite=0 educate=mean income=mean)

poisson: Predictions for volteer

Confidence intervals by delta method

		95% Conf. Interval	
Rate:	.30973	[.27284,	.34662]
Pr(y=0|x):	0.7336	[0.7066,	0.7607]
Pr(y=1|x):	0.2272	[0.2086,	0.2459]
Pr(y=2|x):	0.0352	[0.0281,	0.0423]
Pr(y=3|x):	0.0036	[0.0025,	0.0048]
Pr(y=4|x):	0.0003	[0.0002,	0.0004]
Pr(y=5|x):	0.0000	[0.0000,	0.0000]
Pr(y=6|x):	0.0000	[0.0000,	0.0000]
Pr(y=7|x):	0.0000	[0.0000,	0.0000]
Pr(y=8|x):	0.0000	[0.0000,	0.0000]
Pr(y=9|x):	0.0000	[-0.0000,	0.0000]

	female	nonwhite	educate	income
x=	0	0	13.865226	9.869856

Thus, the expected rate for white males who are at average levels of education and income is 0.310 volunteer activities per year.

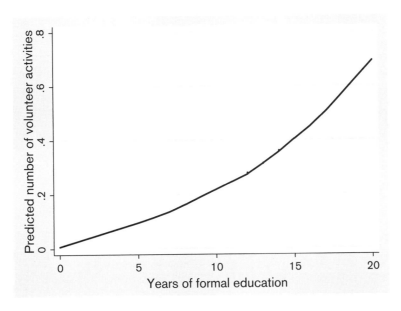

FIGURE 6.5

Another option is to predict the expected rates and then use them in graphs to demonstrate associations. For example, we may wish to predict rates and then look at their association with education.

```
predict rate, n                          * use n or Stata returns probabilities
twoway lowess rate educate
```

The lowess fit line in figure 6.5 is informative in this situation since the log-link results in a nonlinear association between the explanatory and outcome variables. As expected, there is a positive association between years of education and the number of volunteer activities that people report.

Recently, some scholars have argued that the Poisson model is a viable alternative for the log-normal model. Suppose we have a variable that is right-skewed, but with a natural log-transformation it looks normal. Rather than estimating an OLS regression model with a logged version of the variable (as attempted in Chapter 1), we may use a Poisson model as a substitute.[2]

EXAMINING ASSUMPTIONS OF THE POISSON MODEL

Similar to our approach in previous chapters, it is important to consider the assumptions of the Poisson regression model. Use Stata's help menu and search for poisson postestimation. Unfortunately, this shows some limitations. There are only a few post-estimation variables. However, recall that the Poisson regression model is a GLM, so we may use

---

2. See http://blog.stata.com/2011/08/22/use-poisson-rather-than-regress-tell-a-friend/.

Stata's *glm* command to estimate it. The `glm` command has several residuals and influence statistics available and we can predict them to examine the model's assumptions.

First, estimate the model using the `glm` command with a *poisson* family and a *log*-link (see Example 6.6). Next, predict some familiar statistics. (Note: if preferred, the `glm` sub-command `eform` returns exponentiated coefficients.)

Example 6.6

```
glm volteer female nonwhite educate income, family
    (poisson) link(log)
```

```
Generalized linear models                   No. of obs      =       1,944
Optimization     : ML                       Residual df     =       1,939
                                            Scale parameter =           1
Deviance        =   2465.513855             (1/df) Deviance =    1.271539
Pearson         =   4349.349483             (1/df) Pearson  =    2.243089

Variance function: V(u) = u                 [Poisson]
Link function    : g(u) = ln(u)             [Log]

                                            AIC             =    1.739125
Log likelihood   = -1685.429933             BIC             =   -12217.57
```

	Coef.	OIM Std. Err.	z	P>|z|	[95% Conf.	Interval]
female	.2613218	.077849	3.36	0.001	.1087407	.413903
nonwhite	-.2803773	.1083786	-2.59	0.010	-.4927954	-.0679592
educate	.1028018	.0144317	7.12	0.000	.0745161	.1310875
income	.0568278	.0156635	3.63	0.000	.0261278	.0875278
_cons	-3.158302	.2447916	-12.90	0.000	-3.638084	-2.678519

```
predict dev, deviance
predict hat, hat
predict cook, cooksd        * there is no clear threshold, so use 4/(n-k-1)
predict rate, mu            * note that here we use mu rather than n
```

As discussed in the literature on GLMs, deviance residuals and predicted values can be influenced substantially by various conditions, so some statisticians recommend computing adjusted deviance residuals with the following equation:

$$\text{adjusted dev}_i = \text{dev}_i + \frac{1}{6\sqrt{\hat{y}_i}}$$

In Stata we may use the following code to compute these residuals:

```
generate adjdev = dev + (1/(6*sqrt(rate)))
```

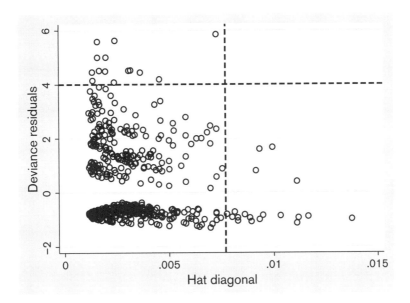

FIGURE 6.6

We may now use the predicted values and the influence statistics to assess the model. For example, look for influential observations using the deviance residuals and the leverage values. Leverage values above $3\bar{h}$ and deviance residuals above 4 are considered as influential.

```
twoway scatter dev hat, yline(4) xline(0.0077)
```

Figure 6.6 shows quite a few points that should attract our concern, including several large leverage values and a number of outliers. Consider also whether the adjusted deviance residuals follow a normal distribution. Given the large sample, we might be safe using the deviance residuals, but we can also examine how the adjusted ones behave.

```
kdensity adjdev, normal
```

It seems clear that we do not have normally distributed residuals; much of the mass is concentrated on the low end (see figure 6.7). Perhaps there is something awry with the Poisson regression model that we should address. One likely issue is that the *volteer* variable is overdispersed; its variance exceeds its mean. There may also be other intricacies that are addressed later. So we should examine some alternatives.

THE EXTRADISPERSED POISSON REGRESSION MODEL

Stata's `glm` command provides a way to estimate an extradispersed Poisson regression model. This approach allows there to be overdispersion or underdispersion, so it is quite flexible. This model is provided in Example 6.7.

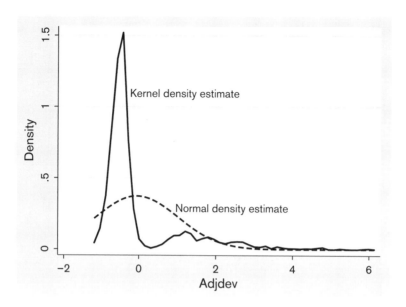

FIGURE 6.7

In the `glm` command we added `scale(dev)`, which asks Stata to scale the standard errors by a transformation of the deviance-based dispersion (1.27) (see the `glm` output). If we compare the standard errors to the previous model, we find that the standard errors from that model have been multiplied in the following manner:

$$\text{SE} \times \sqrt{1.2715}\,.$$

For example, in the previous model the standard error for the *female* coefficient was 0.0778. In this model it is 0.0778 × 1.12715 = 0.0877. It is relatively easy to see the effects of overdispersion (which a deviance-based dispersion greater than 1 indicates): the standard errors increase.

It is interesting to note that the model is not estimated by the ML approach we used to estimate most of the models up until this point. Instead, in the `glm` command we asked for iteratively reweighted least squares (IRLS), which is a form of quasi-maximum likelihood. This is a technical nuance that does not change the utility or the general interpretations of this Poisson model. For more information, see Hardin and Hilbe (2012).

Other than the larger standard errors, the overdispersed model may be used the same way as the normal Poisson model, with exponentiated coefficients interpreted as IRRs. Unfortunately, this model does not solve the problems of influential observations, outliers, or non-normal residuals (check this for yourself). We must therefore consider other alternatives.

Example 6.7

```
glm volteer female nonwhite educate income, family(poisson)
    link(log) scale(dev) irls
```

```
Generalized linear models                    No. of obs      =      1,944
Optimization       : MQL Fisher scoring      Residual df     =      1,939
                     (IRLS EIM)              Scale parameter =          1
Deviance        =    2465.513855            (1/df) Deviance =   1.271539
Pearson         =    4349.279397            (1/df) Pearson  =   2.243053

Variance function: V(u) = u                 [Poisson]
Link function    : g(u) = ln(u)             [Log]

                                            BIC             =  -12217.57
```

volteer	Coef.	EIM Std. Err.	z	P>|z|	[95% Conf. Interval]	
female	.2613218	.0877836	2.98	0.003	.0892691	.4333745
nonwhite	-.2803773	.1222097	-2.29	0.022	-.5199039	-.0408508
educate	.1028018	.0162734	6.32	0.000	.0709065	.1346972
income	.0568278	.0176625	3.22	0.001	.0222099	.0914456
_cons	-3.158302	.2760313	-11.44	0.000	-3.699313	-2.61729

(Standard errors scaled using square root of deviance-based dispersion.)

THE NEGATIVE BINOMIAL REGRESSION MODEL

Another alternative to the standard Poisson regression model when the outcome variable is overdispersed (but not underdispersed) is the negative binomial (NB) regression model. Recall that the NB distribution is concerned with how many trials we must run to get a particular number of successes, given we have an a priori probability of success with each trial. So, for example, we may wish to estimate how many people we must approach to get 10 signatures on a petition if the probability of success (p) for any randomly chosen person is 0.6. Recall from Chapter 2 that the PMF for the NB distribution is

$$p(i) = \binom{i-1}{r-1}(1-p)^{i-r}p^r,$$

where i equals the number of trials and r equals the number of successes we wish to obtain. The NB distribution may be considered a mixture of Poisson random variables, with these mixtures following a gamma (Γ) distribution. However, this complicates the log-likelihood used in NB regression, so it is not presented here (see Hilbe, 2014, Chapter 5, for details).

The NB distribution is particularly useful for overdispersed count variables. In fact, as mentioned in Chapter 2, the variance of the NB distribution typically exceeds the mean by a substantial amount. The results of our volunteer model using Stata's nbreg command are provided in Example 6.8.

Notice that we once again asked for IRRs; thus, the coefficients indicate the ratio of rates for two groups, such as male and female, or the expected increase in the rate for each one-unit increase in the explanatory variable, such as a 1-year increase in education.

The interpretations are thus similar to those in the Poisson model. For example,

·   Adjusting for the effects of sex, race/ethnicity, and income, each 1-year increase in education is associated with an 11.9% increase in the number of volunteer activities in the past year.

But also notice that, unlike the previous Poisson models, the *nonwhite* coefficient is not statistically significant at the $p < 0.05$ level. Thus, it appears that the NB model is the most conservative of the three we have examined thus far.

Example 6.8

```
nbreg volteer female nonwhite educate income, irr
```

```
Negative binomial regression                Number of obs    =      1,944
                                            LR chi2(4)       =      42.71
Dispersion       = mean                     Prob > chi2      =     0.0000
Log likelihood = -1419.7818                 Pseudo R2        =     0.0148
```

volteer	IRR	Std. Err.	z	P>|z|	[95% Conf. Interval]	
female	1.328975	.1648787	2.29	0.022	1.042109	1.694809
nonwhite	.7326605	.1188648	-1.92	0.055	.5330969	1.00693
educate	1.118508	.027145	4.61	0.000	1.06655	1.172996
income	1.053303	.0236049	2.32	0.020	1.00804	1.100599
_cons	.0388758	.0146518	-8.62	0.000	.0185726	.081374
/lnalpha	1.362715	.0979507			1.170735	1.554695
alpha	3.906786	.3826725			3.224362	4.733642

```
LR test of alpha=0: chibar2(01) = 531.30          Prob >= chibar2 = 0.000
```

An important aspect of the NB regression model is found in the rows labeled *lnalpha* and *alpha*. The former is simply the natural logarithm of the latter. If there is no overdispersion in the outcome variable, the alpha is expected to be zero. Stata provides a likelihood ratio test of alpha = 0. In this model, the null hypothesis of alpha = 0 is clearly rejected, with a *p*-value far smaller than any rule of thumb might indicate. Among the three models estimated thus far, it is probably safest and most accurate to go with the NB regression model than with the Poisson or overdispersed Poisson model. If there is still any remaining uncertainty, try estimating both the Poisson and NB models using the glm command and compare the AIC and

BIC values using `estat ic`. These are substantially lower for the NB model (although the PREs are very close—check for yourself). However, it is still important to consider the model diagnostics since, as we learned earlier, there may be problems with outliers, high leverage points, and the distribution of the residuals.

Before doing this, consider what the `margins` command provides. As with the other models estimated thus far, `margins` is handy for computing predicted rates and probabilities. For instance, the following shows the predicted rates for males and females at particular levels of the other explanatory variables.

Example 6.8 (Continued)

```
margins, at(female=(0 1) nonwhite=0 educate=12 income=5)
```

```
Adjusted predictions                          Number of obs    =       1,944
Model VCE    : OIM

Expression   : Predicted number of events, predict()

1._at        : female       =           0
               nonwhite     =           0
               educate      =          12
               income       =           5

2._at        : female       =           1
               nonwhite     =           0
               educate      =          12
               income       =           5
```

	Margin	Delta-method Std. Err.	z	P>|z|	[95% Conf. Interval]	
_at						
1	.1932472	.0298992	6.46	0.000	.1346458	.2518487
2	.2568208	.0346653	7.41	0.000	.1888781	.3247636

It is now helpful to compare these predicted probabilities to those from the Poisson model.

Poisson		Negative binomial	
Male	*Female*	*Male*	*Female*
0.193	0.252	0.193	0.257

They are surprisingly similar, which suggests that the overdispersion does not affect the sex differences much. However, if we examine the probabilities of the expected number of activities using `prvalue`, we can see some of the key differences between the Poisson and NB models.

Example 6.8 (Continued)

```
prvalue, x(female=0 nonwhite=0 educate=mean income=mean)

nbreg: Predictions for volteer

Confidence intervals by delta method

                                95% Conf. Interval
        Rate:              .30667   [ .25216,     .36118]
        Pr(y=0|x):         0.8174   [ 0.7972,     0.8377]
        Pr(y=1|x):         0.1140   [ 0.1076,     0.1204]
        Pr(y=2|x):         0.0390   [ 0.0337,     0.0444]
        Pr(y=3|x):         0.0160   [ 0.0125,     0.0195]
        Pr(y=4|x):         0.0071   [ 0.0050,     0.0092]
        Pr(y=5|x):         0.0033   [ 0.0020,     0.0045]
        Pr(y=6|x):         0.0016   [ 0.0008,     0.0023]
        Pr(y=7|x):         0.0008   [ 0.0004,     0.0012]
        Pr(y=8|x):         0.0004   [ 0.0001,     0.0006]
        Pr(y=9|x):         0.0002   [ 0.0001,     0.0003]

           female   nonwhite    educate      income
    x=           0          0  13.865226    9.869856
```

If we compare these to the Poisson results, it is clear that the probabilities differ mainly in the higher counts. For example, in the Poisson model the probability that the expected number of activities is 4 is 0.0003, whereas in the NB model it is 0.0071.

It is also useful to graph the predicted counts or probabilities. For instance, after the NB model, we may use the `predict` postcommand to get the expected counts (`predict count, n`). These may then be used in a graph. For example, consider figure 6.8. It shows four predicted counts, with white females appearing to have the highest average number of volunteer activities. However, we should also consider the coefficients and standard errors when making conclusions such as this one. Looking back at the NB model results, is this conclusion valid?

```
graph dot (mean) count, over(nonwhite) over(female)
```

CHECKING ASSUMPTIONS OF THE NEGATIVE BINOMIAL MODEL

As shown earlier, search for `help nbreg postestimation`. This indicates that there are a limited number of predicted values available. Once again, though, we can take advantage of the `glm` command and its numerous predicted values. To reestimate the NB model using it, type the code found in Example 6.9.

Example 6.9

```
qui glm volteer female nonwhite educate income,
    family(nbinomial) link(log)
```

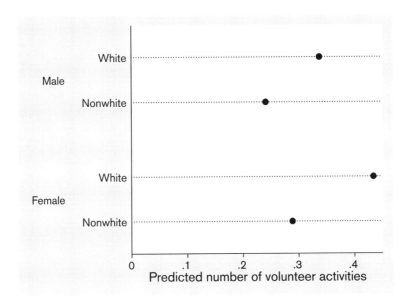

FIGURE 6.8

Next, ask for the same set of predicted values as we did following the Poisson model. These may then be used to test a few of the assumptions of the model.

Example 6.9 (Continued)

```
predict dev, deviance
predict hat, hat
predict cook, cooksd
predict rate, mu
generate adjdev = dev + (1/(6*sqrt(rate)))
twoway scatter dev hat, yline(4) xline(0.0077)
```

Figure 6.9 provides promising results. Even though there are still a number of large leverage values, there are no deviance residuals greater than 4 (although, depending on one's rule of thumb, they could still be problematic). Another graph is suggestive of heteroscedasticity (construct this using the appropriate predicted values). Asking Stata for robust standard errors may be appropriate in this situation.

ZERO-INFLATED COUNT MODELS

A common reason that count variables are overdispersed is because distinct probabilistic processes affect the shift from zero to one and from one to more frequent counts. One way to think about this is that, underlying many count variables, there are actually two

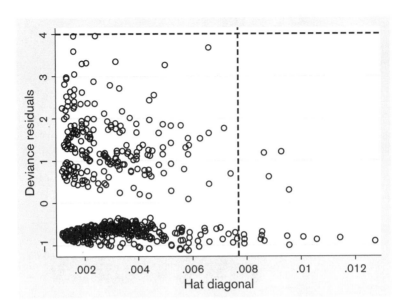

FIGURE 6.9

probability distributions: a binomial distribution that dictates whether there are any events (vs. no events) and another distribution (perhaps Poisson) that dictates the frequency of events occurring. To get a better handle on this, look at the distribution of volunteer activities for those members of the sample who volunteered at least once (see figure 6.10).

```
hist volteer if volteer>0, frequency normal
```

The distribution is clearly not normal; it is more like a Poisson or an extradispersed Poisson distribution. Nevertheless, this provides some important information. Moreover, conceptually speaking, it makes sense to consider that the factors that affect whether individuals volunteer at all may differ from those that affect how many activities they volunteered for in the past year. Examining summary statistics for *volteer* > 0 shows other interesting characteristics.

One classification of regression models that may be used when hypothesizing that there are two distinct distributions is known as *zero-inflated* models. A related model is called a *hurdle model*, since, in essence, we are "hurdling" over some point to reach another distribution (Zorn, 1998). Stata includes zero-inflated models that are designed for Poisson and NB distributions.

The Stata command for a zero-inflated Poisson model is called `zip`. There are two parts to a *zip* model: the inflation part that estimates the probability of moving from zero events to positive events and the Poisson part that estimates the frequency of events for those who are involved in any events. For example, if we hypothesize that education affects whether one volunteers at all, then the education variable should go into the inflation part of the model. But it may also remain in the frequency part of the model as well. This model is shown in Example 6.10.

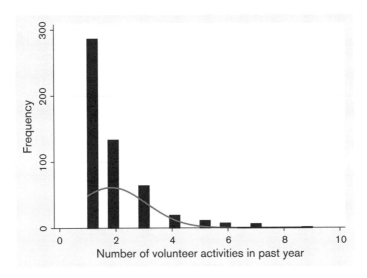

FIGURE 6.10

We may now see that there are two sets of coefficients representing the two parts of the model. Note, though, that the education coefficient in the second panel is not an IRR; rather, it remains logged. Moreover, it is a logistic coefficient, so it may be exponentiated to reveal an odds ratio. The main peculiarity to keep in mind is that it represents the odds of a zero outcome versus a positive outcome.

Example 6.10

```
zip volteer female nonwhite educate income, inflate
   (educate) irr
```

```
Zero-inflated Poisson regression              Number of obs   =      1,944
                                               Nonzero obs     =        376
                                               Zero obs        =      1,568

Inflation model = logit                        LR chi2(4)      =      23.79
Log likelihood   = -1434.377                   Prob > chi2     =     0.0001
```

volteer	IRR	Std. Err.	z	P>|z|	[95% Conf. Interval]	
volteer						
female	1.22903	.1161644	2.18	0.029	1.021196	1.479161
nonwhite	.8179197	.1099613	-1.50	0.135	.628456	1.064502
educate	1.058112	.0226514	2.64	0.008	1.014634	1.103452
income	1.050095	.019763	2.60	0.009	1.012066	1.089553
_cons	.3339494	.1162509	-3.15	0.002	.1687989	.6606809
inflate						
educate	-.0718015	.0276852	-2.59	0.010	-.1260635	-.0175396
_cons	2.011738	.4119309	4.88	0.000	1.204368	2.819108

What do these results suggest about the frequency of volunteer work? First, note that females, those with more education, or those reporting higher incomes tend to volunteer for more activities. For example, the expected number of activities among females is 23% higher than the expected number of activities among males, adjusting for the effects of race/ethnicity, education, and income. Moreover, the predicted odds of any volunteer work are lower for those with fewer years of education (OR = exp(−0.072) = 0.93). The odds ratio suggests that each 1-year increase in education is associated with a 7% increase in the odds of any volunteer work.

What does a zero-inflated NB regression model indicate (see Example 6.11)?

Example 6.11

```
zinb volteer female nonwhite educate income,
    inflate(educate) irr
```

```
Zero-inflated negative binomial regression      Number of obs   =      1,944
                                                Nonzero obs     =        376
                                                Zero obs        =      1,568

Inflation model = logit                         LR chi2(4)      =      18.52
Log likelihood  = -1416.365                     Prob > chi2     =     0.0010
```

volteer	IRR	Std. Err.	z	P>|z|	[95% Conf. Interval]	
volteer						
female	1.300111	.1526977	2.23	0.025	1.032778	1.636643
nonwhite	.7518705	.1173136	-1.83	0.068	.5537719	1.020834
educate	1.070543	.0355067	2.06	0.040	1.003165	1.142447
income	1.054342	.0227653	2.45	0.014	1.010653	1.099919
_cons	.1664952	.0911496	-3.27	0.001	.056938	.486857
inflate						
educate	-.0720809	.0468349	-1.54	0.124	-.1638756	.0197138
_cons	1.283221	.7143871	1.80	0.072	-.1169522	2.683394
/lnalpha	-.0505682	.4623769	-0.11	0.913	-.9568102	.8556739
alpha	.9506891	.4395767			.3841162	2.35296

The results are similar to those from the zip model, except for the standard errors and *p*-values, which provide conflicting information. The main difference is in the education effect: it is not statistically significant in the inflated portion of the NB model. However, also notice that the *p*-value for the log of alpha is not statistically significant. This suggests a lack of overdispersion when considering the zero-inflated nature of volunteer activities. In this situation, it is probably best to rely on the zero-inflated Poisson model. In fact, we may have presumed this would be an issue since, when examining summary statistics for those who had positive events (summarize volteer if volteer>0, detail), the variance

(1.56) was actually smaller than the mean (1.84). Hence, we have underdispersion and the NB regression model is inappropriate.

A statistical test that is recommended when estimating zero-inflated models is called a *Vuong test* (Vuong, 1989). It compares the zero-inflated model to the main model, such as the zip to the Poisson model, and provides a *z*-test. A statistically significant *z*-value suggests that the two models are different and, thus, the zero-inflated model is preferred (Loeys et al., 2012). Adding the word `vuong` as a subcommand with `zip` or `zinb` requests this test. For example, here are the Vuong tests for the two models.

Examples 6.10 and 6.11 (Continued)

```
Vuong test of zip vs. standard Poisson:
   z = 8.00 Pr>z = 0.0000
Vuong test of zinb vs. standard negative binomial:
   z = 1.35 Pr>z = 0.0887
```

These tests suggest that the zip model is preferred to the Poisson model, yet the zinb model is not preferred to the standard NB model. Placing `zip` at the end of the `zinb` command also provides an LR test comparing the models. The Vuong and LR tests support the conclusion that the zero-inflated Poisson model is preferred in this situation.

We next use the zero-inflated Poisson model and examine the predicted rates for males and females. Example 6.12 provides the predicted probabilities for males and females after reestimating the model in Example 6.10 (but note that the model command is modified by including *i.female* rather than *female* as an explanatory variable).

Example 6.12

```
margins female, at(nonwhite=0 educate=12 income=5)
   expression (exp(xb(volteer)))      * This provides exponentiated values
```

```
Adjusted predictions                        Number of obs    =    1,944
Model VCE      : OIM

Expression     : exp(xb(volteer))
at             : nonwhite         =          0
                 educate          =         12
                 income           =          5
```

		Delta-method				
	Margin	Std. Err.	z	P>|z|	[95% Conf.	Interval]
female						
male	.8398523	.1196723	7.02	0.000	.6052989	1.074406
female	1.032204	.1253621	8.23	0.000	.7864983	1.277909

We may then compare these to what the original Poisson model provided.

Poisson	
Male	*Female*
0.193	0.252

Zero-inflated Poisson	
Male	*Female*
0.840	1.032

The differences are quite striking because we now are examining the expected rate for those who participated in any volunteer work, rather than for everyone in the sample who answered the volunteer question. This seems to be a more realistic comparison.

The `prvalue` command also provides some interesting results.

Example 6.12 (Continued)

```
prvalue, x(female=0 nonwhite=0 educate=mean income=mean)

zip: Predictions for volteer

  Expected y:           .06776
  Pr(Always0|z):        0.8782
  Pr(y=0|x,z):          0.9480
  Pr(y=1|x):            0.0388
  Pr(y=2|x):            0.0108
  Pr(y=3|x):            0.0020
  Pr(y=4|x):            0.0003
  Pr(y=5|x):            0.0000
  Pr(y=6|x):            0.0000
  Pr(y=7|x):            0.0000
  Pr(y=8|x):            0.0000
  Pr(y=9|x):            0.0000

x values for count equation

          female    nonwhite     educate       income
x=             0           0   13.865226     9.869856

z values for binary equation

        educate
z=          .5
```

These predicted probabilities show that we expect about 87.8% of adult males to never volunteer, whereas we expect about 3.9% of males to volunteer for one activity. Among women (results not shown), about 4.2% are expected to volunteer for one activity. Note that the predicted probabilities are for white men and women at the means of education and income. We might also vary these values to examine the predicted probabilities for an assortment of groups. For instance, figure 6.11 provides predicted probabilities for men and women at various levels of education.

```
qui margins female, at(nonwhite=0 educate=(10 12 14 16 18)
   income=5) expression (exp(xb(volteer)))
marginsplot, noci
```

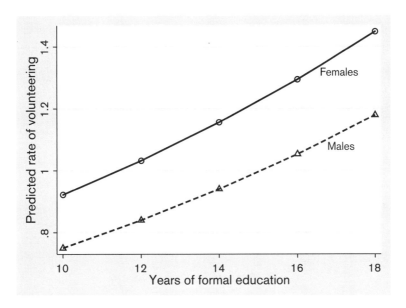

FIGURE 6.11

TESTING ASSUMPTIONS OF ZERO-INFLATED MODELS

There are few predicted values available following Stata's zero-inflated models. Although there is little guidance on this in the literature, we may wish to compute residuals by subtracting the actual count from the predicted count. Taking the z-scores of these residuals using the `egen` command provides standardized residuals (see Example 6.13).

Example 6.13

```
qui zip volteer female nonwhite educate income,
   inflate(educate) irr                    * based on example 6.9
predict rate, n                            * we are back to specifying n
generate residuals = volteer - rate
egen zresid = std(residuals)
```

Follow these steps by constructing a residual-by-predicted scatterplot. It may be best to show this graph only for those who have positive *volteer* values.

```
twoway scatter zresid rate if volteer>0, jitter(5)
```

The results shown in figure 6.12 provide little evidence of heteroscedasticity, but there are a substantial number of outliers shown (notice the large number of standardized residuals above 4, for example). Since this is a long-tailed count variable, having so many large values is probably not surprising. A problem with this approach is that the residuals are heteroscedastic and asymmetric, so other types are needed. It is possible to compute deviance

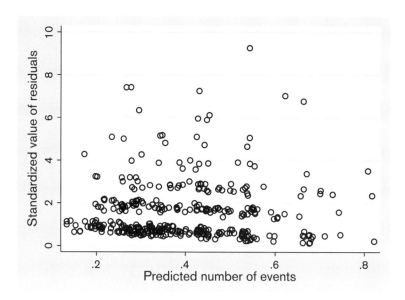

FIGURE 6.12

residuals and adjusted deviance residuals for this model, although we would have to calculate them using Stata's `generate` command.

A simpler alternative—at least in terms of computation—is to estimate *Anscombe residuals*. These tend to be closest to normally distributed residuals in Poisson models and are homoscedastic. Using the *volteer* values and the predicted rates from the previous model, Anscombe residuals may be computed in Stata using the following code (they are also directly available after `glm`):

```
gen anscombe = [1.5 * ((volteer^(2/3)) -
   (rate^(2/3)))] / (rate^(1/6))
```
                                                    * after Example 6.10

Figure 6.13 provides a scatterplot of the Anscombe residuals and the predicted number of events. It shows little evidence of heteroscedasticity, but there remain a number of outliers that still reflect the long tail of the Poisson distribution.

As a final step, look at the distribution of the Anscombe residuals for those who reported any volunteer activities in the past year. Figure 6.14 indicates that the residuals have a positive skew, but it is not as severe as others we have seen recently. Given this information, the zero-inflated Poisson model appears to be the most appropriate model we have estimated thus far. And this modeling exercise supports the notion that two distinct probability distributions underlie the rate of volunteering among adults in the United States.

```
twoway scatter anscombe rate if volteer>0, jitter(5)
kdensity anscombe if volteer>0
```

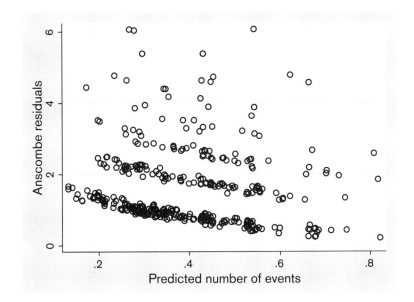

FIGURE 6.13

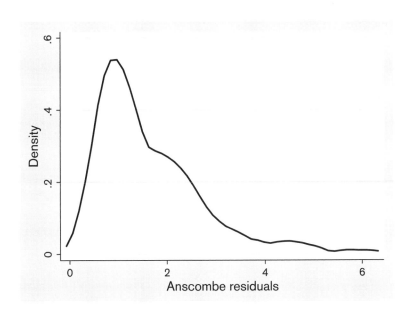

FIGURE 6.14

FINAL WORDS

This chapter has discussed a useful set of regression models appropriate for count variables. Count variables are found throughout the research community, from health studies that examine symptoms of illness to astronomical studies of black holes. Poisson, extradispersed Poisson, negative binomial, and zero-inflated models are especially useful when one is faced with count variables, especially those that measure rare events. They are also useful when estimating models for rates of various outcomes in populations, such as the rate of disease or crime per 100,000 population across cities, counties, states, or other geographic units.

Although these models have become used more frequently in recent years, there has not been sufficient attention to examining their assumptions or other nuances. Just as with the other models presented in previous chapters, count models should be evaluated for independence of the error terms, influential observations, and other characteristics. Moreover, the type of model that is appropriate, such as a regular or a zero-inflated Poisson model, is both a conceptual and an empirical question that must be addressed carefully.

EXERCISES FOR CHAPTER 6

1. Consider the *stress* data set (*stress.dta*). It includes information from about 650 adolescents in the United States who were asked various questions about the number of stressful life events (stress) they had experienced over the previous year. They were also asked about several other issues.
   a. Construct a histogram and estimate summary statistics for the variable *stress*. What do you think is its most likely probability distribution? What evidence suggests this?
   b. Plot a normal probability (Q–Q) plot of *stress*. Comment on its departure from a normal distribution.

2. Estimate a Poisson regression model and a negative binomial regression model, both of which use *stress* as the outcome variable and the following explanatory variables: *cohes*, *esteem*, *grades*, and *sattach*.
   a. Interpret the coefficients from both models associated with the variables *cohes* and *sattach*. Use the percentage change formula to interpret these coefficients.
   b. What are the predicted levels of stress from this model when family cohesion is at the upper quartile versus when family cohesion is at the lower quartile and the other explanatory variables are set at their means? You may use either model to estimate these predicted levels.
   c. Create a graph (your choice of the type) that predicts the rate of stress for those who report low, medium, and high levels of cohesion (at the means of the other variables). What does the graph indicate?

3. Compare the Poisson and negative binomial models estimated in exercise 2. Which model do you prefer and why? Provide details about how you compared the two models.

4. Estimate a zero-inflated Poisson model and a zero-inflated negative binomial model, each of which uses *stress* as the outcome variable, *esteem*, *grades*, and *cohes* as the explanatory variables, and *cohes* as the inflation variable. Request Vuong tests. Which model do you prefer in this situation? Why?

## SEVEN

Event History Models

Thus far, we have been concerned with what might be called "static" variables. These are variables that measure some current condition: scores on a life satisfaction scale, attitudes toward corporal punishment, political ideology, or the number of volunteer activities a person has been involved in over the past year. Whereas these types of variables seem to exhaust the possibilities, there is another type of variable that is hinted at by the use of the term "static." These are variables that explicitly address some time dimension. But what do we do with measures of activities or behaviors that have occurred in the past year? Although these might be thought of as having a time dimension (e.g., number of volunteer activities in the past year), we still tend to measure them by treating time so that it has, at best, only a tangential role in our models.

There are different, though often related, types of models that explicitly address a time dimension, however. One type involves models that use change scores or estimate change in some statistical way. Often, researchers simply use a linear regression model and examine the impact of time 1 variables on changes from time 1 to time 2 for some outcome variable. It is not uncommon for regression models of the following form to be estimated:

$$y_t = \alpha + \beta_1 y_{t-1} + \beta_2 x_1 + \cdots + \beta_k x_{k-1}.$$

For example, we might be interested in determining the impact of stressful life events on changes in adolescent depression from one month to the next. Most books on structural equation models have some discussion of this type of analysis. There are also books and articles on longitudinal data analysis that discuss how to use change scores in regression models and some of the assumptions required to properly estimate them (e.g., Finkel, 2004;

Skrondal and Rabe-Hesketh, 2004). In the next chapter, we will become familiar with several models appropriate for longitudinal data, or data that have been collected repeatedly over time from a sample of individuals or from other units of observation (e.g., a set of companies). Some models that are not covered in this book, however, are designed to predict some value of a variable based on its past values and other variables, but the data are usually limited to one or a small number of units. These are known as *time-series models* when aggregate data are used. Examples of time-series models include examining the dynamics of stock markets, a city's crime rates, or a nation's unemployment rates over time (Shumway and Stoffer, 2010).

EVENT HISTORY MODELS

A second general type of model concerns the timing of events. For instance, medical researchers are often interested in the effects of certain drugs on the timing of recovery among a sample of patients. In fact, these statistical models are most commonly known as *survival models* because biostatisticians, epidemiologists, and other medical researchers often use them to study the time between the diagnosis of a disease and death, or between therapeutic intervention and death or recovery.

Social and behavioral scientists have adapted these types of models for a variety of purposes. Under the term *event history analysis* or *duration analysis*, sociologists, political scientists, and economists are interested in issues such as the following: among those who have divorced, what is their average length of marriage? What is the average length of a war (from declaration to cessation) and can we predict this based on some set of explanatory variables? What is the average length of one's first job? Does this affect the average length of one's second job or the probability that one changes jobs in the future?

Event history models have been developed to explicitly address these and similar questions. They are a particular type of longitudinal analysis that focuses on predicting when or how long until an event, or multiple events, occurs (Allison, 1995). Thus, the outcome variable in an event history model quantifies the duration of time one spends in a state before some episode takes place (Box-Steffensmeier and Jones, 2004). Events include transitions such as those mentioned earlier (e.g., job change) as well as death, birth, heart transplant, marriage, completing graduate school, or one's first delinquent act. Many events, birth and death being notable exceptions, can occur more than once. A person can commit delinquent acts continuously in each month; a person might get married several times; or a person may have several jobs over a 40-year employment career.

A key for event history models is identifying when the event occurred. Sometimes one can be very specific (Samuel died on September 14 at 5:16 p.m.), although often we have to place the event into a particular time interval (Sally had her first baby in 2007). This distinction is important because it gives rise to two types of event history models: *continuous-time models* and *discrete-time models*. In continuous-time models the researcher assumes that the event can occur at any time, but, as with continuous variables, we often have to approximate this. Discrete-time models assume that the event can occur only within distinct time units.

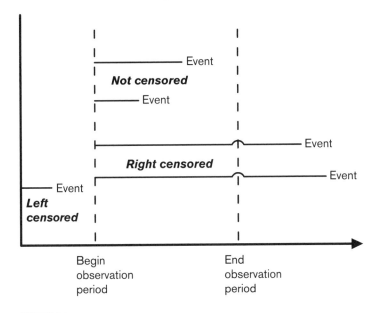

Censoring of Observations

FIGURE 7.1

Practically speaking, discrete-time models are often used to approximate continuous-time models, such as if we wished to model what predicted the length of time until one's first marriage, but we had only annual data on a set of individuals (see Yamaguchi, 1991, Chapters 2 and 3). We return to this issue later.

But why not simply create a continuous variable that measures the time until some event occurs (e.g., 5 months passed between the time Jessie was married and her first child was born) and then plug this variable into a linear regression model? Even if this time variable is not normally distributed (typically it is not), we might simply transform it and then estimate an ordinary least squares or maximum-likelihood linear regression model.

There are two impediments to this approach. First, what happens if, in a sample of first births, we find that 30% of respondents had not had a child by the time we complete the study. What should we do? Drop them from the analysis? This is unwise since it results in a biased sample and ultimately biases any parameter estimates from the model. The general issue of sample members who have not experienced the event during the "observation" period is known as *censoring*. Censoring includes *right censoring*, such as the situation just described. Right censoring also involves subjects who drop out of a study (due to death or utter exhaustion from answering all of our questions). *Left censoring* includes situations when sample members have already experienced the event of interest (e.g., first child born) before the observation period begins. The final type of censoring is *interval censoring*: the sample member experienced the event, but we cannot determine precisely when it occurred. Figure 7.1 illustrates right and left censoring.

A second impediment concerns the variables we wish to use as explanatory variables. Assume we have monthly measures from a sample of married couples and wish to predict the length of time from marriage to first child. Ignoring for now the issue that some may not have had a child during the observation period, suppose the monthly measures included questions about the use of contraceptive devices. This variable may very well change over time. It is known as a *time-varying* or *time-dependent* covariate. If we wished to use it in a linear regression model, it would be virtually impossible to use it in a typical way, even though it is certainly reasonable to argue that contraceptive use affects the timing of the birth of one's first child.

Rather than relying on a linear regression model, a preferred approach is to use a statistical approach that is designed explicitly to deal with these issues. This is where event history models become useful. There are actually a variety of event history or survival models from which to choose. There are, as mentioned earlier, discrete- and continuous-time models. At a more finely tuned level, there are models known as univariate, exponential, Weibull, Gompertz, gamma, competing risks, proportional hazards, as well as several other types. We do not have the space to cover all of these (there are entire books devoted to them; e.g., Blossfeld et al., 2009; Box-Steffensmeier and Jones, 2004), so we focus on only a few. It is important to note that several of these models are generalized linear models (GLMs) since they involve either a transformation to normality via a link function or, in the case of discrete-time models, we may use logistic, probit, or multinomial regression to estimate them. The key distinction is the way the data are set up; the estimation and interpretation of these models are similar to what we learned earlier.

Before seeing how event history models are estimated and interpreted, it is essential that we explore two important functions: *survivor* and *hazard functions*. It is best to learn about these through an example. Consider a variable that gauges the time until a sample of people marry. Imagine that this variable is continuously distributed over some time interval, such as ages 18–70. The survivor function ($S(t)$) gives the probability of "surviving" (a loose term) beyond some specific time point. For example, what is the probability that a person "survives" until age 30 and does not get married? Survivor functions are rarely normally distributed. Rather, they typically have a variety of shapes depending on the outcome and the population.

The hazard function is the instantaneous probability of an event at some specific time point, denoted as t, divided by the probability of not having experienced the event prior to that specific time point. In other words, it is the probability of the event divided by the survivor function. Another way of viewing the hazard function is that it measures the conditional probability of experiencing an event given that one has not experienced the event before this time point. More formally, this relationship may be expressed as

$$h(t) = \frac{f(t)}{S(t)} = \lim_{\Delta t \to 0} \frac{P(t \le T < \Delta t \mid T \ge t)}{\Delta t}.$$

In this expression, $f(t)$ is the probability density function (PDF) of the times to the event and $S(t)$ is the survivor function.[1] The right-hand side of the expression shows that the hazard is a conditional measure. Hence, $h(t)$ is often called the *conditional failure rate*. Less formally, the *hazard rate* may be thought of as the expected number of events per unit of time (or within some small interval of time; note the limit in the right side of the expression of $h(t)$):

$$\text{Average hazard rate} = \frac{\text{number of "failures"}}{\text{sum of observed survival times}}.$$

In this respect, the hazard is very similar to the rate (λ) in the Poisson distribution. The primary difference is that the hazard is conditional upon the person "surviving" until the particular time point (e.g., age 30). A key similarity, though, is that both are interpreted based on some measure or interval of time. For instance, someone may have a predicted count of 1.2 delinquent acts per year; similarly, if the hazard of committing delinquent acts from ages 16–17 is 1.2, then a 16-year-old is expected to commit 1.2 delinquent acts per year during that period of time. If we switched the interval to months, both the rate and the hazard would also change. Returning to our marriage example, we may compute the hazard of marriage for those between the ages of 25 and 30 by examining the conditional probability of marriage among those who have not married by age 25 as we follow them through age 30.

EXAMPLE OF SURVIVOR AND HAZARD FUNCTIONS

It is helpful at this point to examine what survivor and hazard functions look like both numerically and graphically. Perhaps the most common graphical approach for estimating survival curves is with the *Kaplan–Meier (KM) method*. The simplest way to understand the KM method is with data that have no censored observations. When there is no censoring, KM provides the percent of observations with event times greater than some time point, t. In order to see this, consider the following random subsample of data from the National Survey of Family Growth. The *event1* data set (*event1.dta*) limits the sample to only those who married during the observation period. A few observations from the data file are provided in Example 7.1.

---

1. Chapter 2 introduced the notion of probability mass functions for discrete random variables assumed to follow particular distributions (e.g., binomial, Poisson). It also briefly described the PDF. Since time-until-an-event is assumed to follow a continuous distribution, we use the PDF in this situation.

Example 7.1

```
list marriage attend14 cohab educate nonwhite in 10/20
```

	marriage	attend14	cohab	educate	nonwhite
10.	30	1	1	1	1
11.	12	1	0	1	0
12.	1064	1	0	1	0
13.	94	0	0	1	1
14.	1	0	0	1	0
15.	39	0	0	1	0
16.	72	1	1	1	1
17.	39	1	1	1	0
18.	62	1	1	1	0
19.	61	0	0	1	1
20.	117	1	0	1	0

The *marriage* variable shows the length of time in days from when respondents moved out of their parents' home until they married.

In order to use Stata with event history data, we must first tell the program that there is a time dimension by using the stset command. Since the variable *marriage* gauges the "time" variable, we use it in the command structure provided in Example 7.2.

Example 7.2

```
stset marriage
```

```
       failure event:  (assumed to fail at time=marriage)
obs. time interval:  (0, marriage]
 exit on or before:  failure
```

```
      59  total observations
       0  exclusions

      59  observations remaining, representing
      59  failures in single-record/single-failure data
    5635  total analysis time at risk and under observation
                                    at risk from t =          0
                           earliest observed entry t =          0
                                last observed exit t =       1071
```

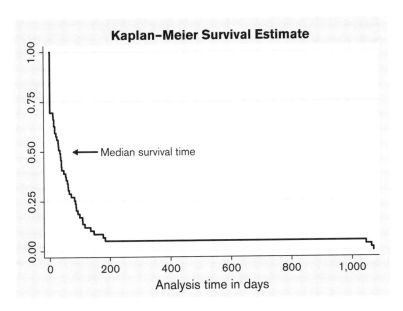

Kaplan–Meier Survival Estimate

← Median survival time

Analysis time in days

FIGURE 7.2

We can see the assumptions that Stata makes. It notices that every observation "failed." In other words, there is no censoring in these data because everyone had married by the end of the observation period. Taking the sum of the marriage variable derives the total analysis time (5,635 days). The Stata printout also shows that the minimum time until marriage is 0 and the maximum time is 1,071 days.

We may examine the estimated survivor distribution function graphically with the KM method by using the following command:

```
sts graph
```

The graphical depiction of the survivor function in figure 7.2 shows that almost all of the sample members married within about 200 days of leaving their parents' homes. Of particular interest in event history models is the median "failure" time, or, in this example, the time point by which 50% of the sample had married. In this graph it appears to be about 100 days, although it is difficult to tell precisely when it occurred.

To find the precise median, as well as other statistics, we may use the commands sts list and stsum. Be careful, though: with large data sets Stata provides a lot of printout when using the first command. Example 7.3 provides an abbreviated version of the output from the sts list command.

Example 7.3

sts list   * *most of the printout is omitted*

```
        failure _d:  1 (meaning all fail)
   analysis time _t:  marriage
```

Time	Beg. Total	Fail	Net Lost	Survivor Function	Std. Error	[95% Conf. Int.]	
1	59	18	0	0.6949	0.0599	0.5604	0.7955
11	41	1	0	0.6780	0.0608	0.5428	0.7810
12	40	1	0	0.6610	0.0616	0.5254	0.7663
14	39	2	0	0.6271	0.0630	0.4910	0.7363
17	37	2	0	0.5932	0.0640	0.4572	0.7058
21	35	1	0	0.5763	0.0643	0.4405	0.6903
24	34	1	0	0.5593	0.0646	0.4240	0.6747
28	33	1	0	0.5424	0.0649	0.4076	0.6590
29	32	1	0	0.5254	0.0650	0.3914	0.6431
30	31	1	0	0.5085	0.0651	0.3753	0.6270
34	30	1	0	0.4915	0.0651	0.3593	0.6109
36	29	1	0	0.4746	0.0650	0.3435	0.5946
37	28	1	0	0.4576	0.0649	0.3278	0.5782
39	27	2	0	0.4237	0.0643	0.2969	0.5449
40	25	1	0	0.4068	0.0640	0.2817	0.5280
47	24	1	0	0.3898	0.0635	0.2666	0.5111
53	23	1	0	0.3729	0.0630	0.2517	0.4939
55	22	1	0	0.3559	0.0623	0.2369	0.4767

Notice that there are many sample members who "failed" at time 1, or within 1 day of leaving their parents' homes they were married. The median survival time for this subset of the data may be found by looking at the column labeled *Survivor Function*. The 50% point occurs at about 34 days. We may also use this column to make some interpretations, such as "The estimated probability that a sample member does not get married for at least 21 days is 0.58." It should be clear that this means that about 58% of respondents were not married within 21 days of leaving their parents' homes.

The stsum command also provides important information (see Example 7.4).

Example 7.4

stsum

```
        failure _d:  1 (meaning all fail)
   analysis time _t:  marriage
```

	time at risk	incidence rate	no. of subjects	Survival time 25%	50%	75%
total	5635	.0104703	59	1	34	86

The estimate of the median in this output is precise: 34 days. The *incidence rate* shown in this printout is the hazard rate, or the "rate" of marriage per one unit of time (in days). The

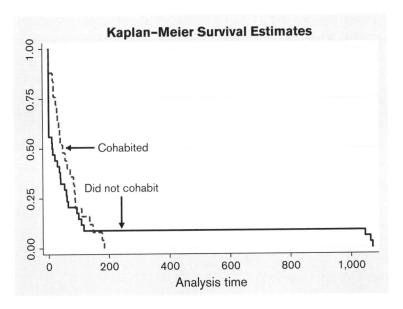

FIGURE 7.3

reciprocal of this number is 95.5, so the average sample member is expected to "survive" about 96 days before getting married. The mean length of time and the median length of time until marriage differ substantially. This occurs in many situations because survival curves, and the lengths of time they measure, are rarely distributed normally. Given the long tail of the marriage curve, the median is preferred as a measure of central tendency in this situation.

So far, this is not very interesting. In the next step, we consider differences by potential explanatory variables. For instance, we might suspect that cohabitation before marriage affects the length of time until marriage. In Stata, we may ask for separate survivor functions (see figure 7.3).

```
sts graph, by(cohab)
```

Notice that all the cohabiters had married by day 200, whereas a group of non-cohabiters remained unmarried for a substantial length of time. More importantly, though, the median length of time suggests that cohabiters "survived," on average, slightly longer than non-cohabiters. It is difficult to tell how much higher the median is for cohabiters, though. It may also be important to determine whether the difference is statistically significant.

There are two basic tests for comparing survival curves: the log-rank test and the Wilcoxon test. Both are based on ranks, so they are nonparametric. The key difference is that the Wilcoxon test is less sensitive to differences that occur later in time, as we see in figure 7.3.

Both tests are implemented using Stata's `sts test` command, although the Wilcoxon test seems best suited for our data.

Example 7.5

```
sts test cohab, wilcoxon
```

```
            failure _d:  1 (meaning all fail)
      analysis time _t:  marriage
```

```
     Wilcoxon (Breslow) test for equality of survivor functions
```

cohab	Events observed	Events expected	Sum of ranks
0	34	31.54	267
1	25	27.46	-267
Total	59	59.00	0

```
            chi2(1) =      4.27
            Pr>chi2 =    0.0389
```

It appears that there is a statistically significant difference in the time until marriage. And, whereas we expect about 27.5 of the cohabiters to marry by the overall median days, only 25 had done so. Combine this information with the distributions shown in figure 7.3 and it should be clear that cohabiters last a bit longer than the others before marriage occurs. It is also helpful to consider output from the following command: `stsum, by(cohab)`. What are the median survival times for cohabiters and non-cohabiters?

CONTINUOUS-TIME EVENT HISTORY MODELS WITH CENSORED DATA

What we have done so far is unrealistic. The original data include substantial proportion of the respondents who did not marry during the observation period; they were right censored. In addition, we have not considered any potential explanatory variables besides cohabitation. In this section and the next, we extend the model by (a) looking at the full sample and (b) using event history models to estimate the hazard of or survival time until marriage.

There are numerous event history models from which to choose. However, we limit our examination to four models that are designed for continuous-time data: the lognormal, the exponential, the Weibull, and Cox proportional hazards (PH) regression. The first three are parametric models since they make distinct assumptions about the shape of the error terms.

Before considering each of these models, we should discuss how log-likelihood functions might be developed for parametric models. In general, we first must define the *cumulative hazard function*; it integrates the hazard rates over a time period:

$$\Lambda(t) = \int_0^t h(t)dt = -\ln(S(t)).$$

Generally speaking, this is as if we summed each of them from when observation begins until time t (what role does censoring play?). Consider also that this function has a specific relationship to the survivor function. To develop a log-likelihood function that includes censoring we must consider those who experience the event during the observation period and those who do not. The former contributes information to the hazard, whereas the latter does not. Yet both must be in the function:

$$\ln L = \sum_{i=1}^{N} \left(v_i \log(\lambda(t_i)) - \Lambda(t_i) \right).$$

The v_i denotes whether the event occurred during the observation period (0 = no, 1 = yes), $\lambda(t_i)$ is the hazard at time t, and $\Lambda(t_i)$ represents the cumulative hazard function. The term λ serves a similar purpose here as it does in Poisson regression: it gauges the (hazard) rate of events occurring during a fixed unit of time. Note that when the event does not occur ($v_i = 0$), we are left to sum the contribution from the cumulative hazard function, or those who "survived" during the entire observation period. In this way, censored observations are included in the log-likelihood. This general version of the log-likelihood function may be customized for each presumed distribution that follows (Zhou, 2016).

Lognormal Event History Model

The PDF of the lognormal distribution is

$$f(t) = \frac{1}{\sigma\sqrt{2\pi}} t^{-1} \exp\left[-\frac{1}{2}\left(\frac{\log(t) - \mu}{\sigma} \right)^2 \right] \text{ if } t > 0.$$

This PDF looks different from the PDF for the normal distribution (see Chapter 3) because we are explicitly modeling the time until an event, t. The "shape" parameter, σ, estimates the variance of the error term (Box-Steffensmeier and Jones, 2004). We may then modify this function to accommodate a model with explanatory variables and estimated coefficients (**x′B**), as well as identify the log-likelihood function.

Since the lognormal model uses the log link, we may exponentiate the coefficients and come up with something tangible: the ratio of the expected survival times of two or more groups. First, switch to the *event2* data set (*event2.dta*) so we may examine those who married and those who did not during the observation period. Second, before estimating the model, we must identify two things for Stata: the time variable and the failure variable. In this data set, *marriage* remains as the time variable and *evermarr* (marriage during the observation period: 0 = no; 1 = yes) gauges whether a respondent "failed." The Stata command `table evermarr` shows that 458 individuals never married (and were thus right censored) and 1,048 individuals married during the observation period. The Stata command `stset` is used again (see Example 7.6).

Example 7.6

```
stset marriage, failure(evermarr)
```

```
    failure event:  evermarr != 0 & evermarr < .
obs. time interval:  (0, marriage]
 exit on or before:  failure
```

```
   1506  total observations
      0  exclusions
```

```
   1506  observations remaining, representing
   1048  failures in single-record/single-failure data
 754457  total analysis time at risk and under observation
                                      at risk from t =         0
                                earliest observed entry t =    0
                                   last observed exit t =      1500
```

Once this step is taken, we may use the `streg` command to estimate the event history model. Notice that we tell Stata that the presumed distribution is the lognormal (`tr` provides *time ratios*; see Example 7.7).

Example 7.7

```
streg educate age1sex attend14 cohab nonwhite,
    dist(lognormal) tr
```

```
Lognormal regression -- accelerated failure-time form
```

```
No. of subjects =       1,506          Number of obs   =       1,506
No. of failures =       1,048
Time at risk    =     754457
                                       LR chi2(5)      =      127.43
Log likelihood  =  -3268.9873          Prob > chi2     =      0.0000
```

_t	Time Ratio	Std. Err.	z	P>|z|	[95% Conf. Interval]	
educate	1.130854	.0410046	3.39	0.001	1.053276	1.214146
age1sex	.929208	.0326014	-2.09	0.036	.867458	.9953537
attend14	2.971875	.7118385	4.55	0.000	1.858431	4.752418
cohab	1.466572	.2943135	1.91	0.056	.9896534	2.173319
nonwhite	10.23181	2.540452	9.37	0.000	6.289384	16.64551
_cons	16.05421	11.70171	3.81	0.000	3.847324	66.99139
/ln_sig	1.290994	.0237682	54.32	0.000	1.244409	1.337579
sigma	3.6364	.0864306			3.470884	3.809808

As mentioned earlier, the exponentiated coefficients reveal the expected ratio of "survival" times. For example, the coefficient for the *nonwhite* variable indicates that nonwhites are expected to "survive" about 10 times longer than whites, after adjusting for the effects of the other variables in the model. Each additional year of education is associated with a 13% increase in the time until marriage, adjusting for the effects of the other variables in the model. Age at first sexual experience is negatively associated with time until marriage.

The shape parameter, sigma (σ), listed at the bottom of the printout is the estimate of the variance of the error term. It is fairly large, thus indicating substantial variability in the error distribution or the residual variation.

Exponential Event History Model

We now examine the exponential model. This model assumes that the random component follows an exponential distribution. Its probability distribution is commonly used in product quality assessments to determine how long an electronic device will operate before it needs to be replaced (this is a key concern of reliability engineering), so, in practice, it is explicitly interested in time to "failure." The PDF is

$$f(t) = \lambda \exp(-\lambda t), \text{ where } t > 0; \lambda \geq 0.$$

As noted earlier, not unlike the Poisson distribution we use λ to designate the hazard of the event occurring. The mean of the exponential distribution is simply $1/\lambda$.

An important assumption of this model is that the hazards are assumed to be proportional over time. In other words, in our model it assumes that the hazard rate of marriage is the same at 45 days as it is at 500 days. So the coefficients measure only the expected differences in hazard rates. Simply examining the survivor curves displayed earlier suggests that this is not a wise assumption to make.

Example 7.8

```
streg educate age1sex attend14 cohab nonwhite,
    dist(exponential)
```

Exponential regression -- log relative-hazard form

No. of subjects =	1,506		Number of obs	=		1,506
No. of failures =	1,048					
Time at risk =	754457					
			LR chi2(5)	=		323.16
Log likelihood =	-5188.0854		Prob > chi2	=		0.0000

_t	Haz. Ratio	Std. Err.	z	P>|z|	[95% Conf. Interval]	
educate	.9901144	.0103996	-0.95	0.344	.9699399	1.010708
age1sex	1.060299	.0097238	6.38	0.000	1.041411	1.07953
attend14	.5031306	.0368606	-9.38	0.000	.4358323	.5808205
cohab	1.321993	.0826226	4.47	0.000	1.169581	1.494266
nonwhite	.3513487	.0303799	-12.10	0.000	.2965773	.4162353
_cons	.0011274	.0002358	-32.45	0.000	.0007483	.0016987

The exponential model shown in Example 7.8 is interested in the "hazard" of marriage, which is the inverse of the lognormal model where we examined the survival time until marriage. Most researchers are interested in what are known as *hazard ratios*. These are similar to odds or rate ratios: a value greater than 1 indicates a positive association between increasing values of the explanatory variable and the hazard of the outcome; a value between zero and one indicates a negative association.

For instance, the hazard of marriage among cohabiters is about 1.32 times the hazard of marriage among non-cohabiters, after adjusting for the effects of the other variables in the model. However, recall that this model assumes that the hazard rates are constant over time. This is a dubious assumption to make, especially when considering cohabitation. The hazard rates may increase or decrease over time.

Weibull Event History Model

One way to determine whether or not the hazard rate is constant is with the *Weibull model*. This model assumes that the error terms have a similar shape as the exponential model, but it does not assume that hazard rates are the same over time. Rather, it estimates changes in the hazard rate based on values of $1/p$.[2] If this value is greater than 1, then the hazard is decreasing over time; if it is between 0.5 and 1, the hazard is increasing; and if it is between 0 and 0.5, it is increasing at an increasing rate (i.e., the second derivative is positive). The PDF of the Weibull distribution is

$$f(t) = \alpha \lambda t^{\alpha-1} \exp\left(-\lambda t^{\alpha}\right), \text{ where } t > 0; \lambda \geq 0.$$

The Weibull PDF is similar to the exponential PDF, but it includes the extra term α. If $\alpha = 1$, then the function is simply the exponential PDF. In a Weibull model, α is estimated from the data, although Stata reports it as p.

Example 7.9 shows the event history model that assumes a Weibull distribution. There are substantial differences between the exponential and the Weibull models. This is because the estimated dispersion parameter at the bottom of the printout ($1/p = 3.25$) is highly statistically significant; we may conclude that the hazard of marriage is decreasing substantially over time. Hence, it seems that the Weibull model is a better choice than the exponential model in this situation.

---

2. Some representations of the Weibull model use σ for this quantity. However, here we use the Stata parlance so as not to confuse the results of the Weibull model with the results of the exponential model.

Example 7.9

```
streg educate age1sex attend14 cohab nonwhite,
    dist(weibull)
```

```
Weibull regression -- log relative-hazard form
```

No. of subjects =	1,506		Number of obs	=	1,506	
No. of failures =	1,048					
Time at risk =	754457					
			LR chi2(5)	=	145.93	
Log likelihood =	-3409.4673		Prob > chi2	=	0.0000	

_t	Haz. Ratio	Std. Err.	z	P>|z|	[95% Conf. Interval]	
educate	.9795059	.0105745	-1.92	0.055	.958998	1.000452
age1sex	1.035775	.0101711	3.58	0.000	1.016031	1.055903
attend14	.6572003	.048196	-5.72	0.000	.5692128	.7587887
cohab	1.080057	.0678723	1.23	0.220	.9548958	1.221623
nonwhite	.458035	.0396917	-9.01	0.000	.3864884	.5428263
_cons	.1673049	.0369348	-8.10	0.000	.1085412	.257883
/ln_p	-1.179021	.0258007	-45.70	0.000	-1.22959	-1.128453
p	.3075796	.0079358			.2924125	.3235333
1/p	3.251191	.0838829			3.090872	3.419827

This is an awful lot to digest. It is often easier to look at the distribution of the survival or hazard curves and decide which model to use. As a more rigorous alternative, we may also rely on the Akaike's information criterion (AIC) and Bayesian information criterion (BIC) to compare the models. We may find these by using the estat ic command after estimating each model. The following table provides this information for each of the three models.

Model	AIC	BIC
Lognormal	6,551.98	6,589.20
Exponential	10,388.17	10,420.07
Weibull	6,832.94	6,870.16

It is quite clear that the exponential model is not appropriate for estimating time until marriage. The lognormal model appears to be a much better option. But there are also graphical techniques that are useful to judging the appropriateness of event history models. Recall that in the past we plotted some form of the residuals against the predicted values or against a normal distribution. In event history models, we may plot what are called *Cox–Snell residuals* against a transformation of the *empirical survival curve* (the KM estimator we saw earlier). If the assumptions of the model are met, the residuals should match up with a variant of the KM estimator along a diagonal line. After estimating the lognormal model, use the predict command to request Cox–Snell residuals (see Example 7.10).

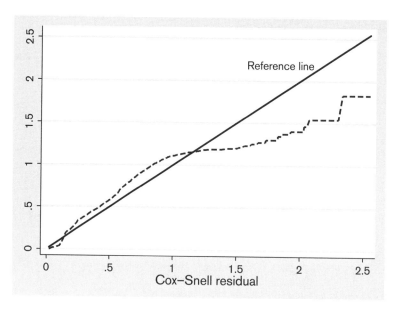

FIGURE 7.4

Example 7.10

```
qui streg educate age1sex attend14 cohab nonwhite,
  dist(lognormal) tr
qui predict double cs, csnell
```

We must then tell Stata that the new failure time variable is the Cox–Snell residuals.

```
qui stset cs, failure(evermarr)
```

Subsequently, use the `sts generate` command to compute the KM estimator. The key is to plot the negative logged values of this estimator by the Cox–Snell residuals to look for any variations from a diagonal line.

```
qui sts generate km=s
qui generate double H = -ln(km)
line H cs cs, sort            * the second cs provides a 45° reference line
```

Figure 7.4 shows that there is substantial variation at the high end of the residuals, thus calling into question whether we have met the assumptions of the model. As an exercise, create versions of this graph for each of the other distributions. In addition, test a model that uses the log-logistic distribution, which is similar to the lognormal distribution, and construct a

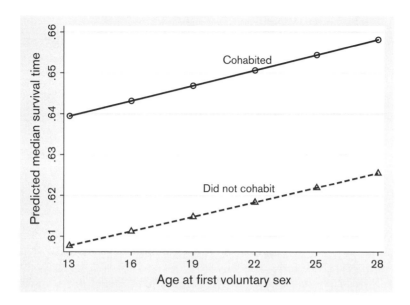

FIGURE 7.5

similar graph. Which distribution appears to fit best? Is this different from our conclusions based on the AICs and BICs?

Before considering another type of residual, assume we have decided that the lognormal distribution provides the best fitting model. How might we present some of its results? As shown in earlier chapters, the `margins` post-estimation command is useful. Considering that cohabitation and age at first sex appear to have an association with time until marriage, we will plot predicted values for these two variables. Figure 7.5 provides this graph.

```
qui margins cohab, at(nonwhite=0 age1sex=(13 16 19 22 25
   28)) atmeans
```
         * *after Example 7.7 with* i.cohab *entered as an explanatory variable*
```
marginsplot, noci
```

The graph shows the predicted median time for each group at different ages at first sex. The lines are roughly parallel, as expected, since we assume—based on the model setup—that cohabitation and age at first sex do not interact (although they might). One message from the graph is that we expect cohabiters to "survive" longer until they marry.

Another useful set of predicted values is the deviance residuals. These may be used like other deviance residuals discussed in previous chapters to check for unusual observations. However, the key is to plot them against the survival or ranked survival time variable. For instance, after reestimating the lognormal event history model, predict the deviance residuals and plot them against the *marriage* variable.

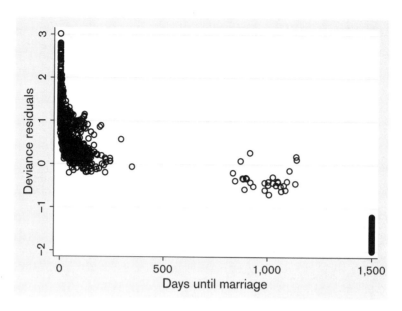

FIGURE 7.6

```
predict dev, deviance                    * after Example 7.7
twoway scatter dev marriage
```

Figure 7.6 shows some of the peculiarities in the data better. The marriage times are clumped together at two key points: less than about 200 days and around 1,000 days (the 1,500 days are the censored observations). However, the deviance residuals are not particularly extreme, so there do not appear to be any outliers of concern in the lognormal model. But also notice that the smallest residuals occur at later points in the time-until-marriage distribution.

We have not yet discussed how to include time-varying covariates in event history models. These are discussed later when we review discrete-time models.

THE COX PROPORTIONAL HAZARDS MODEL

The final option discussed for event history data is perhaps the most widely used model. It is known as the *Cox PH model*. Its popularity rests on the fact that the Cox PH model does not require a particular distribution for the survival times (or, more precisely, the random component). Rather, it is known as a *semi-parametric model*. As with the other event history models, we are interested in estimating conditional hazards, but use the following function to do so:

$$h(t|\mathbf{x}) = h_0(t)\exp(\mathbf{x'\beta}).$$

The β represents a vector of regression coefficients, with **x** representing the explanatory variables (covariates). As suggested earlier, the baseline hazard (h_0)—which is similar to the intercept in a linear regression model—is not assumed to follow any particular distribution; rather, it is estimated from the data in a nonparametric manner. The log-likelihood used in model estimation is thus

$$\ln L\left(\mathbf{x}'\boldsymbol{\beta}|t\right) = \sum_{i=1}^{N}\delta_i\left(x_i'\boldsymbol{\beta}\right) - \sum_{i=1}^{N}\delta_i \ln\left\{\sum_{j\in R(t_i)}\exp\left(x_j'\boldsymbol{\beta}\right)\right\}.$$

The Greek letter δ (lower case delta) represents a vector of censoring identifiers. $R(t_i) = (j: t_j \geq t_i)$ indicates the risk set at time t_i. Recall that people fall out of the risk set once they experience the event; so they no longer contribute to the likelihood function (see Zhou, 2016, for more details).

A key advantage of this model is that it easily allows time-varying covariates and it can accommodate both continuous- and discrete-time outcomes. Since even the best fitting parametric model—the lognormal—had some problems, it may be best to use—or at least consider—the Cox PH model with our time-to-marriage variable.

An important issue for the Cox PH model is its *proportional hazards assumption*. This was described briefly earlier. Moreover, it is helpful to think about how this assumption is similar to the assumption made with a dummy variable in a linear regression model: we assume the slopes are the same; only the intercepts differ. However, as we learn later, the proportional hazards assumption may be relaxed in a straightforward manner.

The model is set up as before. First, use `stset` to describe the time and event variables and then use `stcox` to estimate the model. The subcommand `efron` tells Stata what to do when there are ties in the time dimension (see Example 7.11).

Example 7.11

```
stcox educate age1sex attend14 cohab nonwhite, efron
```

```
Cox regression -- Efron method for ties

No. of subjects =        1,506          Number of obs    =        1,506
No. of failures =        1,048
Time at risk    =      754457
                                        LR chi2(5)       =       118.63
Log likelihood  =   -7106.8978          Prob > chi2      =       0.0000
```

_t	Haz. Ratio	Std. Err.	z	P>|z|	[95% Conf. Interval]	
educate	.9676438	.0104215	-3.05	0.002	.9474321	.9882867
age1sex	1.023474	.0101849	2.33	0.020	1.003705	1.043632
attend14	.6956732	.0509935	-4.95	0.000	.6025754	.8031546
cohab	.9507317	.0600379	-0.80	0.424	.8400504	1.075996
nonwhite	.4794834	.0415583	-8.48	0.000	.4045733	.5682636

Rather than specifying the output with no hazard ratios, we allow the program to present them (note that the subcommand `nohr` is omitted). The coefficients may be interpreted the same way as before; the percent change interpretation may also be used. It might be useful to compare these results to those we saw earlier: where are the differences and why do they occur?

Here are some interpretations from the Cox PH model:

- The hazard of marriage is expected to decrease by about (0.968 − 1) × 100 = 3% for each 1-year increase in education, adjusting for the effects of the other variables in the model.

- Attending religious services at age 14 is associated with about a 30% decrease in the hazard of marriage, adjusting for the effects of the other variables in the model.

An important issue to examine is whether the model meets the proportional hazards assumption. Stata provides a simple approach for testing this assumption with its `phtest` post-estimation command. This involves estimating a set of residuals and then determining whether they indicate proportional hazards.

The null hypothesis is that the hazards are proportional. As shown in the proportional hazards test (`phtest`), we may reject the null for all the explanatory variables except religious service attendance. In this situation, we should explore the associations further to see if we can identify the nonproportional hazards. For instance, we might hypothesize that cohabitation and length of time until marriage are related in a nonproportional manner. A Cox model that includes an interaction term designed to test this hypothesis is provided in Example 7.12.

Example 7.11 (Continued)

estat phtest, rank detail   * *null hypothesis: hazards are proportional*

Test of proportional-hazards assumption

Time: Rank(t)

	rho	chi2	df	Prob>chi2
educate	0.25785	62.40	1	0.0000
age1sex	0.13899	16.74	1	0.0000
attend14	-0.02638	0.73	1	0.3914
cohab	0.38091	149.51	1	0.0000
nonwhite	0.07837	6.53	1	0.0106
global test		223.00	5	0.0000

Example 7.12

```
stcox educate age1sex attend14 cohab c.cohab#c.marriage, efron
```

```
Cox regression -- Efron method for ties

No. of subjects =        1,506           Number of obs    =        1,506
No. of failures =        1,048
Time at risk    =       754457
                                         LR chi2(5)       =       763.76
Log likelihood  =   -6784.3294           Prob > chi2      =       0.0000
```

_t	Haz. Ratio	Std. Err.	z	P>|z|	[95% Conf. Interval]	
educate	.9624289	.0109374	-3.37	0.001	.941229	.9841062
age1sex	1.041585	.0105982	4.00	0.000	1.021019	1.062565
attend14	.7694989	.0564091	-3.57	0.000	.6665145	.8883957
cohab	3.398678	.3212429	12.94	0.000	2.823934	4.090398
c.cohab#c.marriage	.9931206	.0009458	-7.25	0.000	.9912685	.9949762

Note first that both cohabitation and the interaction term have statistically significant coefficients. To understand better what is going on, use the stphplot command with cohabitation as the covariate.

```
stphplot, by(cohab)
```

Figure 7.7 provides what is termed a *log cumulative hazard function*. It indicates that the hazards over time change for the two groups. As suggested by the Cox PH model with the interaction term, the hazard of marriage for those who cohabitated begins high, but then, proportionally speaking, it decreases much more rapidly than among those who have not cohabited. In fact, the expected hazards cross at around 150 days (exp(5)).

We may also adjust the graph for the other explanatory variables. The following variation of the previous command adjusts the graph in figure 7.7 for education, age at first sexual experience, and race/ethnicity. However, the results are roughly the same (the graph is omitted; try producing it yourself).

```
stphplot, by(cohab) adj(educate age1sex nonwhite)
```

Finally, it is a simple step to include time-varying covariates in the Cox model by using the tvc subcommand. For example, suppose that cohabitation and education are time varying; that is, their values change over the observation period (alas, they do not in this data set). The Cox PH model shown in Example 7.13 could then be estimated.

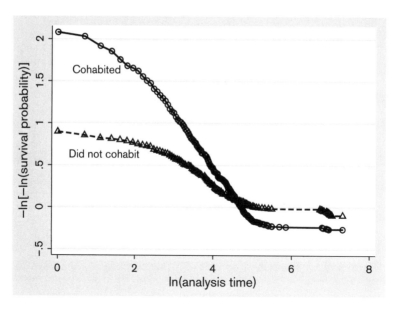

FIGURE 7.7

```
stcox age1sex attend14 nonwhite, efron tvc(educate cohab)
   texp(ln(_t))
```

The last part of the command asks Stata to assume that the changes in the hazards are log-normal rather than the default of normally distributed changes. This is often more reasonable in longitudinal models with time-varying effects. The main implication is that the effects of education and cohabitation, if they are time varying, are allowed to interact with time (log-time here), thus allowing their hazard rates to change over the observation period. This is often a more realistic assumption than presuming constant hazard rates.

DISCRETE-TIME EVENT HISTORY MODELS

A nice aspect of several of the GLMs discussed in previous chapters is that they may be adopted for use as discrete-time event history models. As shown later, the Cox PH model may also be used for discrete-time models. The key for using GLMs is to set up the data properly. We need to treat each observation as a time point for each individual rather than just treating the respondents as observations. In other words, the unit of analysis is now person-time, rather than just the individual.

In discrete-time models we assume that each person's involvement in the event of interest occurs within some distinct time period. For instance, longitudinal data are often collected on a periodic basis—such as annually—using questionnaires that ask respondents about events that occurred in the last year, month, or week. The data setup for discrete-time models requires us to treat each time period as a separate observation for each respondent. Whether or not the event occurred in a particular time period is then estimated with a logistic regression model, a multinomial logistic regression model (if the event includes more than two possibilities), or a complementary log-log model. Although we have not discussed this last option, it comes in handy if one is willing to assume that the events occur in continuous time (Allison, 1995). Moreover, it is available in Stata's `glm` command `[link(cloglog)]`.

Example 7.14 provides an example of discrete-time event history data. The *event3* data (*event3.dta*) are from a longitudinal study of adolescent behaviors. Each year a sample of adolescents filled out questionnaires that asked about drug use, delinquency, stressful life events, family relations, and many other issues. The variable *delinq1* is coded as 0 = no delinquent acts in that year and 1 = the respondent's first delinquent act was committed in that year. Notice that each respondent contributed up to 4 years of information.

Example 7.14

```
list newid year age stress delinq1 in 121/132
```

	newid	year	age	stress	delinq1
121.	403	1	14	1	1
122.	403	2	15	1	0
123.	403	3	16	7	0
124.	403	4	17	5	0
125.	423	1	13	1	1
126.	423	2	14	0	0
127.	423	3	15	0	0
128.	423	4	16	0	0
129.	433	1	12	0	0
130.	433	2	13	0	1
131.	433	3	14	2	0
132.	433	4	15	0	0

Thus, respondent 403 reported his first delinquent act (at least under the observation period) at age 14 during year 1. Respondent 433 committed his first delinquent act at age 13 during year 2. If respondents did not report a delinquent act during the 4 years of observation, they are considered as right censored.

In Stata, once the data are set up in this manner,[3] we may use the `logit` or `glm` command to estimate a discrete-time event history model with logistic regression. However, we

---

3. Stata has several convenient commands for transforming data sets into structures suitable for discrete-time event history models or other types of analysis. Type `help reshape` in Stata's command window for more information.

also need to tell Stata that the *year* variable should be treated as a categorical indicator. A practical way to do this is with Stata's i. option (as shown earlier, i.*variable* specifies an indicator variable), which creates a set of dummy variables from a categorical variable (see Example 7.15).

Example 7.15

```
logit delinq1 stress cohes i.year, or
```

```
Logistic regression                             Number of obs   =       2,604
                                                LR chi2(5)      =      238.43
                                                Prob > chi2     =      0.0000
Log likelihood = -995.98366                     Pseudo R2       =      0.1069
```

delinq1	Odds Ratio	Std. Err.	z	P>|z|	[95% Conf. Interval]	
stress	1.130257	.0346213	4.00	0.000	1.064398	1.200192
cohes	.9715197	.0048293	-5.81	0.000	.9621004	.9810313
year						
2	.3858035	.0549758	-6.68	0.000	.2917915	.5101051
3	.2154061	.0353694	-9.35	0.000	.1561319	.2971832
4	.1547798	.0285326	-10.12	0.000	.1078453	.2221403
_cons	1.471793	.4421556	1.29	0.198	.8168246	2.651944

The year variables (year is now represented with dummy variables, with *year* = 1 as the reference category) show the effect of time on the odds of committing one's first delinquent act. We see that the odds decrease over time, thus indicating that the odds are highest during the first observation period (could this be affected by left censoring?). The other explanatory variables are typically the focus of interest in this type of model. For instance, the stress coefficient suggests that, adjusting for the effects of family cohesion, each one-unit increase in stressful events is associated with a 13% increase in the odds of committing an initial delinquent act. On the other hand, family cohesion is negatively associated with the odds of committing an initial delinquent act, adjusting for the effects of stress and time.

This model may also be estimated with the Cox PH model. We first must tell Stata that these are event history data using the stset command. The main difference for these data is that we also must inform Stata that there is an ID variable so it can identify the individuals in the data set (see Example 7.16).

Example 7.16

```
stset year, id(newid) fail(delinq1)
```

```
                id:  newid
     failure event:  delinq1 != 0 & delinq1 < .
obs. time interval:  (year[_n-1], year]
 exit on or before:  failure
```

```
    2604  total observations
     858  observations begin on or after (first) failure
```

```
    1746  observations remaining, representing
     651  subjects
     399  failures in single-failure-per-subject data
    1746  total analysis time at risk and under observation
                                      at risk from t =            0
                               earliest observed entry t =        0
                                   last observed exit t =         4
```

We may then estimate the Cox PH model as before.

Example 7.16 (Continued)

```
stcox stress cohes, efron
```

```
Cox regression -- Efron method for ties

No. of subjects =          651           Number of obs     =        1,746
No. of failures =          399
Time at risk    =         1746
                                         LR chi2(2)        =        122.39
Log likelihood  =    -2364.3714          Prob > chi2       =        0.0000
```

_t	Haz. Ratio	Std. Err.	z	P>|z|	[95% Conf. Interval]	
stress	1.156791	.0292854	5.75	0.000	1.100793	1.215637
cohes	.9659971	.0039772	-8.40	0.000	.9582333	.9738238

Notice that the hazard ratios for stress and family cohesion are similar to the odds ratios in the logistic regression model. This is normally the case, so that either model may be used to estimate the associations between the outcome and the explanatory variables. An advantage of the logistic model is that it is useful for examining the effects of discrete time explicitly. In the logistic model, we saw that the odds of initial delinquency decreased over time.

Using `margins`, the predicted values from the model that focuses on stressful life events are represented as relative hazards, with an increasing rate as stress increases (see figure 7.8).

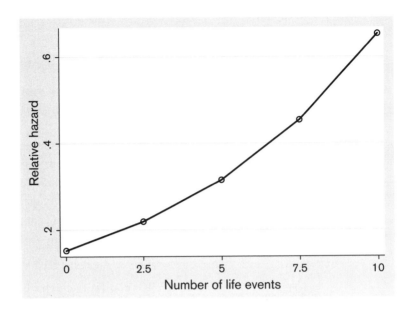

FIGURE 7.8

```
qui margins, at(stress=(0 2.5 5 7.5 10)) atmeans
```
                                                        * after Example 7.16
```
marginsplot, noci
```

We may now check the proportional hazards assumption using the same procedure as earlier.

Example 7.16 (Continued)

```
estat phtest, rank detail
```

Time: Rank(t)

	rho	chi2	df	Prob>chi2
stress	-0.01150	0.05	1	0.8157
cohes	-0.05334	0.91	1	0.3391
global test		0.91	2	0.6331

From these results it seems that there is little concern that the model violates the proportional hazards assumption. The relationships involving stressful life events and family cohesion do not involve nonproportional hazards.

However, these data do suffer from a lurking problem. Think about the way delinquency is measured. Many adolescents may have committed their first delinquent act before the first observation period. Hence, we likely have left censoring. There are methods for handling this issue, but they are beyond the scope of this presentation.

Another discrete-time model is to imagine delinquency—or some other outcome—as repeatable. In other words, we examine whether delinquent acts occurred in more than 1 year and what factors predict their occurrence. The *event4* data set (*event4.dta*) provides data that show repeated events. Review this data set and observe that delinquency may occur in multiple years for each individual.

To estimate this event history model we utilize the same commands as before, with *year* specified as a set of dummy variables using the `i.variable` name convention. Example 7.17 provides the Stata commands and output for a logistic regression model.

The coefficients for the year variables indicate the odds of reporting delinquent events for each year; we see that in each year the odds increase relative to the odds in year 1. Moreover, each one-unit increase in stressful life events is associated with a 22% increase in the odds of delinquency, adjusting for the effects of time and family cohesion.

Example 7.17

```
logit delinq stress cohes i.year, or
```

```
Logistic regression                             Number of obs   =      2,604
                                                LR chi2(5)      =     252.53
                                                Prob > chi2     =     0.0000
Log likelihood = -1552.5174                     Pseudo R2       =     0.0752
```

delinq	Odds Ratio	Std. Err.	z	P>|z|	[95% Conf. Interval]	
stress	1.219323	.0307588	7.86	0.000	1.160503	1.281125
cohes	.9580394	.0037072	-11.08	0.000	.9508009	.9653331
year						
2	1.37116	.1706464	2.54	0.011	1.074367	1.749942
3	1.301327	.1626341	2.11	0.035	1.018608	1.662516
4	1.41845	.1781966	2.78	0.005	1.108867	1.814464
_cons	2.43191	.580454	3.72	0.000	1.523283	3.882525

As before, this model may also be estimated using the Cox PH model. We first tell Stata that these are event history data and then use `stcox` to estimate the model. However, we also change the subcommand from `efron` to `exactp` because there are so many ties (not surprising since there are only four observation points). The results of the model are shown in Example 7.18.

Example 7.18

```
stset year, id(newid) fail(delinq)

               id:  newid
    failure event:  delinq != 0 & delinq < .
obs. time interval:  (year[_n-1], year]
 exit on or before:  failure
```

```
   2604  total observations
    858  observations begin on or after (first) failure
```

```
   1746  observations remaining, representing
    651  subjects
    399  failures in single-failure-per-subject data
   1746  total analysis time at risk and under observation
                                  at risk from t =           0
                            earliest observed entry t =        0
                                 last observed exit t =        4
```

Once again, the results are similar to those from the logistic regression model. Moreover, if we wish to model the explanatory variables as time-varying covariates, the `tvc` subcommand may be used. The Stata help menu for `stcox` provides additional details on introducing time-varying effects.

Example 7.18 (Continued)

```
stcox stress cohes, exactp
```

Cox regression -- exact partial likelihood

No. of subjects =	651	Number of obs =	1,746
No. of failures =	399		
Time at risk =	1746		
		LR chi2(2) =	127.50
Log likelihood =	-839.53968	Prob > chi2 =	0.0000

_t	Haz. Ratio	Std. Err.	z	P>|z|	[95% Conf. Interval]	
stress	1.203654	.040596	5.50	0.000	1.12666	1.285909
cohes	.9569676	.0050658	-8.31	0.000	.9470901	.9669481

FINAL WORDS

Event history models are valuable empirical tools for studying many phenomena, literally from birth to death. If data are available, continuous-time models are especially valuable for estimating how long it takes until some event occurs. Discrete-time models can also be useful. However, an important assumption that these models make is that there is no dependence among the observations. However, it seems clear that getting involved in delinquency in one year probably affects involvement in subsequent years. Therefore, we have violated a key assumption of the regression model. Perhaps the best method for addressing dependence is with *random-* or *fixed-effects regression models* and a more general set of models known as *generalized estimating equations.* These models are discussed in the Chapter 8.

Another problem involves what is known as *unobserved heterogeneity* (Blossfeld et al., 2009, Chapter 10). This concept addresses, in a very basic sense, the fact that people bring all sorts of baggage along with them when they "decide" to engage in some type of event or simply experience an event. When we study adolescent drug use, for example, we must admit that there are many unobserved factors—psychological, neurological, sociological—that influence it. We cannot expect to include all of the factors that affect this outcome in our study. Recall that this is the problem of omitted variable bias that plagues regression models.

However, this issue may be especially problematic for event history models because omitted variables may affect the likelihood of entering or exiting some state (e.g., marriage), but also affect (obviously) the hazard of the event (Yamaguchi, 1991). Unobserved heterogeneity tends to bias the hazard rate estimates. People with truly high hazards tend to drop out of the study early (or they may experience left censoring), so the remaining lower hazard people make it seem as if the hazard is low. This is a particular problem when the event can occur only once. For instance, in our time-to-marriage example, not only was the hazard particularly high during the first part of the observation period, but we also found a sharply decreasing hazard for those who cohabited. Yet, this may be due to some unmeasured trait—common among those who cohabit—that led many to marry quickly. Hence, our estimate of the hazard rate may be biased. This problem is most acute when the hazard rates are decreasing. When they increase, we may be satisfied that the true hazard is increasing for at least some groups (Allison, 1995). Stata provides random- and fixed-effects models that are appropriate for adjusting for unobserved heterogeneity (see the next chapter for an introductory overview of these models). Frailty models (also available in Stata and other statistical software) are also useful as a control for unobserved heterogeneity (see Therneau and Grambsch, 2000).

EXERCISES FOR CHAPTER 7

1. The *firstsex* data set (*firstsex.dta*) contains information from a sample of more than 1,700 young men who participated in the National Survey of Adolescent Males. The variable *firstsex* measures the number of months from the participant's 12th birthday until first sexual intercourse. At the end of the observation period, about 10% of respondents had not had sexual intercourse. Right censoring is identified by the variable *eversex*. After examining the coding of the variables, consider the following exercises.

   a. Determine the following statistics: total time at risk, the median survival time until first sexual intercourse, and the average hazard rate per month.

   b. Construct a graph of the survival curve.

   c. Construct a graph of the survival curve stratified by *pstudent*. Comment about what the graph shows about the differences between these two groups.

   d. Determine the median survival time for the two groups distinguished by *pstudent*. Is this difference statistically significant? What do you conclude about the difference between the two groups?

2. Estimate two event history (survival) models: a lognormal and a Weibull. Use the following explanatory variables: *famsize, pareduc, momdad, black, othrace, reborn,* and *pstudent.*

   a. Compare the two models statistically to determine which fits the data better.

   b. From your best fitting model, interpret the coefficients (transformed if you prefer) associated with the variables *momdad, black,* and *pstudent.*

   c. Discuss why the exponential model is not appropriate for this analysis.

3. Estimate a Cox proportional hazards model using the same set of explanatory variables as in exercise 2.

   a. Interpret the coefficients associated with the variables *famsize* and *pstudent.*

   b. Construct a graph (your choice of the type) that shows the relative hazards for the following family sizes: three, six, and nine. Set the other explanatory variables at their means or modes (depending on the type of variable).

   c. Test the proportional hazards assumption. Indicate which, if any, variables do not follow this assumption. If you find any, discuss why they do not appear to meet this assumption. Feel free to use graphical or other techniques to support your answer.

## EIGHT

Regression Models for Longitudinal Data

Discrete-time event history models are appropriate for some types of longitudinal data, such as when we are interested in predicting whether or how long it takes for an event or repeated events to occur. However, these types of data are not limited to identifying and predicting particular events. Longitudinal or panel data, as they are often called, are also used to study behaviors, attitudes, or other outcomes that can change over time. For example, we may be interested in not only whether adolescents are involved in any delinquency, but also their level of involvement during particular time periods. Similarly, some researchers are interested in changes in levels of self-esteem, community involvement, test scores over time, or physical growth during infancy. Thus, longitudinal data allow us to study the dynamics of various outcomes, which is often superior to merely examining an outcome at one time point, as we do with cross-sectional data.

Regression models for longitudinal data may be considered as relatively straightforward extensions of the generalized linear models (GLMs) examined in previous chapters. For instance, consider the following regression model:

$$y_{it} = \alpha_i + \beta x_{it} + \varepsilon_{it}.$$

Note the differences between this model and those from earlier chapters. We now have extra subscripts in each term: the outcome and explanatory variables, as well as the error term, are subscripted with t to indicate that they vary not only across individuals but also across time. The intercept also varies across individual units (i) because each unit can have a mean estimated over the periods of observation (the same issues can also apply to data collected over spatial units).

For instance, consider the *event4* data set that was mentioned in Chapter 7. As shown in Example 8.1, delinquency is measured each year over the 4-year observation period, so there is mean involvement in delinquency across the 4 years for each individual (e.g., *newid* = 403's mean delinquency is 4/4 = 1, whereas *newid* = 433's mean delinquency is 1/4 = 0.25). Similarly, age and stress are measured each year; all of these variables vary over time.

Example 8.1

```
list newid year age stress delinq in 121/132
```

	newid	year	age	stress	delinq
121.	403	1	14	1	1
122.	403	2	15	1	1
123.	403	3	16	7	1
124.	403	4	17	5	1
125.	423	1	13	1	1
126.	423	2	14	0	1
127.	423	3	15	0	1
128.	423	4	16	0	0
129.	433	1	12	0	0
130.	433	2	13	0	1
131.	433	3	14	2	0
132.	433	4	15	0	0

There are several advantages of longitudinal data relative to cross-sectional data. First, the causal order is specified better. We may see, for example, that some conditions came before some outcome, thus making is easier to establish causal order among variables (although other conditions also play a role). Second, problematic issues such as age, period, and cohort effects may be sorted out with longitudinal data. We may ask, for instance, whether changes in political party membership occur as people get older or whether people born during particular periods (e.g., the 1960s) are more apt to join particular parties. Third, given appropriate statistical procedures, longitudinal data are useful for determining *changes within individuals* and *differences across individuals* (Bijleveld and van der Kamp, 1998). The ability to distinguish these *within-unit* and *between-unit* differences also gives us an opportunity to control for unobserved heterogeneity. Since we are observing changes within individuals over time, we can adjust for unobserved differences across individuals.

Most of the models shown thus far are amenable to longitudinal analysis. However, using the methods discussed in Chapter 7 for discrete-time event history analysis forces us to assume that the observations are independent. This is rarely a reasonable assumption for repeated events. Committing a delinquent act in one year is likely to be associated with committing a delinquent act in another year. Similarly, a personality trait measured at age 13 is likely to be positively correlated with the same trait at ages 14 and 15.

FIXED- AND RANDOM-EFFECTS REGRESSION MODELS

A popular approach when analyzing longitudinal data with repeated events or measures involves *fixed-* and *random-effects regression models*. The choice between these two models is primarily conceptual, but, unfortunately, there are several definitions of these models that cause confusion among users. One definition is that fixed-effects models are used when the analyst wishes to generalize only to the sample members in the data set. Random-effects models are more appropriate when sample members are drawn from a larger population. Another definition indicates that the coefficients from fixed-effects models are constant across individuals or some other unit (e.g., schools, neighborhoods), whereas coefficients from random-effects models are allowed to vary. For instance, we may use a random-effects model to examine whether the association between, say, family relations and delinquency differs across various types of neighborhoods (see Chapter 9).

We rely mainly on this latter definitional approach to these models. However, the "random effect" in the following random-effects models tends to focus on the intercept: it is allowed to vary randomly across units (α_i). Thus, given longitudinal data, we allow the baseline of the outcome variable to vary across individuals. A nice feature of random-effects regression is that it may accommodate both time-variant and time-invariant covariates. Fixed-effects models are appropriate when we wish to examine what affects changes within-individual units (e.g., people, organizations). In particular, they remove the effects of time-invariant characteristics so we may observe the effects of time-varying explanatory variables on a time-varying outcome (Allison, 2009). Thus, they are not as useful when we wish to estimate the effects of both time-invariant and time-variant explanatory variables on some outcome since the former drop out of the analysis.

As shown earlier, the fixed-effects regression model in general may be represented as

$$y_{it} = \alpha_i + \beta x_{it} + \varepsilon_{it} .$$

Note that the intercept includes a subscript i that denotes that there are potentially different individual-specific baseline values for each individual unit in the data. Perhaps the most advantageous quality of the fixed-effects model is that individuals serve as their own controls, thus allowing us to get closer to a causal effect than with other models (Angrist and Pischke, 2009). Another way of looking at this is that the fixed-effects model controls for all the stable variables that are not observed directly. However, as suggested earlier, two assumptions are particularly important for identifying this model:

1. The unobserved factors that simultaneously influence the outcome and the explanatory variables are time-invariant.

2. The individual-specific effects are correlated with the explanatory variables.

The first assumption is essential since we must presume that variables not appearing in the model do not affect the association between the time-varying explanatory variables and the time-varying outcome variable. Moreover, whatever effects these omitted variables have on the individual are assumed to be the same regardless of when they occur (Allison, 2009).[1]

The random-effects model may be characterized by

$$y_{it} = \alpha + \beta x_{it} + \mu_{it} + \varepsilon_{it} \, .$$

In this model, the error is divided into between-individual (μ_{it}) and within-individual (ε_{it}) components. A key assumption is that the individual-specific effects are not correlated with the explanatory variables, thus time-invariant variables may be included in the model. As with many other types of regression models, the omitted variables should not affect the variables included in the model; otherwise, there is omitted variable bias.

As indicated by the second assumption listed earlier, a key difference between the fixed- and random-effects models involves the assumption they make about the correlation between what is not observed and the explanatory variables in the model. As already implied, the random-effects model assumes a zero correlation between the unobserved heterogeneity—represented by the within-individual errors—and the explanatory variables. The fixed-effects model, on the other hand, allows for a nonzero correlation (Allison, 2009; Greene, 2000). For instance, one way to think about which model is most appropriate is to consider the answer to the following inquiry: when studying the impact of changes in marijuana use on changes in wages, can we assume that physiological or environmental factors that are constant within persons are not correlated with marijuana use? The answer to this type of question helps determine whether one relies on a fixed- or a random-effects model.

A statistical procedure known as the Hausman test is often used to adjudicate between these two regression models, but it is only directly appropriate for linear models. If the null hypothesis for the Hausman test is rejected (analysts typically use $p < 0.05$ as the decision rule), then the fixed-effects model is preferred to the random-effects model.

---

1. It is interesting to note that the fixed-effects regression model may also be estimated by including a set of dummy variables for the individual identifiers in a linear regression model. Thus, another way to estimate this model is with what is called a *least squares dummy variable* model (Allison, 2009). For example, if we wished to use this model to estimate changes in self-esteem (see Example 8.2), we could use the Stata code `regress esteem x-vars i.newid`. Be careful, though, since this model requires a large matrix for estimation and will produce a long output. An alternative is to use Stata's `areg` command. Nonetheless, this approach to fixed-effects regression is instructive for understanding how the model operates.

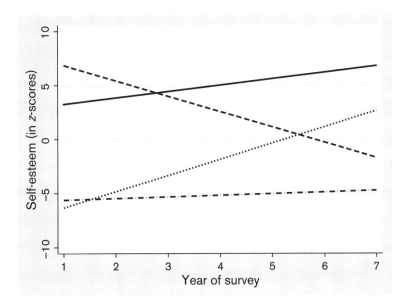

FIGURE 8.1

Stata includes several commands to estimate these two models, including those appropriate for outcome variables that are continuous (`xtreg`), binary (`xtlogit`, `xtprobit`), and count (`xtpoisson`, `xtnbreg`). As discussed in Chapter 9, Stata also has a set of procedures for mixed-effects models, which may include both fixed- and random-effects models.

As an example of these two models, we'll now use a more extensive set of data from the same source as the *event4* data set. The *esteem* data set (*esteem.dta*) includes 7 years of longitudinal data with measures of self-esteem from each year. Suppose we wish to predict the level of self-esteem with information about stressful life events, family cohesion, sex, and race/ethnicity. Before seeing these regression models, though, examine a small portion of the data by considering self-esteem among four respondents followed over the 7-year period.

Figure 8.1 shows what it means to have random intercepts. Notice that each individual's baseline self-esteem score differs, with the lowest at a little less than −5 and the highest at about 6. Random-effects models allow for these differing intercepts. But we often wish to determine whether a set of explanatory variables predict self-esteem as it changes over time.

Example 8.2 furnishes the steps used in Stata to estimate a fixed-effects linear regression model. This is followed by a random-effects model and the Hausman test to compare the two. The models predict changes in self-esteem with stressful life events and family cohesion.

```
xtset newid year          * this command is used to tell Stata that these are
                            longitudinal data with newid as the unit
                            identifier and year as the time marker

xtreg esteem stress cohes male nonwhite, fe    * fixed-effects model
```

```
Fixed-effects (within) regression          Number of obs      =       5,250
Group variable: newid                      Number of groups   =         750

R-sq:                                      Obs per group:
    within  = 0.0766                                    min =           7
    between = 0.2569                                    avg =         7.0
    overall = 0.1677                                    max =           7

                                           F(2,4498)          =      186.45
corr(u_i, Xb)   = 0.1795                    Prob > F           =      0.0000
```

esteem	Coef.	Std. Err.	t	P>|t|	[95% Conf. Interval]	
stress	-.0589644	.0163352	-3.61	0.000	-.0909893	-.0269394
cohes	.1894794	.0101819	18.61	0.000	.1695179	.2094409
male	0	(omitted)				
nonwhite	0	(omitted)				
_cons	2.21e-10	.0659641	0.00	1.000	-.1293221	.1293221
sigma_u	4.5849682					
sigma_e	4.7795564					
rho	.47922967	(fraction of variance due to u_i)				

```
F test that all u_i=0: F(749, 4498) = 6.19              Prob > F = 0.0000
```

```
estimates store fixed          * stores estimates for the Hausman test
```

The results of the fixed-effects model show how time-invariant explanatory variables are treated: their effects are not estimated since we assume that time-invariant influences are fixed by the model and do not affect the association between changes in stress or cohesion and changes in self-esteem.

The interpretation of coefficients for the fixed-effects model focuses on the effect of changes in, say, stressful life events on changes in self-esteem. One interpretation of this particular coefficient is the following:

- Statistically adjusting for changes in family cohesion, a one-unit change in stressful events is associated with a −0.059 decrease in self-esteem relative to an individual's average self-esteem score.

Note that we are comparing self-esteem scores among the same individuals over time. Thus, we assume that as an individual experiences an increase in stressful life events, we expect that individual will also experience a decrease in self-esteem.. In general, this model focuses on changes within individuals rather than differences across individuals.

The printout also provides the estimated correlation between the within-individual errors—which, you may recall, represent unobserved heterogeneity—and the explanatory variables (0.18). The bottom panel shows *sigma_u* and *sigma_e*, which furnish the variance due to differences between individuals (*sigma_u*) and the variance due to differences within individuals (*sigma_e*). The statistic *rho* provides the proportion of the overall variance that is attributable to between-individual differences. Hence, about 48% of the variability is due to changes between individuals. This suggests that self-esteem is highly variable across adolescents over time, even after accounting for stressful events and family cohesion.

The random-effects model using the same general set of explanatory variables is estimated with the Stata code provided in Example 8.3.

Example 8.3

xtreg esteem stress cohes male nonwhite, re

* random-effects model

```
Random-effects GLS regression              Number of obs      =       5,250
Group variable: newid                      Number of groups   =         750

R-sq:                                      Obs per group:
     within  = 0.0764                                    min =           7
     between = 0.2990                                    avg =         7.0
     overall = 0.1918                                    max =           7

                                           Wald chi2(4)       =      652.47
corr(u_i, X)    = 0 (assumed)              Prob > chi2        =      0.0000
```

esteem	Coef.	Std. Err.	z	P>|z|	[95% Conf. Interval]	
stress	-.0828434	.0154533	-5.36	0.000	-.1131313	-.0525555
cohes	.2159616	.009186	23.51	0.000	.1979573	.2339659
male	2.064147	.3168456	6.51	0.000	1.443141	2.685153
nonwhite	.0632558	.495001	0.13	0.898	-.9069284	1.03344
_cons	-1.033907	.2311983	-4.47	0.000	-1.487047	-.5807662
sigma_u	3.915267					
sigma_e	4.7795564					
rho	.40156982	(fraction of variance due to u_i)				

estimates store random   ** stores the next estimates for the Hausman test*

As mentioned earlier, the random-effects model allows time-invariant explanatory variables. Note that the coefficient for stressful life events is different in this model than in the fixed-effects model. Not surprisingly, its interpretation is also different. In this example, the coefficient estimates the *average* effect of changes in stressful life events over time and between individuals on changes in self-esteem. Moreover, as assumed by this type of model, the correlation between the unobserved heterogeneity and the explanatory variables is zero. We can test whether this assumption is valid by using the aforementioned Hausman test.

Examples 8.2 and 8.3 (Continued)

hausman fixed random   * *requests the Hausman test*

| | —— Coefficients —— | | | |
| | (b) | (B) | (b-B) | sqrt(diag(V_b-V_B)) |
	fixed	random	Difference	S.E.
stress	-.0589644	-.0828434	.023879	.0052947
cohes	.1894794	.2159616	-.0264822	.0043917

```
                        b = consistent under Ho and Ha; obtained from xtreg
            B = inconsistent under Ha, efficient under Ho; obtained from xtreg

    Test:  Ho:  difference in coefficients not systematic

            chi2(2) = (b-B)'[(V_b-V_B)^(-1)](b-B)
                    =        49.85
            Prob>chi2 =      0.0000
```

The Hausman test suggests that we should reject the null hypothesis and assume that the unobserved heterogeneity is correlated with the explanatory variables. Thus, the fixed-effects results are preferred for this model. Nonetheless, it is important to note what happened to the time-invariant covariates in the fixed-effects model: since it directly considers only time-varying effects, the variables (*male* and *nonwhite*) are dropped from consideration. However, their effects are not completely absent. Rather, they are subsumed under the *sigma_u* part of the model. This should cause us to pause and consider whether or not our conceptual model treats time-invariant explanatory variables as important. If, say, the relationship between self-esteem and sex is something we wish to focus on, then the fixed-effects model may be disadvantageous since we cannot determine their statistical association. However, if the goal is to estimate changes within individuals and adjust for unobserved heterogeneity, then the fixed-effects model offers a preferred approach.

There are situations where we can use a time-invariant covariate in a fixed-effects model; however, it is actually part of a time-variant explanatory process. Suppose we hypothesize that the association between changes in stressful life events and changes in self-esteem varies by sex, with males and females demonstrating different patterns. In this case, we might consider including an interaction term between *male* and *stress*, as in Example 8.4.

Example 8.4

```
qui xtreg esteem c.stress#i.male cohes, fe
margins male, at(stress=(0 10 20 30 40 50)) atmeans
marginsplot, noci
```

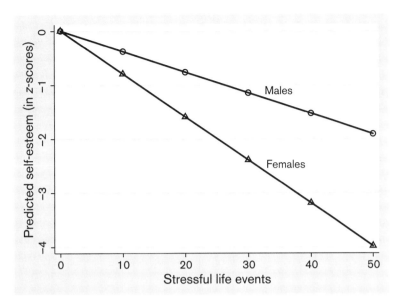

FIGURE 8.2

The graph shown in figure 8.2 suggests that changes in stressful events are more consequential for females than for males. Although this supports previous studies of adolescent
self-esteem (Baldwin and Hoffmann, 2002), consider the model upon which the figure is
based. What does it show regarding the interaction term?

Fixed- and random-effects linear models make the same assumptions as other linear
regression models, so, although testing for unobserved heterogeneity is critical, other tests
should also be considered. One complication is that the residuals have two components:
within-individual and between-individual components. Thus, we have two types of residuals
to examine for violations of assumptions such as heteroscedastic errors and non-normality.
Examining the post-estimation options available for Stata's xtreg command shows how to
save these residuals, as well as the predicted values. One common issue with longitudinal
data is autocorrelation among the errors (see Chapter 1 and the discussion later in this chapter), so some analysts recommend using the cluster option with the person-level identifi-
cation variable to guard against this problem.

As an exercise, save the combined residual, which Stata labels as *ue*, and examine its distribution. Does it follow a normal distribution? What implications does the answer to this
question have for the model? Should it be modified in some way?

GENERALIZED ESTIMATING EQUATIONS FOR LONGITUDINAL DATA

A more general form of longitudinal analysis that offers greater flexibility than the random-
effects models falls under a class of models known as *generalized estimating equations*

(*GEEs*).[2] GEEs may be used to estimate most GLMs, such as those available in Stata's `glm` command. An advantage of GEEs over other methods of longitudinal data analysis is that they allow different correlation structures to be assumed for the within-unit portion of the model. In other words, we may impose a correlation structure on the within-unit errors and compare different types to arrive at the best fitting model.

To see how this operates, return to our model of delinquency in the *event4* data set. Recall that some adolescents reported no delinquent activities across the 4 years, some reported activities only during 1 year, and so forth. When we estimated a model of delinquency using discrete-time event history procedures, we were forced to assume that the observations were uncorrelated within individuals over time. Clearly this is an unrealistic assumption. Delinquency is bound to be correlated over time within individuals, even after considering several explanatory variables. The GEE approach allows us to consider several possible correlation structures. As with the fixed- and random-effects models, the `xtset` command is used to identify the id variable and the time variable. The `xtgee` command is then invoked (see Example 8.5).

Example 8.5

```
xtset newid year
xtgee delinq stress cohes, family(binomial) link(logit)
    corr(independent) eform      * eform provides exponentiated coefficients
```

```
GEE population-averaged model                Number of obs     =       2,604
Group variable:                       newid  Number of groups  =         651
Link:                                 logit  Obs per group:
Family:                            binomial                       min =       4
Correlation:                    independent                       avg =     4.0
                                                                   max =       4
                                             Wald chi2(2)      =      216.43
Scale parameter:                          1  Prob > chi2       =      0.0000

Pearson chi2(2604):                 2592.09  Deviance          =     3114.66
Dispersion (Pearson):               .995427  Dispersion        =    1.196106
```

delinq	Odds Ratio	Std. Err.	z	P>|z|	[95% Conf. Interval]	
stress	1.204178	.0298121	7.50	0.000	1.147142	1.264049
cohes	.9575221	.0036913	-11.26	0.000	.9503145	.9647843
_cons	3.251671	.7081729	5.41	0.000	2.121904	4.982962

This model asks for a binomial model with a logit link (logistic regression) and an independent correlation structure. To see what this latter aspect means, we may use the `xtcorr` command after the `xtgee` command (`estat wcorrelation` also works).

---

2. As shown later in the chapter, the random-effects model is actually a special case of a GEE.

Example 8.5 (Continued)

xtcorr

```
Estimated within-newid correlation matrix R:

        c1      c2      c3      c4
r1   1.0000
r2   0.0000  1.0000
r3   0.0000  0.0000  1.0000
r4   0.0000  0.0000  0.0000  1.0000
```

Although there are modest differences, these results are very close to those from the discrete-time event history model using the `logit` command (see Example 7.17). This is because the correlation structures of the errors are identical; both assume zero correlations across years within individuals. For example, after accounting for the variation in delinquency due to stressful life events and cohesion, we assume that the residual variance at, say, time 1 is not correlated with the residual variance at time 2 or 3.

This is an unreasonable assumption for longitudinal data, so consider the correlation structure in Example 8.6.

Example 8.6

xtgee delinq stress cohes, family(binomial) link(logit)
    corr(exchangeable) eform

```
GEE population-averaged model              Number of obs      =     2,604
Group variable:                    newid   Number of groups   =       651
Link:                              logit   Obs per group:
Family:                         binomial                min =         4
Correlation:                exchangeable                avg =       4.0
                                                         max =         4
                                           Wald chi2(2)       =    118.79
Scale parameter:                       1   Prob > chi2        =    0.0000
```

delinq	Odds Ratio	Std. Err.	z	P>|z|	[95% Conf. Interval]	
stress	1.161748	.0288896	6.03	0.000	1.106483	1.219773
cohes	.9662708	.0039744	-8.34	0.000	.9585125	.974092
_cons	2.203419	.5097001	3.42	0.001	1.400216	3.467361

xtcorr

```
Estimated within-newid correlation matrix R:

        c1      c2      c3      c4
r1   1.0000
r2   0.3458  1.0000
r3   0.3458  0.3458  1.0000
r4   0.3458  0.3458  0.3458  1.0000
```

Here we assume that the correlations are identical regardless of which years are being compared. Note that the coefficients and standard errors are different in this model: both coefficients are closer to zero and the standard errors differ slightly. Using an exchangeable correlation structure is similar to using a random-effects model, such as xtlogit.

Next, assume that the correlations are larger for time points that are closer together. Example 8.7 shows an autoregressive correlation structure.

Example 8.7

```
xtgee delinq stress cohes, family(binomial) link(logit)
    corr(ar1) eform
```

```
GEE population-averaged model              Number of obs      =     2,604
Group and time vars:          newid year   Number of groups   =       651
Link:                               logit   Obs per group:
Family:                          binomial                min =         4
Correlation:                        AR(1)                avg =       4.0
                                                         max =         4
                                             Wald chi2(2)      =    119.54
Scale parameter:                        1    Prob > chi2       =    0.0000
```

delinq	Odds Ratio	Std. Err.	z	P>|z|	[95% Conf. Interval]	
stress	1.157053	.0289306	5.83	0.000	1.101717	1.215168
cohes	.9660144	.0039698	-8.41	0.000	.9582649	.9738265
_cons	2.220481	.5141274	3.45	0.001	1.410461	3.495693

```
    xtcorr
```

Estimated within-newid correlation matrix R:

```
        c1        c2        c3        c4
r1   1.0000
r2   0.4181   1.0000
r3   0.1748   0.4181   1.0000
r4   0.0731   0.1748   0.4181   1.0000
```

The coefficients and standard errors are similar to those from the previous model. However, the correlation structure is different since we assume that the correlations are smaller as more years separate the adolescents. This *decay*, as it is often called, follows an autoregressive(1) pattern: $\{(t-1)^{q+1}, (t-2)^{q+2}, \ldots\}$. In the model we just estimated, the correlations are $0.418^2 = 0.175$ and $0.418^3 = 0.073$. This structure seems more realistic for our model of delinquent behavior.

Although there are several other possibilities, we examine just one more correlation structure: unstructured. This simply allows the data and the estimated model to dictate the correlations. Thus, they can differ across each pair of years (see Example 8.8).

Example 8.8

```
xtgee delinq stress cohes, family(binomial) link(logit)
    corr(unstructured) eform
```

```
GEE population-averaged model              Number of obs     =        2,604
Group and time vars:          newid year   Number of groups  =          651
Link:                               logit  Obs per group:
Family:                          binomial                    min =          4
Correlation:                 unstructured                    avg =        4.0
                                                             max =          4
                                           Wald chi2(2)      =       112.79
Scale parameter:                      1    Prob > chi2       =       0.0000
```

delinq	Odds Ratio	Std. Err.	z	P>|z|	[95% Conf. Interval]	
stress	1.15793	.0289391	5.87	0.000	1.102577	1.216061
cohes	.9671169	.0039891	-8.11	0.000	.9593299	.9749671
_cons	2.092143	.4861042	3.18	0.001	1.326835	3.298875

```
    xtcorr
```

```
Estimated within-newid correlation matrix R:

        c1       c2       c3       c4
r1  1.0000
r2  0.3858  1.0000
r3  0.2851  0.4589  1.0000
r4  0.2159  0.3233  0.4118  1.0000
```

It appears we have converged to a consistent set of results. All three of the models that relax the assumption of no correlation within units display very similar results in terms of odds ratios and z-values. According to the unstructured model, there is some decay in the correlations, but not as extreme as the autoregressive(1) structure indicates. Thus, it appears the correlation structure of the within-unit residual variation is important for specifying the correct model.

Assuming we have arrived at a consistent and reasonable model, we may then consider what it indicates about the association between the explanatory variables and the probability of delinquency. For instance, after reestimating the GEE model with the unstructured correlation matrix, use the margins command to estimate predicted values. This is followed by marginsplot to graph the results. We examine the association between family cohesion and the probability of delinquency.

```
    qui margins, at(cohes=(20 40 60)) atmeans      * after Example 8.8
    marginsplot, noci
```

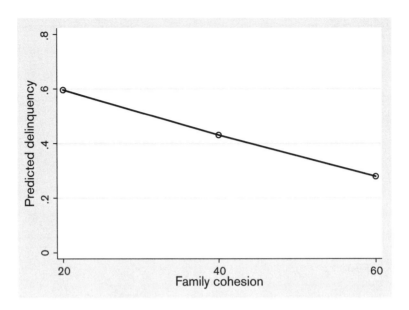

FIGURE 8.3

The graph in figure 8.3 shows a substantial association between levels of family cohesion and the probability of engaging in delinquent activities. Note that predicted probability decreases by about 0.3 units from the lowest to the highest levels of cohesion.

Consider that each of the `xtgee` models is labeled as a *GEE population-averaged model*. A population-averaged estimator differs from a random-effects estimator in the way they are interpreted. For instance, compare two longitudinal models using the *esteem* data set. Use a logistic regression model to estimate a dichotomous measure of depressive symptoms. In the data set this is the variable *bi_depress* (type `codebook bi_depress`). First, estimate an `xtgee` population-averaged model (see Example 8.9).

Example 8.9

```
xtgee bi_depress male, family(binomial) link(logit) eform
```

```
GEE population-averaged model          Number of obs     =      5,250
Group variable:                 newid  Number of groups  =        750
Link:                           logit  Obs per group:
Family:                      binomial                     min =          7
Correlation:              exchangeable                     avg =        7.0
                                                           max =          7
                                       Wald chi2(1)      =      32.68
Scale parameter:                    1  Prob > chi2       =     0.0000
```

bi_depress	Odds Ratio	Std. Err.	z	P>|z|	[95% Conf. Interval]	
male	.5797514	.0552868	-5.72	0.000	.4809155	.6988998
_cons	.785521	.0509893	-3.72	0.000	.6916797	.8920938

Since *male* is coded as {0 = female; 1 = male), the interpretation of the coefficient is that the expected odds that the *average* male is depressed is 0.58 times the expected odds that the *average* female is depressed. Another way of saying this is that it represents the expected odds that a male picked at random is depressed compared to the expected odds that a female picked at random is depressed.

The alternative is the random-effects model that we used earlier (fixed-effects does not make sense since the variable *male* is time-invariant). In this model (see Example 8.10), the interpretation involves the odds of a person being depressed if a male compared with the odds *of the same person* being depressed if a female. Thus, we would have to claim from this model that the odds of being depressed if male is 0.46 times the odds of being depressed if female for the same person. This does not make sense. (Actually, we normally interpret the random-effects coefficient in terms of changing conditions for the individual, which is conceivable if a gender transition occurs.)

Example 8.10

```
xtlogit bi_depress male, re or
```

```
Random-effects logistic regression          Number of obs      =        5,250
Group variable: newid                       Number of groups   =          750

Random effects u_i ~ Gaussian               Obs per group:
                                                          min =            7
                                                          avg =          7.0
                                                          max =            7

Integration method: mvaghermite             Integration pts.   =           12

                                            Wald chi2(1)       =        32.17
Log likelihood  = -3061.8922                Prob > chi2        =       0.0000
```

bi_depress	OR	Std. Err.	z	P>|z|	[95% Conf. Interval]	
male	.4644421	.0627999	-5.67	0.000	.3563163	.6053793
_cons	.7052378	.0662572	-3.72	0.000	.586631	.8478247
/lnsig2u	.8670089	.094022			.6827291	1.051289
sigma_u	1.542654	.0725217			1.406866	1.691548
rho	.4197405	.0228999			.3756347	.4651668

LR test of rho=0: <u>chibar2(01) = 741.46</u>   Prob >= chibar2 = 0.000

It is more sensible to use the random-effects model for explanatory variables that can change (even if they might not in the sample). In this data set, for instance, the variable *fstruct* indicates whether or not the adolescent lived with both parents. Suppose we estimate the association between this variable and depressive symptoms using a random-effects logistic regression model (see Example 8.11).

Example 8.11

```
xtlogit bi_depress fstruct, re or
```

```
Random-effects logistic regression        Number of obs     =       5,250
Group variable: newid                     Number of groups  =         750

Random effects u_i ~ Gaussian             Obs per group:
                                                        min =           7
                                                        avg =         7.0
                                                        max =           7

Integration method: mvaghermite           Integration pts.  =          12

                                          Wald chi2(1)      =        7.48
Log likelihood  = -3074.1835              Prob > chi2       =      0.0062
```

bi_depress	OR	Std. Err.	z	P>|z|	[95% Conf. Interval]	
fstruct	.6854874	.0946232	-2.74	0.006	.5229997	.8984576
_cons	.595356	.0615295	-5.02	0.000	.4861898	.7290338
/lnsig2u	.9156763	.0931057			.7331925	1.09816
sigma_u	1.580653	.0735839			1.442815	1.731659
rho	.431638	.0228413			.3875423	.4768447

```
LR test of rho=0: chibar2(01) = 787.44          Prob >= chibar2 = 0.000
```

Here we may claim that the odds of being depressed if living with both parents is 0.69 times the odds of being depressed if one does not live with both parents. In this situation, since adolescents can live with both parents or not at different times of their lives, the interpretation is reasonable.

As mentioned earlier in the chapter, most of the assumptions of the various regression models discussed in earlier chapters also apply to the fixed- and random-effects models. Similarly, they apply to GEE models as well. The key assumption that is relaxed in these models is independence of observations: they are designed for data in which observations are dependent. Yet assumptions involving other issues such as homoscedastic errors, linearity, measurement error, collinearity, and model specification remain important. Moreover, influential observations can create the same difficulties with these models as with any other regression models.

Unfortunately, Stata's xtgee and related xt commands offer few post-estimation options designed for checking these assumptions. Fortunately, there are ways to compute post-estimation residuals and leverage values in Stata that are not overly taxing from a programming perspective (see Hardin and Hilbe, 2013). And these may be used to examine the assumptions inherent in these models. As an exercise, estimate an xtgee model with *esteem* as the outcome variable and *stress, cohes, male,* and *nonwhite* as explanatory variables. Then, generate standardized residuals and explore them. Notice that there are some large negative outliers. What are some steps you might take to consider these further? What consequences might they have for the model and its results?

FINAL WORDS

Many research projects in various scientific disciplines collect data over time. Researchers study the growth of vocabulary words among toddlers, changes in test scores among students, shifts in criminal activity among adults, changes in health conditions among the elderly, and the dynamics of family relations among parents and children, to name only a few interests. Yet, because observations are not independent in longitudinal data, regression models require some modification to estimate these outcomes. Fixed-effects, random-effects, and GEE models offer flexible tools for these types of data. However, we merely scratched the surface of the statistical possibilities when longitudinal data are available. Chapter 9 describes another approach, but books on longitudinal modeling provide much more depth (e.g., Rabe-Hesketh and Skrondal. 2012; Skrondal and Rabe-Hesketh, 2004).

EXERCISES FOR CHAPTER 8

The *workdata* data set (*workdata.dta*) has information from a large sample of adults about their hourly wages and a few other aspects of their education and work life. The data cover a 20-year period. We wish to predict hourly wages using a set of explanatory variables. Use a GEE to estimate the model. The hourly wage variable (ln_wage) is logged and adjusted for inflation. Thus, use an appropriate GEE model to predict this hourly wage variable with the following explanatory variables: collgrad, union, black, otherrace, age, and age-squared (note that we are using white as the reference category for race/ethnicity).

1. Explain which type of GEE model you used and why you selected it.

2. Interpret the coefficients associated with the following variables: *collgrad* and *union.*

3. Save the predicted values from the model and construct a graph that has average predicted logged hourly wages on the *y*-axis and *age* on the *x*-axis. Explain what the graph indicates about the association between wages and age.

## NINE

Multilevel Regression Models

A number of scientific disciplines utilize data that are quantitative and hierarchical. While the term quantitative is clearly understood, the term hierarchical is appreciated less often. *Hierarchical data* refer to information that is collected at multiple units of observation. Units of observation can include things such as informal social groups, schools, organizations, neighborhoods, counties, or nation-states. This can also include individuals who are followed over time in a longitudinal design, as discussed in Chapter 8. In this situation, time is nested within different individuals, so there are two levels. Some additional examples may help clarify this general type of data structure.

First, sociologists and education researchers often collect information from students, teachers, and school administrators. Since such data collection is usually expensive and time-consuming, researchers typically collect data on samples of students in particular classrooms, the teachers who direct these classrooms, and the administrators to whom these teachers report. Hence, students are grouped or nested within classrooms and teachers are grouped or nested within schools. In practice, the researcher has data on individual characteristics from students, classroom characteristics from teachers, and school characteristics from administrators.

Second, clinical psychologists frequently conduct research on the effectiveness of various therapies to treat mental disorders such as depression, anxiety, or phobias. Data are collected on patients who are undergoing specific types of therapies so that these regimens may be compared directly. To make studies efficient, the patients are often drawn from lists of specific therapists, some of whom provide each type of therapeutic regimen. Consider how this is similar to the data collection effort described in the previous paragraph: patients are nested within types of therapies, but also by specific therapists. Any one therapist may, for example, treat several of the patients who are participating in the study. Yet these therapists also have particular characteristics (perhaps some are more empathetic than others) that may affect treatment and, ultimately, the outcome of interest (which may be recovery or relapse, depending on the project).

Third, urban anthropologists interested in cultural and social aspects of city life often combine rich descriptive data on individual lives, behavioral outcomes, social exchanges and interactions, and community characteristics in various urban areas. While perhaps not lending themselves directly to quantitative analysis, advances in qualitative and network analysis allow these data to be quantified and analyzed with statistical techniques. Once again, though, these types of data are hierarchical in the sense that individuals are nested within groups or social networks and these networks interact in different community contexts. Anthropologists might be interested in whether characteristics of these networks or communities affect the behaviors of individuals, or in how community factors affect interactions among groups.

In order to provide a concrete example of hierarchical or *multilevel data*, consider the table below. It shows a small selection of data from a study that sampled residents of 99 towns in an unidentified province. The data are from the *multilevel_data12* data set (*multilevel_data12.dta*). The variable *idcomm* references the particular towns in the data set, whereas the *id* variable indicates individuals who live in these towns. The variable *income* is a categorical variable that gauges annual household income. It is simple to see that there are multiple survey respondents who are nested within each town. In the parlance of research that utilizes multilevel data, the town level is referred to as *level-2* and the individual level is referred to as *level-1*. If we were fortunate enough to get data on towns from multiple provinces, we could also have a level-3: the province level.

	idcomm	id	income
1.	25	10001	2
2.	25	10004	4
3.	25	10005	3
4.	25	10008	2
5.	25	10009	8
100.	30	20008	6
101.	30	20009	2
102.	30	20010	4
103.	30	20011	3
104.	30	20012	2

There are numerous other examples that may be drawn from a variety of disciplines. Although theory often suggests, sometimes rather vigorously, that this data hierarchy is important in determining relationships among elements of the "groups," a common approach when facing such data has been to ignore the hierarchy or treat it as an analytic nuisance that must be corrected through some statistical procedure.

Unfortunately, using quantitative techniques that ignore the hierarchical nature of data may lead to incorrect inferences and conclusions. As we know from previous chapters, an important assumption of many regression models is that the observations (or, more precisely, the errors) are statistically independent; that is, that observations do not influence one another or share unobserved characteristics. Yet there is clear evidence that students grouped in classrooms, to use a well-known example, both influence one another and share characteristics that affect numerous outcomes. Using an ordinary least squares (OLS) regression model, or most of the generalized linear models (GLMs) discussed thus far, to determine whether a student-level characteristic, such as self-esteem, affects an outcome, such as being disruptive in the class, is inappropriate because students who share a classroom are clearly not independent. The result of using such a statistical model might be incorrect inferences about the true effect of the explanatory variables on the outcome variable. Thus, any conclusions drawn from this type of statistical exercise may be misleading.

A simple approach to consider when data are hierarchical and the analyst does not care about the potential effects of group-level variables or variability across groups is to identify a variable that gauges the clustering structure, such as the *idcomm* variable shown earlier. This variable is then used to adjust the standard errors in a regression model. Details of this type of approach are provided in texts that concern survey sampling (e.g., Heeringa et al., 2010). In Stata, the group identifier (variable that identifies how the data are clustered or nested) is used along with the `cluster` command to adjust the standard errors for the non-independence of observations. For example, suppose we are interested in predicting the *income* variable shown earlier using as explanatory variables *male* and *married*. An OLS regression model that ignores the clustering by town is provided in Example 9.1.

Example 9.1

```
regress income male married
```

Source	SS	df	MS		
Model	7937.50228	2	3968.75114		
Residual	35914.122	9,856	3.64388413		
Total	43851.6243	9,858	4.4483287		

Number of obs =	9,859
F(2, 9856) =	1089.15
Prob > F =	0.0000
R-squared =	0.1810
Adj R-squared =	0.1808
Root MSE =	1.9089

income	Coef.	Std. Err.	t	P>|t|	[95% Conf. Interval]	
male	.5187255	.039416	13.16	0.000	.4414621	.5959889
married	1.74758	.0422214	41.39	0.000	1.664817	1.830342
_cons	2.930973	.0363492	80.63	0.000	2.859721	3.002225

The standard errors in this model are likely biased because of the nonindependence of observations. However, the `cluster` command may be used to adjust the standard errors for the possible dependence due to sample members who live in the same town.

Example 9.2

```
regress income male married, cluster(idcomm)
```

```
Linear regression                          Number of obs   =      9,859
                                           F(2, 98)        =    1091.79
                                           Prob > F        =     0.0000
                                           R-squared       =     0.1810
                                           Root MSE        =     1.9089
```

(Std. Err. adjusted for 99 clusters in idcomm)

income	Coef.	Robust Std. Err.	t	P>|t|	[95% Conf. Interval]	
male	.5187255	.0420724	12.33	0.000	.4352342	.6022168
married	1.74758	.0441666	39.57	0.000	1.659933	1.835227
_cons	2.930973	.0406193	72.16	0.000	2.850365	3.01158

In Example 9.2, the standard errors have been adjusted for the clustering of individuals within the 99 clusters (towns) identified by *idcomm*. Notice that the standard errors are larger in the model that takes into account the hierarchical data structure. This is often the case. However, if there is no clustering effect, the standard errors are the same in both models. Stata can also deal with complex survey designs, some of which have multiple levels of data, with its set of survey commands (see `help svy` and `help svy estimation`). Nonetheless, correct model specification is still needed regardless of how the clustering is addressed. Moreover, when there is a sparse number of level-2 units, using a clustering approach to correct the standard errors can lead to biases.

But there are many studies that are designed to examine not only individual effects of variables on one another, but also how these effects may vary across groups defined by schools, organizations, neighborhoods, and so forth. Moreover, researchers are often interested in the effects of group-level variables on individual-level outcomes. Fortunately, there are statistical models that allow researchers to take advantage of hierarchical or multilevel data structures. These are known as *multilevel models*.

THE BASIC APPROACH OF MULTILEVEL MODELS

Multilevel models—also known as *hierarchical models* and *mixed-effects models*—stem from developments in agricultural statistics, educational statistics, biometry, and econometrics. It was recognized long ago in agricultural studies of crop growth that an efficient design called for growing sets of plants in shared environments. Yet if one wished to extrapolate findings to the broader "population" of plants, one needed to be able to consider that the relationships found in an analysis might be "random." Hence, as mentioned in Chapter 8, a model that sees these relationships as different, or random, rather than fixed across units of, for example, agricultural plots are called *random-effects models*. Not only do the characteristics of a plant vary (since we have many plants to examine), but the relationships between these characteristics (e.g., how much water the plant receives) and some outcome of interest (e.g., plant growth) also vary.

In general, a multilevel model separates the variability of the outcome variable into two components: the variability that is due to differences across individual units (e.g., plants, people) and the variability that is due to differences across groups (e.g., plots of land, schools, neighborhoods) (Snijders and Bosker, 2012). Recall that random- and fixed-effects models, as well as generalized estimating equation (GEE) models, did the same thing: they also consider the clustering of data and partition the variability. However, their interest tends to focus on changes within individuals over time (see Chapter 8).[1]

As an example, reexamine the *multilevel_data12* data set. Recall that it includes individual respondents grouped within towns. We continue to focus on the variable *income*. This variable has an overall mean of 4.36 (recall that it is measured in categories), a standard deviation of 2.11, and a variance of 4.45. However, think about the towns that the respondents reside in. Each town may have a different mean, reflecting higher or lower average incomes among respondents. Thus, whereas there is overall variability among the individuals in the data set, there is also variability across towns. Another way of stating this is that, although there is an overall fixed mean in the data set, there is also random variation in the means across towns. This is where the term *random-effects* comes from.

For instance, examine the Stata results in Example 9.3 that list means of the *income* variable for several of the towns in the data set. Notice that town 25 has a mean of 3.74 and town 30 has a mean of 3.88, but town 35 has a mean of about 4.75. At least upon first glance, it appears that there is variation in means across towns that should be taken into account.

Example 9.3

`mean income, over(idcomm)`   * note: the output is edited

Over	Mean	Std. Err.	[95% Conf.	Interval]
income				
25	3.739583	.210599	3.326766	4.1524
30	3.877778	.2028682	3.480115	4.275441
35	4.747368	.2024311	4.350562	5.144175
45	4.339623	.1914088	3.964422	4.714823
50	4.375	.2164309	3.950751	4.799249
55	4.358974	.2400011	3.888523	4.829426
80	3.412371	.1903789	3.03919	3.785553
105	5.294118	.2299821	4.843306	5.74493
135	4.106796	.2087523	3.697599	4.515993
225	5.540541	.2109934	5.12695	5.954131
240	4.152381	.1997164	3.760896	4.543866
285	4.241071	.2010452	3.846982	4.635161

1. Fixed- and random-effects models are also used with multilevel data collected over spatial units, such as schools, counties, or countries (Arceneaux and Nickerson, 2009). Moreover, a subsequent section of this chapter shows how to use multilevel models with longitudinal data.

Identifying the variability across individuals and communities is similar to conducting a one-way analysis of variance (ANOVA) model. Just as we wish to compare variances within units and across units in a one-way ANOVA model (or in a fixed- or random-effects model; see Chapter 8), here we compute and compare the variance that is due to individual-level variation and that is due to community-level variation. We may also set up a regression model using the following equation:

$$y_{ij} = \alpha_j .$$

In this notation, which you may recognize as an intercept-only model, the y is subscripted to indicate that there are multiple observations (i), but there are also multiple groups (designated by the j subscript) within which these observations are nested. The α is subscripted with a j because, as suggested by Example 9.3, there are multiple means across the groups. In a multilevel model, we estimate three parameters: the overall mean (α_0, or *grand mean*, as it is often called), the between-group variance (σ_u^2), and the within-group variance (σ_e^2).[2] In Stata, one way to estimate these parameters is by using the `mixed` command.

This command requires two things: an outcome variable and the variable that identifies how individuals are grouped. As we learned earlier, in the *multilevel_data12* data set this latter variable is *idcomm*. The Stata code in Example 9.4 returns a one-way ANOVA for the variable *income*. The double bar separates the command into the fixed-effects component and the random-effects component (hence the term mixed-effects). The random effects are allowed to vary across communities that are designated by the *idcomm* variable.

The results indicate that there are 9,859 individual observations nested within 99 groups (towns). The minimum number of observations in a town is 70 and the maximum is 119, with a mean of almost 100 per town. The first panel indicates the grand or overall mean of income, which is about 4.36. This is known as the *fixed-effects estimate* for this model.

---

2. These may also be referred to as *between-group* and *between-individual* variability. However, these terms are dependent on the type of hierarchical data. As shown later in the chapter as well as in Chapter 8, with longitudinal data time periods are nested within individuals so the terms *between-individual* and *within-individual* variability are often used.

Example 9.4

`mixed income || idcomm:`

```
Mixed-effects ML regression                        Number of obs     =      9,859
Group variable: idcomm                             Number of groups  =         99

                                                   Obs per group:
                                                                 min =         70
                                                                 avg =       99.6
                                                                 max =        119

                                                   Wald chi2(0)      =          .
Log likelihood = -21234.631                        Prob > chi2       =          .
```

income	Coef.	Std. Err.	z	P>|z|	[95% Conf. Interval]	
_cons	4.361973	.0463907	94.03	0.000	4.271049	4.452897

Random-effects Parameters	Estimate	Std. Err.	[95% Conf. Interval]	
idcomm: Identity				
var(_cons)	.1697445	.030322	.1196033	.2409065
var(Residual)	4.279162	.0612567	4.160769	4.400923

LR test vs. linear model: <u>chibar2(01) = 223.21</u>   Prob >= chibar2 = 0.0000

The panel labeled *Random-effects Parameters* shows the estimates for the variability of *income*, or the random-effects portion of the model. As described earlier, there are two estimates. The first is the variation that is due to differences between communities (`var(_cons)` or σ_u^2). The second is variation that is due to differences within communities (between individuals; `var(Residual)` or σ_ε^2). The latter is larger than the former; this is common in community studies since variability in attitudes, behaviors, and so forth is typically larger across individuals than it is across groups. One way to characterize this variability is with a ratio measure of the community-level variance to the total variance. For this model, we may compute this as

$$\rho(\text{ICC}) = \frac{\sigma_u^2}{\sigma_u^2 + \sigma_\varepsilon^2} = \frac{0.170}{1.70 + 4.28} = 0.038 \ .$$

This quantity is called the *intraclass correlation*, or ICC (Stata computes it, along with standard errors and confidence intervals, with the following post-estimation command `estat icc`). It measures the proportion of variability in the outcome that is between groups (towns). Thus, about 3.8% of the variability in *income* is between towns; most of the variability occurs within towns. The ICC is also called the *cluster effect*. Note that it is the same as the quantity *rho* that is estimated in the random-effects model discussed in Chapter 8.[3]

The ICC is useful because if there is no effect of clustering, then a regression model that does not account for it may be used. If it is a substantial number, then it is important to take into account the clustering since this suggests that the observations are not independent. Although there is no standard rule of thumb for the size of the ICC, some analysts recommend that it be used to compute the *design effect*, which is the ratio of the sampling variance for the statistic computed using the multilevel design divided by the sampling variance that would have been obtained assuming a simple random sample. A design effect of two or more is usually indicative that the clustering is substantial enough to warrant a multilevel model (or some other statistical adjustment). However, there are important exceptions to this, such as when analysts are interested in the effects of level-2 explanatory variables or when cluster sizes are small (Lai and Kwok, 2015). The ICC may be combined with the average group size to yield the design effect:

$$d^2 = 1 + \left(\widehat{\text{group size}} - 1\right) \times \text{ICC} = 1 + (99.6 - 1) \times 0.038 = 4.75$$

The 99.6 portion of the equation is the average group size given in the Stata multilevel printout. Notice that the larger the average group size, the larger the design effect. Since the design effect is substantially greater than two, it seems best to rely on a multilevel model for this outcome variable.

Stata also provides a likelihood ratio χ^2 test at the bottom of the printout that compares a linear regression model to the multilevel model. The null hypothesis is that they are identical, statistically speaking. Here we may reject the null hypothesis with confidence and conclude that there is a substantial hierarchical effect for this outcome. However, as Stata warns us, this χ^2 test is conservative because its reported significance level is an estimated upper bound of the actual significance level.

An issue that is raised often with multilevel models is how many level-1 units there should be per level-2 unit. For instance, suppose we have, on average, only four or five individuals per school or neighborhood in our data set. Can we still use a multilevel model? The answer to this question depends on several issues, such as how many level-2 units are available, ICC, the number of explanatory variables in the model and their interrelationships, and whether the analyst is interested in individual-level or aggregate-level associations. Several studies have shown that having a sufficient number of level-2 units is especially important (see Snijders and Bosker, 2012), but definitive rules of thumb are difficult to come by.

---

3. As shown later in the chapter, the random-effects model described in Chapter 8 is virtually identical to the multilevel model with a random intercept term that is shown here.

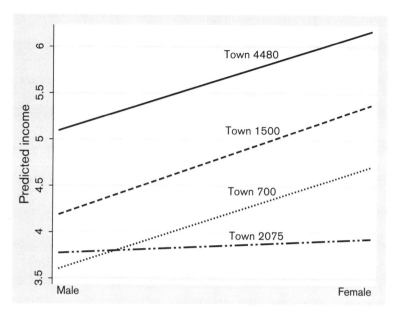

FIGURE 9.1

NOTE: Of course, this is a poor example of a graph since the variable *male* is a dummy variable.

THE MULTILEVEL LINEAR REGRESSION MODEL

Now consider a hypothesis that suggests a link between income and another variable in the data set. One hypothesis we may wish to consider is whether males have higher incomes than females across the various towns. Hence, we explore the association between the variables *male* and *income*. Although we already know that mean income varies across communities, our concern now changes: we ask whether the regression slopes that indicate the association between *male* and *income* vary across communities. In other words, rather than investigating only the random intercept, we also examine a random slope.

One way to visualize this is with a graph. Consider figure 9.1 that shows linear slopes for four different towns: 700, 1500, 2075, and 4480. In other words, we have four distinct regression models that display the linear association between sex and income among respondents in the particular towns. Notice that there are differences in the slopes for these four towns. The slopes for three of the towns appear to be roughly parallel, although the intercepts are different. But town 2075 has a noticeably flatter slope than the other towns. As with the means, we may also consider slopes for each of the 99 towns and assess their differences.

```
graph twoway lfit income male if idcomm==700 ||       ///
lfit income male if idcomm==1500 ||       ///
lfit income male if idcomm==2075 ||       ///
lfit income male if idcomm==4480
```

In general, then, we may examine two questions:

1. Do the mean levels of income differ across towns?
2. Do the linear associations between the sex of residents and their incomes differ across towns?

First, though, note that the OLS regression equation that does not take the hierarchical structure into account is as follows:

$$\widehat{\text{income}} = 3.989 + 0.834(\text{male})$$

Thus, with a very small p-value, we may say that, on average, males report higher annual household incomes than females. Females report, on average, annual incomes of about 3.99 units, whereas males report average household incomes about 0.83 units higher (4.82). But, again, this is an overall effect that does not take into account differences across towns.

Using Stata's `mixed` command, Example 9.5 illustrates a random-effects model that simply takes into account the clustering of individuals within towns to estimate the variability within and across groups. As noted earlier, this is commonly called a *random-intercept model* since the intercept is allowed to vary randomly across the groups.

The `xtreg` command we saw in the last chapter may also be used to estimate the random-intercept model. The main difference is that Stata's `mixed` command uses restricted maximum-likelihood (REML) estimation and its `xtreg` command uses generalized least squares (GLS) estimation. Both models (see the `xtreg` model in Example 9.6) indicate that the average intercept—or predicted income for females—is about 4.00. Moreover, males, on average, report about 0.8 units more on this income variable than do females. However, as shown by the variance of the intercept in the random-effects panel of the `mixed` model (and the `sigma_u` coefficient in the random-effects GLS model, which is in standard deviation units), the intercept varies across the towns. Note that the ICC (*rho*) that may be computed from the `mixed` model is almost identical to the value `rho` in the `xtreg` printout:

$$\texttt{mixed}\ rho\ (\rho) = 0.16/(0.16 + 4.12) = \textbf{0.037}.$$

Thus, as mentioned earlier, the `rho` value in Stata's `xtreg` model is the ICC.

Example 9.5

```
mixed income male || idcomm:
```

```
Mixed-effects ML regression              Number of obs     =       9,859
Group variable: idcomm                   Number of groups  =          99

                                         Obs per group:
                                                        min =          70
                                                        avg =        99.6
                                                        max =         119

                                         Wald chi2(1)      =      389.09
Log likelihood = -21043.833              Prob > chi2       =      0.0000
```

income	Coef.	Std. Err.	z	P>|z|	[95% Conf. Interval]	
male	.814822	.0413082	19.73	0.000	.7338594	.8957846
_cons	3.998739	.0487078	82.10	0.000	3.903273	4.094205

Random-effects Parameters	Estimate	Std. Err.	[95% Conf. Interval]	
idcomm: Identity				
var(_cons)	.1596286	.0286575	.1122788	.2269466
var(Residual)	4.117453	.058942	4.003535	4.234614

LR test vs. linear model: chibar2(01) = 216.21   Prob >= chibar2 = 0.0000

Example 9.6

```
qui xtset idcomm
xtreg income male, re
```

```
Random-effects GLS regression            Number of obs     =       9,859
Group variable: idcomm                   Number of groups  =          99

R-sq:                                    Obs per group:
    within  = 0.0378                                    min =          70
    between = 0.1091                                    avg =        99.6
    overall = 0.0386                                    max =         119

                                         Wald chi2(1)      =      389.08
corr(u_i, X)   = 0 (assumed)             Prob > chi2       =      0.0000
```

income	Coef.	Std. Err.	z	P>|z|	[95% Conf. Interval]	
male	.8150066	.0413181	19.73	0.000	.7340246	.8959886
_cons	3.998649	.0479742	83.35	0.000	3.904621	4.092677

sigma_u	.39050637					
sigma_e	2.0292391					
rho	.0357106	(fraction of variance due to u_i)				

A more interesting model examines whether the slopes also vary randomly across towns. This extends the model estimated in Example 9.5 by telling Stata that we wish to test a *random-slope, random-intercept model.*

In Example 9.7, we test the hypothesis that the slope gauging the linear association between sex and income varies across towns (the `idcomm: male` portion of the Stata code allows the *male* slope to vary across the level-2 units). As shown in the Random-effects panel, there is variability in the *male* coefficient ($\sigma = 0.015$). A straightforward way of representing this variability is by computing 95% confidence intervals for the slope using the standard formula:

$$95\% \text{ CI} = 0.814 \pm 1.96 \times \sqrt{0.0152} = \{0.572, 1.056\}.$$

Thus, we claim that we expect about 95% of the slopes to fall within the interval 0.572 and 1.056. In other words, in some towns the slope of the relationship between sex and income (i.e., the average income difference between males and females) is much larger than in other towns. But it is still positive across the towns represented in the data set.

Example 9.7

```
mixed income male || idcomm: male
```

```
Mixed-effects ML regression                    Number of obs      =       9,859
Group variable: idcomm                         Number of groups   =          99

                                               Obs per group:
                                                              min =          70
                                                              avg =        99.6
                                                              max =         119

                                               Wald chi2(1)       =      356.59
Log likelihood = -21043.572                    Prob > chi2        =      0.0000
```

income	Coef.	Std. Err.	z	P>|z|	[95% Conf. Interval]	
male	.8143975	.0431272	18.88	0.000	.7298699	.8989252
_cons	3.998561	.0484022	82.61	0.000	3.903694	4.093427

Random-effects Parameters	Estimate	Std. Err.	[95% Conf. Interval]	
idcomm: Independent				
var(male)	.0151873	.0223215	.000852	.270728
var(_cons)	.1567337	.0288689	.1092393	.2248773
var(Residual)	4.113672	.0591081	3.999439	4.231169

```
LR test vs. linear model: chi2(2) = 216.73          Prob > chi2 = 0.0000
```

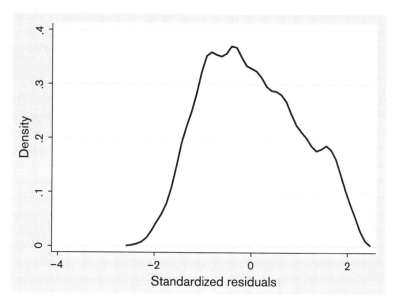

FIGURE 9.2

CHECKING MODEL ASSUMPTIONS

As with any regression model, we should check the assumptions. To test the normality assumption we may request standardized residuals after estimating the `mixed` model and then determine if they follow a normal distribution (see figure 9.2).

> **Example 9.7 (Continued)**
>
> ```
> predict rstandard, rstandard
> kdensity rstandard
> ```

It appears that the residuals do not strictly follow a normal distribution, but they are close (a `qnorm` plot confirms this). In addition, consider the following:

> **Example 9.7 (Continued)**
>
> ```
> predict fitted_i, fitted
> ```

The subcommand `fitted` provides predicted values at the individual level that take into account the random variation among individuals (specifying simply `predict pred, xb` requests the predicted values for the fixed-effects portion of *income*). We may then check for heteroscedasticity at the individual-level by plotting the standardized residuals by the predicted values. For this model a residual-by-fitted scatterplot that uses the predicted values with random variation shows an odd pattern that is likely indicative of the categorical nature of the income variable (check this for yourself).

At the group level, a number of different post-model predictions are available. For instance, an important test is to compute what are known as the *best linear unbiased predictions (BLUPs)* of the random effects. These may be treated as group-level residuals. The Stata code to compute these is

```
predict u1 u0, reffects
```

By default, this results in estimates of the random effects for the slope ($u1$) and intercept ($u0$) for each group. These may then be used to estimate slopes and intercepts for each town in the data set:

```
generate slope_t = _b[male] + u1
generate intercept_t = _b[_cons] + u0
```

One way to consider whether there are influential observations and normality at the group level is to use the following command:

```
statsby, by(idcomm) clear: mean u1 u0 slope_t intercept_t
```

This asks for the mean values of the BLUPs of the random effects, the town-specific slopes, and the town-specific intercepts. But it also creates a group-level data set of these estimates. Be careful because using statsby along with the clear subcommand replaces the data in Stata's memory.

Quantile plots of the slopes and intercepts (using qnorm or kdensity) indicate slight departures from normality for the group-level random effects (which are treated here as residuals). A separate investigation—using summary statistics, boxplots, dot plots, and scatterplots—indicates that town 285 has a particularly large slope relative to the rest of the towns, whereas two others (towns 225 and 4480) have relatively large intercepts. Figure 9.3 shows the scatterplot of the town-level intercepts and slopes. According to a lowess fit line, there is a modest positive association between the model's slopes and intercepts. The analyst might then wish to explore what it is about particular towns that leads to greater than average slopes (mean differences between males and females) and intercepts (higher mean incomes among female residents); or why the towns with larger income differences also tend to have higher average household incomes in general.

```
twoway scatter _b_slope_t _b_intercept_t, mlabel(idcomm)
```

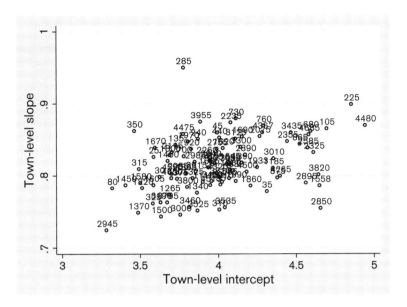

FIGURE 9.3

GROUP-LEVEL VARIABLES AND CROSS-LEVEL INTERACTIONS

A key advantage of multilevel models is that they may include not only individual-level variables, such as sex and marital status, but also group-level variables. In the *multilevel_data12* data set, for example, there are variables that measure the towns' population and level of economic disadvantage. We might be interested in whether, say, household income is lower, on average, in larger towns or in more disadvantaged areas. Example 9.8 illustrates a model that tests this. Consider that we simply add the group-level variables to the fixed portion of the model command.

The results indicate that towns with more people tend to have individuals who report higher household incomes. Surprisingly, however, those counties that have greater economic disadvantage have sample members from towns within them who report higher household incomes, on average. Although unexpected, this is the sort of statistical association that can be revealed in a multilevel analysis.

Example 9.8

```
mixed income male pop2000 disadvantage || idcomm: male
```

Mixed-effects ML regression	Number of obs	=	9,859
Group variable: idcomm	Number of groups	=	99

```
                                            Obs per group:
                                                       min =        70
                                                       avg =      99.6
                                                       max =       119

                                            Wald chi2(3)    =    379.71
Log likelihood = -21036.416                 Prob > chi2     =    0.0000
```

income	Coef.	Std. Err.	z	P>|z|	[95% Conf. Interval]	
male	.8150767	.0427344	19.07	0.000	.7313189	.8988346
pop2000	.0000756	.0000209	3.61	0.000	.0000346	.0001167
disadvantage	.094317	.0420807	2.24	0.025	.0118403	.1767936
_cons	3.856731	.0598709	64.42	0.000	3.739386	3.974075

Random-effects Parameters	Estimate	Std. Err.	[95% Conf. Interval]	
idcomm: Independent				
var(male)	.0118878	.0213813	.0003501	.4037099
var(_cons)	.1274867	.0246717	.0872444	.1862913
var(Residual)	4.115085	.059123	4.000822	4.232611

```
LR test vs. linear model: chi2(2) = 163.63              Prob > chi2 = 0.0000
```

Researchers often wish to present some measure of the proportion of variance explained by a regression model. For example, the well-known R^2 statistic is usually presented for an OLS regression model. In multilevel models there are analogous measures. Perhaps the most common is variance explained at the lowest level of aggregation (level-1: individuals in this model). This requires information from the within-group variances (σ_e^2).[2] from the one-way ANOVA (intercept-only) model and the hypothesized model. For the previously estimated models (Examples 9.4 and 9.8), this is computed as

$$\frac{\sigma_e^2(\text{ANOVA}) - \sigma_e^2(\text{hypothesized model})}{\sigma_e^2(\text{ANOVA})} = \frac{4.28 - 4.12}{4.28} = 0.037 .$$

This indicates that about 3.7% of the variance in income within the towns, and hence between individuals, is explained by the model that includes sex, community-level disadvantage, and population size.

We may also estimate the proportion of variance explained between groups using information from the between-group variances (σ_u^2) in the one-way ANOVA model and the hypothesized model. For the previous model, this is

$$\frac{0.170 - 0.127}{0.170} = 0.25 .$$

Therefore, the hypothesized model accounts for about 25% of the variability in income between towns. Given that the male slope also has a variance component, we could determine the proportion of variance explained between groups due to the random-slope component.

Another advantage of multilevel models is their ability to test interesting hypotheses about contextual effects on individual-level associations. Utilizing what are termed *cross-level interactions*, we may determine, for instance, whether the association between sex and income or between age and income differs depending on characteristics of the town. Example 9.9 examines if the association between age and income depends on whether a town is high or low on the measure of economic disadvantage. The results suggest that age is negatively associated with income (perhaps due to the older average age of this sample), but that this negative association is slightly stronger in counties with higher levels of economic disadvantage (see Example 9.9 and figure 9.4). It is often simpler to interpret these effects if the variables are first centered and then included in the model.[4]

Example 9.9

```
mixed income c.age##c.disadvantage || idcomm: age
```

Mixed-effects ML regression   Number of obs   =   9,859
Group variable: idcomm   Number of groups =   99

                                           Obs per group:
                                                         min =   70
                                                         avg =   99.6
                                                         max =   119

                                           Wald chi2(3)   =   1312.81
Log likelihood = -20619.96   Prob > chi2   =   0.0000

income	Coef.	Std. Err.	z	P>|z|	[95% Conf. Interval]
age	-.0415394	.0011495	-36.14	0.000	-.0437924	-.0392863
disadvantage	.2444003	.0778165	3.14	0.002	.0918828	.3969177
c.age#c.disadvantage	-.003115	.0011766	-2.65	0.008	-.005421	-.0008089
_cons	6.716011	.0757538	88.66	0.000	6.567536	6.864485

Random-effects Parameters	Estimate	Std. Err.	[95% Conf. Interval]	
idcomm: Independent				
var(age)	1.24e-14	5.24e-12	0	.
var(_cons)	.1095606	.021119	.075089	.1598573
var(Residual)	3.786733	.0542094	3.681961	3.894486

LR test vs. linear model: chi2(2) = 144.00   Prob > chi2 = 0.0000

---

4. There is a substantial and important literature on the use of centering in multilevel models. One of the key issues is that one may center individual-level variables within group or overall. The type of centering has implications for the interpretation of model results. Books and articles on multilevel modeling provide more information (e.g., Paccagnella, 2006; Snijders and Bosker, 2012).

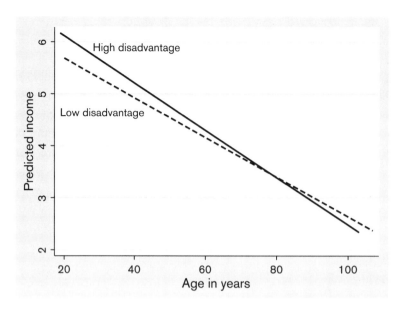

FIGURE 9.4

```
predict pred, xb                              * after Example 9.9
graph twoway lfit pred age if disadvantage < -0.57,
   lpattern(dash) ||lfit pred age if disadvantage > 0.57,
   lpattern(solid)
```

As mentioned earlier, it is important to test the assumptions of multilevel models. Although the residuals from this model appear to follow a normal distribution, a scatterplot of the standardized residuals by the predicted values reveals an odd pattern (determine this for yourself). As noted earlier, this probably reflects the categorical nature of the income variable.

MULTILEVEL GENERALIZED LINEAR MODELS

The mixed command is appropriate for outcome variables that are continuous and normally distributed. Stata also has commands for binomial outcomes (melogit or meqrlogit), count variables (e.g., mepoisson, menbreg), and a variety of other GLMs (meglm) (type help me in Stata's command window for a list of options). Moreover, a user-written Stata program called gllamm (generalized linear latent and mixed models) offers substantial flexibility in estimating multilevel models with data that are measured in various ways. Other software packages designed for multilevel analysis such as MlWin and HLM also offer many options for estimating multilevel models.

In the *multilevel_data12* data set, there is a variable *trust* that is coded as 0 = low trust of others and 1 = high trust of others. Given that this is a binomial outcome variable, we should not use a linear regression model. It is better to rely on a logistic or probit model for this outcome (see Chapter 3). Example 9.10 uses the `meglm` command—which is similar to the `glm` command—to estimate a multilevel logistic regression model designed to predict trust in others, with *age* and *educate* as explanatory variables (the `melogit` command may also be used to estimate this model). The subcommand `eform` requests exponentiated coefficients, or odds ratios for a logistic regression model.

Example 9.10

```
meglm trust age educate || idcomm: age, link(logit)
     family(binomial) eform
```

```
Mixed-effects GLM                        Number of obs    =      9,859
Family:              binomial
Link:                   logit
Group variable:        idcomm            Number of groups =         99

                                         Obs per group:
                                                      min =         70
                                                      avg =       99.6
                                                      max =        119

Integration method: mvaghermite          Integration pts. =          7

                                         Wald chi2(2)     =     296.47
Log likelihood = -4266.0002              Prob > chi2      =     0.0000
```

trust	exp(b)	Std. Err.	z	P>|z|	[95% Conf. Interval]	
age	1.030649	.0018348	16.96	0.000	1.027059	1.034252
educate	1.062582	.0204861	3.15	0.002	1.023179	1.103502
_cons	.0250249	.0040474	-22.80	0.000	.0182264	.0343592
idcomm						
var(age)	3.51e-36	2.66e-21			.	.
var(_cons)	.1710669	.0364447			.1126739	.2597221

```
LR test vs. logistic model: chibar2(01) = 106.39      Prob >= chibar2 = 0.0000
```

This model provides confirmatory evidence that there is random variation in trust across individuals after considering its association age, and education (see the `var(_cons)` estimate). Moreover, higher values of age and education are both associated with a greater odds of reporting high trust of others. For example, each 1-year increase in age is associated with a 3% increase in the odds of high trust, after adjusting for the effects of education. Figure 9.5 provides a visual representation of this association.

```
predict p_trust                          * after Example 9.10
twoway lowess p_trust age
```

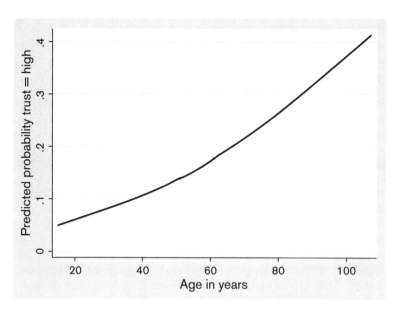

FIGURE 9.5

The graph demonstrates that the probability of high trust is about 0.2 units higher among 80-year-olds than among 20-year-olds in this data set.

As a final step for this model (though there are several others that might be taken), it is a good idea to examine the residuals and predicted probabilities to determine whether there are any unusual patterns in the data. Save the Anscombe residuals and examine whether they are close to a normal or logistic distribution (see Stata's help menu for `meglm` for more information). Figure 9.6 displays a kernel density plot.

```
predict anscombe, anscombe                    * after Example 9.10
kdensity anscombe
```

The distribution of the residuals displays substantial positive skew. Thus, there is substantially more overdispersion in the residuals than might otherwise be expected for a logistic regression model. It is apparent that we should continue to examine the data and model to explore this problem further. It should be noted that there are also specific tests that have been developed for these types of multilevel models that are designed to check for heteroscedasticity, influential observations, and related issues (see, e.g., Snijders and Berkhof, 2008). We have examined only simplified techniques thus far in this chapter.

MULTILEVEL MODELS FOR LONGITUDINAL DATA

Multilevel models also offer an alternative to the GEE models covered in Chapter 8. In particular, they may be used to model what are often called *growth curves* of a variety of

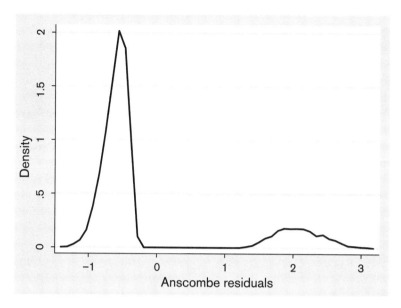

FIGURE 9.6

outcomes.[5] For instance, imagine that we are interested in changes over time in self-esteem. Similar to what we learned in Chapter 8, multilevel data can involve time periods that are nested within individuals; thus, measures of self-esteem are time specific and nested within individual respondents.

Like several of the longitudinal models covered earlier, a key advantage of multilevel models, as we have already learned, is that they are built on the assumption that the observations are dependent in some way. When analyzing longitudinal data with these models, we explicitly examine the dependence within units, so that we no longer have to consider, say, self-esteem scores as independent within individuals across time points. By taking into account the dependence of observations, the standard errors of coefficients exhibit less bias.

Multilevel models may be used to examine different trajectories for the outcome variable across individuals. Figure 9.7 illustrates one way of viewing this (a different version of this graph is in figure 8.1). It provides linear fit lines demonstrating the association between age and self-esteem scores for four individuals using the *esteem* data set (*esteem.*

---

5. The term *growth curve analysis* has been adopted to identify multilevel models (or other longitudinal models) that estimate changes in some outcome—such as a behavior, test score, or personality trait—among individuals. The term *growth* comes mainly from longitudinal studies that focused on physical growth of some species or changes in vocabulary among toddlers. However, it now tends to be used more generally to identify "growth" in a variety of outcomes (see Duncan et al., 2006). Yet using this term is often a misnomer since we might also be interested in decreases in some behaviors, or the ebbs and flows of some outcomes.

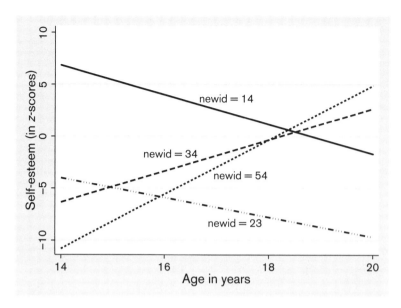

FIGURE 9.7

dta).[6] Recall that this data set has information from several hundred youth over a 7-year period. The graph demonstrates that two of the age–self-esteem slopes are positive and two are negative. But imagine if we estimated different slopes for each individual in this data set; the fixed effect is the overall average of these slopes. The random effect assesses the variation in the individual-level slopes. If we are interested only in random variation of the outcome, and not in the slope, then a random-intercept model is used.

But suppose were ask only whether the association between family cohesion and self-esteem is the same across individuals, thus justifying a random-intercept model. We could instead assume that the effect of family cohesion varies across individuals. Perhaps those experiencing high family cohesion tend to have increasing levels of self-esteem, whereas those who experience poor family cohesion tend to have decreasing levels of self-esteem across the observation period. Moreover, the slope gauging the association of family cohesion and self-esteem may differ by some third variable, such as gender or stressful life events.

```
twoway lfit esteem age if newid==14 || lfit esteem age if
    newid==23 || lfit esteem age if newid==34 || lfit esteem
    age if newid==53
```
                                                                    * *see figure 9.7*

We now examine a multilevel model with longitudinal data. We continue to use the *esteem* data set since it includes an extensive range of years and variables. The `mixed` command is

---

6. Note that we used year as the time dimension in the graph provided in Chapter 8, whereas here we use age as the time dimension. Which one to use is normally a conceptual question that researchers need to address.

used here to estimate a multilevel model for these longitudinal data. As shown earlier in the chapter, a first step is to determine if the outcome variable varies between units (between individuals in this case). Thus, we begin by estimating an intercept-only model (one-way ANOVA).

Example 9.11

```
mixed esteem || newid:
```

```
Mixed-effects ML regression                    Number of obs      =        5,250
Group variable: newid                          Number of groups   =          750

                                               Obs per group:
                                                            min =            7
                                                            avg =          7.0
                                                            max =            7

                                               Wald chi2(0)       =            .
Log likelihood = -16629.106                    Prob > chi2        =            .
```

esteem	Coef.	Std. Err.	z	P>|z|	[95% Conf. Interval]	
_cons	-1.33e-09	.1887915	-0.00	1.000	-.3700246	.3700246

Random-effects Parameters	Estimate	Std. Err.	[95% Conf. Interval]	
newid: Identity				
var(_cons)	23.19924	1.382425	20.64198	26.07332
var(Residual)	24.72703	.5212916	23.72614	25.77015

LR test vs. linear model: <u>chibar2(01)</u> = 1956.38   Prob >= chibar2 = 0.0000

Example 9.11 indicates that the overall mean of self-esteem across all individuals is very close to zero (this is because it was constructed as a centered variable). Moreover, as with the models shown earlier in the chapter, this model has two important variance components: the variability due to between-unit differences (var(_cons) or σ_u^2) and the variability due to within-unit differences (var(Residual) or σ_ε^2). The Random-effects Parameters panel provides these two quantities. They may be used to compare the proportion of variability within units versus between units, or the ICC: 23.2/(23.2 + 24.73) = 0.48. In other words, 48% of the variability in self-esteem is due to between-individual differences. This is identical to the proportion of variance between units that we find in a random-effects model estimated with xtreg. Recall that *rho* captured this (see Chapter 8).

As with any intercept-only regression model, there are obvious limitations to just examining the expected value of the outcome. This is usually true even if one is interested in the decomposition of variance between units and within units. For example, research indicates that self-esteem differs by age across the adolescent years (e.g., Baldwin and Hoffmann, 2002). Consistent with this literature, the next set of multilevel models examines the association between age and self-esteem in two contexts. First, a random-intercept model examines the linear association between age and self-esteem as a fixed effect (i.e., we assume that the age–self-esteem slope is constant across individuals), but with variability in initial levels of self-esteem. Second, a random-slope model examines age's linear association with self-esteem as a random effect (i.e., we assume that the age–self-esteem slope varies across individuals).

The random-intercept model shown in Example 9.12 suggests a weak linear association between age and self-esteem in general. Even though, at some point early in the age distribution, self-esteem varies, it does not appear to change with age in any statistically recognizable way (at least in a linear fashion).

The random-slope model provided in Example 9.13 suggests that the age slope does vary between individuals, however. Even though the average age–self-esteem slope does not differ from zero (*fixed-effects slope* = −0.026, p = 0.444), it is highly variable across individuals [var(age) = 0.066, SE = 0.009; 95% CIs = {0.050, 0.086}]. The standard deviation of the age slope (0.257) may be used to calculate 95% confidence intervals: this slope varies across individuals from about −0.529 to 0.477. In other words, among some youth the association between age and self-esteem is negative, whereas among others it is positive.

It is important to ask whether the random-slope model offers a significantly better fit statistically than the random-intercept model. Since the random-intercept model is nested within the random-slope model, we may use a likelihood ratio χ^2 test to compare them.

Example 9.12

Random-intercept model
mixed esteem age || newid:

Mixed-effects ML regression

Group variable: newid

Number of obs	=	5,250
Number of groups	=	750

Obs per group:

min =		7
avg =		7.0
max =		7

Log likelihood = −16628.945

Wald chi2(1)	=	0.32
Prob > chi2	=	0.5708

esteem	Coef.	Std. Err.	z	P>|z|	[95% Conf. Interval]	
age	−.018798	.0331639	−0.57	0.571	−.0837982	.0462021
_cons	.2986628	.5596835	0.53	0.594	−.7982968	1.395622

Random-effects Parameters	Estimate	Std. Err.	[95% Conf. Interval]	
newid: Identity				
var(_cons)	23.17688	1.381822	20.62081	26.04979
var(Residual)	24.72868	.5213589	23.72766	25.77194

LR test vs. linear model: chibar2(01) = 1950.08   Prob >= chibar2 = 0.0000

Random-slope model

mixed esteem age || newid: age

```
Mixed-effects ML regression                  Number of obs     =        5,250
Group variable: newid                        Number of groups  =          750

                                             Obs per group:
                                                           min =            7
                                                           avg =          7.0
                                                           max =            7

                                             Wald chi2(1)      =         0.59
Log likelihood = -16597.133                  Prob > chi2       =       0.4437
```

esteem	Coef.	Std. Err.	z	P>|z|	[95% Conf. Interval]	
age	-.0261429	.0341344	-0.77	0.444	-.0930451	.0407593
_cons	.430668	.5323684	0.81	0.419	-.6127548	1.474091

Random-effects Parameters	Estimate	Std. Err.	[95% Conf. Interval]	
newid: Independent				
var(age)	.0656951	.0088471	.0504549	.0855388
var(_cons)	7.075266	2.174392	3.873905	12.9222
var(Residual)	24.20633	.5155501	23.21667	25.23818

```
LR test vs. linear model: chi2(2) = 2013.71              Prob > chi2 = 0.0000
```

After estimating the random-intercept model, store the estimates using the Stata command estimates store randint, followed by storing the estimates from the random-slope model: estimates store randslope (note that randint and randslope are user-defined names). Then, ask Stata for a likelihood ratio χ^2 test:

Example 9.12 (Continued)

lrtest randslope randint

```
Likelihood-ratio test                        LR chi2(1)   =        63.62
(Assumption: randint nested in randslope)    Prob > chi2  =       0.0000
```

The test strongly suggests that the random-slope model is preferred, in a statistical sense, over the random-intercept model.

We may also introduce additional explanatory variables to explore if we can account for the variation between individuals, as well as the variation within individuals. Example 9.13 adds two time-varying and two time-invariant effects. In this model, we are interested in the fixed-effects coefficients (slopes) for stressful life events, family cohesion, gender, and race/ethnicity. But we also tell Stata that the *newid* variable specifies the units and that the age, stressful life events, and family cohesion slopes are allowed to vary across individuals. Implicit to this is that we continue to allow the intercept to vary across individuals.

The model indicates that stressful life events are negatively associated with self-esteem, whereas family cohesion and sex (*male*) are positively associated with self-esteem. Moreover, the random-effects panel indicates that the slope for age, stress, and family cohesion, as well as the intercept, vary across individuals. For instance, among some individuals there is a negative association between stressful life events and self-esteem, whereas among others the association is weak or perhaps even positive. As shown earlier, we may compute 95% confidence intervals for the variation of the estimated slopes. Examining this variation, we expect that the stressful life events slope varies randomly across individuals between {−0.077 − (1.96 × 0.144 =} −0.359 and {−0.077 + (1.96 × 0.144 =} 0.205. Therefore, even though, on average, there is a negative association between stressful life events and self-esteem ($\beta = -0.077$), among some youth in the sample there is a positive association. Moreover, even after accounting for random variation in the age, stress, and cohesion slopes, there is still variation in self-esteem across individuals (consider the variance estimate for the constant and its accompanying standard error and confidence intervals).

Example 9.13

```
mixed esteem age stress cohes male || newid: age stress cohes
```

| Mixed-effects ML regression | | | | Number of obs | = | 5,250 |
| Group variable: newid | | | | Number of groups | = | 750 |

Obs per group:

		min =	7
		avg =	7.0
		max =	7

| | | Wald chi2(4) | = | 448.60 |
| Log likelihood = -16247.656 | | Prob > chi2 | = | 0.0000 |

esteem	Coef.	Std. Err.	z	P>|z|	[95% Conf. Interval]	
age	-.0291349	.0325176	-0.90	0.370	-.0928683	.0345985
stress	-.0769964	.0169193	-4.55	0.000	-.1101576	-.0438351
cohes	.2251742	.0115445	19.50	0.000	.2025473	.2478011
male	1.806049	.317693	5.68	0.000	1.183382	2.428716
_cons	-.5573048	.5303808	-1.05	0.293	-1.596832	.4822225

Random-effects Parameters	Estimate	Std. Err.	[95% Conf. Interval]	
newid: Independent				
var(age)	.0528221	.007218	.0404111	.0690446
var(stress)	.0207774	.0093586	.0085939	.0502333
var(cohes)	.0315802	.0047145	.023569	.0423144
var(_cons)	2.188969	1.755035	.4547552	10.53663
var(Residual)	20.28641	.4808481	19.36552	21.25109

LR test vs. linear model: chi2(4) = 1589.72   Prob > chi2 = 0.0000

As discussed earlier, researchers often present some measure of the proportion of variance explained by the multilevel regression model. This requires knowledge of the within-unit variances (σ_ε^2) from the one-way ANOVA model and the hypothesized model. For the previous model, this is

$$\frac{24.73 - 20.29}{24.73} = 0.18 .$$

Thus, a model that includes age, stressful life events, family cohesion, and sex explains about 18% of the variance in self-esteem within individuals.

CROSS-LEVEL INTERACTIONS AND CORRELATIONAL STRUCTURES IN MULTILEVEL MODELS FOR LONGITUDINAL DATA

As with the earlier multilevel model that considered cross-level interactions using the *multilevel_data12* data set, we may also consider cross-level interactions using longitudinal data. For example, we might be interested in whether the age–self-esteem association is different for males and females. The following model considers this issue by including an interaction term between age and sex.

The results displayed in Example 9.14 suggest that age has a negative association among females (coded as zero), but a positive association among males (use the output to compute the slopes for each to verify this statement). Thus, it appears that considering sex differences in self-esteem, as well as whether these sex differences change across the adolescent years is important.

Example 9.14

```
mixed esteem age stress cohes male c.age#c.male || newid:
    age stress cohes
```

```
Mixed-effects ML regression                 Number of obs     =       5,250
Group variable: newid                       Number of groups  =         750

                                            Obs per group:
                                                         min =           7
                                                         avg =         7.0
                                                         max =           7

                                            Wald chi2(5)      =      457.58
Log likelihood = -16243.786                 Prob > chi2       =      0.0000
```

esteem	Coef.	Std. Err.	z	P>|z|	[95% Conf. Interval]	
age	-.1194318	.0459003	-2.60	0.009	-.2093948	-.0294688
stress	-.0755836	.0168788	-4.48	0.000	-.1086653	-.0425018
cohes	.2261328	.0115383	19.60	0.000	.2035182	.2487474
male	-.8587658	1.008216	-0.85	0.394	-2.834833	1.117302
c.age#c.male	.1810757	.0650326	2.78	0.005	.0536142	.3085372
_cons	.7753793	.7139508	1.09	0.277	-.6239385	2.174697

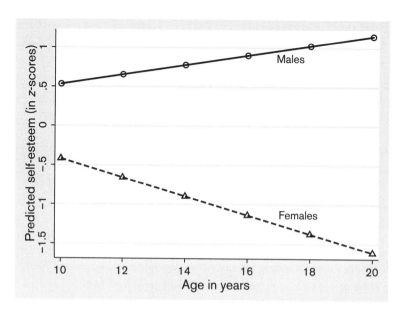

FIGURE 9.8

Random-effects Parameters	Estimate	Std. Err.	[95% Conf. Interval]	
newid: Independent				
var(age)	.0527587	.0071887	.0403936	.068909
var(stress)	.0202308	.0092977	.0082189	.0497979
var(cohes)	.0315105	.0047041	.0235171	.0422209
var(_cons)	2.144316	1.748945	.4335433	10.60584
var(Residual)	20.27289	.4806198	19.35244	21.23712

LR test vs. linear model: chi2(4) = 1581.85   Prob > chi2 = 0.0000

It is often easier to understand the results of the model using predicted values. Figure 9.8 shows an example of how we might use the `margins` command followed by the `marginsplot` command to visualize the association between age, sex, and self-esteem.

```
qui margins, at(age=(10(2)20) male=(0 1)) atmeans
marginsplot, noci
```

Similar to what we saw with the GEE models in Chapter 8, it is often a good idea to relax the assumption of no correlation among the random effects and see if there are any changes in the model. This may be done by adding `cov(unstructured)` to the `mixed` command. Stata will then estimate the correlation between, say, the family cohesion and stress slope coefficients (other correlation structures are also allowed; see `help mixed`).

As shown in Example 9.15, we now note that there is a negative association between the age slope and the cohesion slope $\{r = -0.033/\sqrt{0.662 \times 0.0256} = -0.253\}$. There is also a negative association between the age slope and the intercept ($r = -0.951$). This latter correlation suggests that individuals with higher baseline self-esteem scores tend to have flatter or perhaps negative age slopes relative to those with lower baseline self-esteem scores. In other words, those youth with higher self-esteem in early adolescence often find their self-esteem decreasing as they get older. This may reflect regression towards the mean. Finally, the `wcorrelation` post-command output provides the estimated within-unit correlations.

Example 9.15

```
mixed esteem age stress cohes male nonwhite || newid: age
    stress cohes, cov(unstructured)
```

Mixed-effects ML regression	Number of obs	=	5,250
Group variable: newid	Number of groups	=	750

Obs per group:

min =		7
avg =		7.0
max =		7

	Wald chi2(5)	=	461.51
Log likelihood = -16158.726	Prob > chi2	=	0.0000

esteem	Coef.	Std. Err.	z	P>|z|	[95% Conf. Interval]	
age	-.0223167	.041769	-0.53	0.593	-.1041825	.0595491
stress	-.0677487	.0156312	-4.33	0.000	-.0983852	-.0371122
cohes	.2236237	.0111738	20.01	0.000	.2017235	.2455238
male	1.791212	.3187585	5.62	0.000	1.166456	2.415967
nonwhite	-.1224198	.4985099	-0.25	0.806	-1.099481	.8546416
_cons	-.6161161	.6717916	-0.92	0.359	-1.932804	.7005713

Random-effects Parameters	Estimate	Std. Err.	[95% Conf. Interval]	
newid: Unstructured				
var(age)	.6617192	.0693886	.5387855	.8127025
var(stress)	.0073156	.0067462	.0012003	.0445858
var(cohes)	.0255781	.0043511	.0183261	.0356999
var(_cons)	149.1282	16.76096	119.6439	185.8783
cov(age,stress)	.0516376	.0171702	.0179846	.0852907
cov(age,cohes)	-.0329681	.0118332	-.0561608	-.0097755
cov(age,_cons)	-9.448577	1.059899	-11.52594	-7.371214
cov(stress,cohes)	-.0032702	.004048	-.0112041	.0046636
cov(stress,_cons)	-.7792392	.2626243	-1.293973	-.2645051
cov(cohes,_cons)	.3479258	.1830214	-.0107896	.7066411
var(Residual)	17.9206	.4521471	17.05596	18.82906

LR test vs. linear model: chi2(10) = 1766.95   Prob > chi2 = 0.0000

estat wcorrelation   ** provides correlation structure*

Standard deviations and correlations for newid = 13:

Standard deviations:

obs	1	2	3	4	5	6	7
sd	6.618	6.199	6.126	5.989	6.400	7.087	7.557

Correlations:

obs	1	2	3	4	5	6	7
1	1.000						
2	0.556	1.000					
3	0.405	0.418	1.000				
4	0.370	0.393	0.506	1.000			
5	0.259	0.300	0.505	0.509	1.000		
6	0.123	0.182	0.468	0.484	0.580	1.000	
7	0.050	0.120	0.434	0.459	0.574	0.659	1.000

Given that longitudinal models are usually affected by autocorrelation, reestimate the model with an AR(1) correlation structure. Does this make a difference in the findings?

CHECKING MODEL ASSUMPTIONS

Multilevel models for longitudinal data make many of the same assumptions as other regression models, with the important exception that they explicitly assume dependence of observations among individuals. As suggested earlier, another difference is that these assumptions may be examined both within and across units (individuals in longitudinal analyses). Books on multilevel analysis discuss how to test these assumptions (see, e.g., Snijders and Bosker, 2012). To be consistent with what we learned earlier, examine the predicted values and residuals to determine if there are any unusual patterns. For example, use the `predict` command after the `mixed` model and perform some exploratory analyses.

Example 9.15 (Continued)

predict fitted_i, fitted   ** computes fitted values that take into account the random variation across units*

predict pred, xb   ** computes fitted values for fixed portion only*
predict rstandard, rstandard   ** computes standardized residuals*

We may now use a kernel density graph to check for normality of the residuals (see figure 9.9).

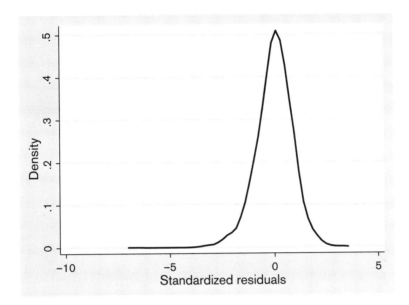

FIGURE 9.9

kdensity rstandard

There is a slightly elongated left tail to the distribution, but it is not overly severe. This is followed by a scatterplot of the residuals by the predicted values for the fixed portion only.

twoway scatter rstandard pred

Figure 9.10 suggests that there is little, if any, problem with heteroscedasticity in the model; the main problem appears to be a few outliers at the bottom end of the distribution of the residuals. This supports the results provided in figure 9.9: the self-esteem residuals are not normally distributed and may require some type of adjustment.

It is also useful to check the slopes and intercepts across individuals. However, when using the post-estimation command to request predicted values with random effects, there are now as many random effects as there are random slopes. For example, if we revert to a simpler model that does not include the covariances among the random effects, but retains the random slopes for age, stressful life events, and family cohesion, the commands in Example 9.16 could be used.

Example 9.16

```
qui mixed esteem age stress cohes male nonwhite || newid:
  age stress cohes
predict u1 u2 u3 u0, reffects
```

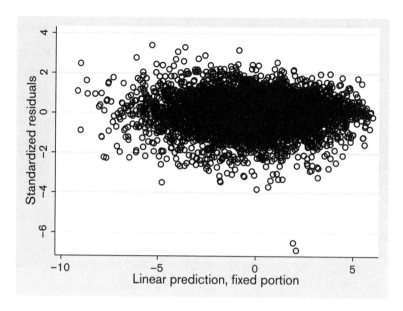

FIGURE 9.10

Next, the slopes and intercepts for each individual in the data set may be computed:

```
generate slope_tage = _b[age] + u1
generate slop_tstress = _b[stress] + u2
generate slop_tcohes = _b[cohes] + u3
generate intercept_t = _b[_cons] + u0
```

Finally, as earlier, consider collapsing the data set using the `statsby` command and examining the slopes, intercepts, and random effects across individuals. Scatterplots for these various measures suggest a negative association between the intercept and the age slope. This supports the point made earlier. However, there is little, if any, association between the intercept and the slopes for stressful life events or family cohesion. Normal quantile plots (`qnorm`) for the random effects of the slopes indicate little deviation from normality of the random age effects, but some departure in the tails for the random portion of the stress and cohesion slopes (verify these claims). Now, consider what other steps an analyst should take.

A MULTILEVEL POISSON REGRESSION MODEL FOR LONGITUDINAL DATA

Earlier, we saw an example of a multilevel model for a binomial outcome variable. In this section, we explore a multilevel model for a count variable, but focus on longitudinal data. As shown in Chapter 8, the *event4* data set (*event4.dta*) is based on 4 years of longitudinal data and includes information on stress, delinquency, and other variables. Although we have used the variable *stress* to predict some outcomes, suppose we wish to predict it using a multilevel regression model. Consider its distribution: it ranges from 0 stressful events to 10 stressful events experienced in the last year. It is therefore a count variable that is a good candidate for a Poisson or negative binomial regression model.

Consider the multilevel Poisson regression model in Example 9.17 that uses `age` as an explanatory variable. The `mepoisson` command is used to estimate this model. As with the Poisson models estimated in Chapter 6, the `irr` subcommand requests incidence rate ratios.

Example 9.17

```
mepoisson stress age || newid:, irr
```

```
Mixed-effects Poisson regression            Number of obs     =      2,604
Group variable:           newid             Number of groups  =        651

                                            Obs per group:
                                                          min =          4
                                                          avg =        4.0
                                                          max =          4

Integration method: mvaghermite             Integration pts.  =          7

                                            Wald chi2(1)      =      65.47
Log likelihood = -4555.4184                 Prob > chi2       =     0.0000
```

stress	IRR	Std. Err.	z	P>|z|	[95% Conf. Interval]	
age	.9161058	.0099208	-8.09	0.000	.8968663	.935758
_cons	6.375324	.9707713	12.17	0.000	4.730311	8.592406

newid						
var(_cons)	.3302106	.0286531			.2785674	.3914278

```
LR test vs. Poisson model: chibar2(01) = 783.59      Prob >= chibar2 = 0.0000
```

The results suggest a negative association with age, with each 1-year increase in age associated with about an 8% decrease in the number of stressful life events experienced. Similar to earlier models, there is substantial variation in life events across individuals (`var(_cons)` = 0.33, SE = 0.03, 95% CIs = {0.28, 0.39}). Figure 9.11 provides a visual representation of the association between age and the predicted values of stress.

> **predict n_stress**   * after Example 9.17
> **twoway mband n_stress age**   * requests median band graph

What does this graph suggest about changes in stressful life events by age? Consider the range of values for the predicted values of stress (0.5–5.9), as well as the range of stress (0–10). How does this inform our understanding of the differences by age shown in figure 9.11? Conceptually speaking, why would these life events be lower for older adolescents?

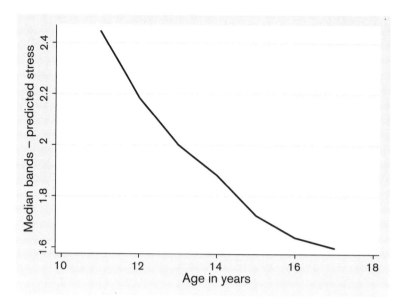

FIGURE 9.11

As with the earlier multilevel GLM, following this example it is a good idea to examine the predicted (fitted) values (both fixed and random components) and the residuals, as well as check for other assumptions of the model. It might also be good practice to compare these results to those from a multilevel negative binomial regression model, especially since stressful life events appear to manifest a slight degree of overdispersion. This is left as an exercise.

FINAL WORDS

Multilevel models offer a useful method for examining how social structures or other conceptually interesting aggregates are associated with individual-level behaviors, experiences, attitudes, and other outcomes. Educational researchers, sociologists, political scientists, and ecologists are particularly fond of these models because they are often faced with data that include a combination of aggregate-level data (e.g., schools, cities, states, ecological niches) and individual-level data. Moreover, they may be extended to accommodate more than two levels, as well as longitudinal and event history data. In fact, some statisticians have argued that multilevel models should be the default starting point for regression models in general because there is little to lose, but potentially much to gain from their use. Yet, these models can become quite complex and lead to convergence problems where the statistical algorithm cannot find a unique solution. Bayesian multilevel models provide one way to overcome these prospective problems, though (Gelman and Hill, 2007).

1. The *neighborhood* data set (*neighborhood.dta*) includes information from more than 2,300 families nested within 524 neighborhoods from a large city in Europe. The variable neighed identifies the neighborhood. We are interested in examining unemployment among the fathers, which is gauged with the variable *dadunemp*. Check its coding as well as what the other variables are designed to measure. Then, use an appropriate regression model with *dadunemp* as the outcome variable.

   a. As a first step, estimate an intercept-only model. Does there appear to be random variation in the intercept? What does the ICC indicate?

   b. Consider a multilevel model with the following explanatory variables: *familywealth, socialcapital, dadocc, dadeduc,* and *deprive*. Interpret the coefficients associated with the variables *familywealth* and *deprive*.

2. Reestimate the model from exercise 1b, but include a random slope for the variable *familywealth*.

   a. What is the 95% confidence interval for the random slope?

   b. What does it suggest about the relationship between *familywealth* and father's unemployment across these neighborhoods?

3. Recall that in the exercises at the end of Chapter 8, you used a data set called *workdata* (*workdata.dta*) to examine wages over time for a sample of adults. In those exercises, you used a GEE model to examine patterns in wages. Here, we use a multilevel model to examine these patterns.

   a. Estimate an intercept-only model with *ln_wage* as the outcome variable. What is the ICC for this model? How is it interpreted?

   b. Extend the model by including the following explanatory variables: *collgrad, union, black, otherrace, age,* and *age*-squared. How does this model compare to the GEE model estimated in the exercise in Chapter 7?

   c. Construct a graph (your choice of the type) that shows the association between age and the predicted values of wages (logged and unlogged). What does the graph suggest about this association?

   d. Check the residuals from the multilevel model using a kernel density graph. What does it suggest about the distribution of the residuals?

4. Using the *workdata* data set, estimate a multilevel model with *ln_wage* as the outcome variable and the following explanatory variables: *collgrad, age,* and *tenure* (years working at the same job). Request a random slope for the *tenure* effect. What does the random component of the tenure slope suggest about the association between tenure and wages?

## TEN

Principal Components and Factor Analysis

We now take on a new task that has important differences and similarities to the models we have learned about in the previous chapters. Up to this point we have been concerned with determining whether a set of explanatory variables predicts some outcome variable. We are now familiar with some powerful statistical tools that may be used to interpret regression coefficients, predict values, compare models, check assumptions, and, hopefully, understand more about associations among variables.

However, another popular task in statistical modeling falls under the general category of *multivariate analysis or data reduction techniques*. This is concerned generally with summarizing large amounts of data in an exploratory or a confirmatory sense. Although there are popular methods such as *data mining* for exploring patterns in data (Hastie et al., 2009), the specific interest we have here is to summarize a large amount of data that are classified by variables into some smaller component(s). The primary concern is in what are termed *variable-directed techniques*. These are often used to create new variables, in particular *latent variables*. As you may recall from discussions of measurement error, latent variables are unmeasured *constructs* that are thought to underlie or account statistically for some set of *manifest* (observed) variables. They are sometimes called *phantom* variables because they are not measured directly, but rather are based on statistical similarities among a set of manifest variables.

These techniques are similar to the regression models examined in earlier chapters since they are designed to explain shared variability among a set of variables. However, unlike linear, logistic, Poisson, or other regression techniques, the models introduced in this chapter are typically concerned with assessing shared variability among a set of variables rather than the unique shared variability among, say, an explanatory variable and a single outcome variable. As we shall see, a helpful way to think about the models in this chapter is to visualize a set of outcome variables that are explained by one or two explanatory variables, which are designated as latent variables or constructs.

There are a large number of techniques available (see Johnson, 1998), but we focus on two of the most popular—and closely related—multivariate analysis tools: *principal components analysis* (PCA) and *factor analysis* (FA). Both are concerned with summarizing a set of variables and describing their shared variability in an efficient way. They are also often used in the development of new or hypothesized latent variables.

For example, suppose we have a few variables that are based on questionnaire data. One of our goals is to assess each participant's level of self-esteem. Yet we know, since we read the vast literature on self-esteem, that it is not possible to measure it with one question. An inquiry such as "How's your self-esteem?" is not sufficient since respondents may have different understandings of this term. Thus, our study utilizes a series of 10 questions that are designed to assess various aspects of self-esteem (e.g., "Do you view yourself positively?" and "Do you have good qualities?"). Now, suppose we wish to determine whether various life events and demographic characteristics affect self-esteem in our sample (or in the target population in an inferential sense). What should we do? Perhaps we could estimate many regression analyses with each question serving as an outcome variable. But this seems unwise since research has shown that self-esteem is a multidimensional construct that is not measured well with a single survey question. Moreover, perhaps we are uncertain whether all of the survey questions are equally or even marginally useful in assessing self-esteem.

Although these issues may appear nettlesome, PCA and FA offer some solutions. In this chapter, we address the issue of what to do when faced with many variables and we are not sure which are best for developing a measure of a multidimensional construct, such as self-esteem. Or, perhaps we wish to create a new measure of some concept, such as optimism. These types of investigations often involve *exploratory (factor) analysis* (EFA).[1] We then discuss what to do to actually develop a measure of an underlying construct. This falls under the term *confirmatory factor analysis* (CFA). There are entire books on CFA (Harrington,

---

1. There is some debate among social and behavioral scientists about whether there is a place for EFA. Some psychometricians, for example, argue that EFA is atheoretical, whereas the behavioral sciences, and the empirical models that are at their core, must be guided by solid theoretical or conceptual models (Mulaik, 2009). Thus, they contend, we should at least have confirmatory models at the heart of our analysis. This is consistent with the hypothetico-deductive model of scientific inquiry (Westland, 2015). One way to keep our eyes focused on confirmatory models, for instance, is to randomly split data sets into two parts: a training set and a test set. The training set is used in an exploratory manner (though still guided by theory or rough hypotheses), but the models are then evaluated on the test data set in a confirmatory manner. Another version of this general approach is called *cross-validation* (Knafl and Grey, 2007).

2008) and a particular estimation approach typically used for CFA known as structural equation models (SEMs; see, e.g., Bollen, 1989; Raykov and Marcoulides, 2012), so this chapter provides merely a brief introduction to these measurement techniques. The interested reader should consider a more thorough treatment, though, by considering some of the books listed in the reference section.

The mathematics underlying FA and PCA can be daunting (including substantial matrix algebra, for instance), so we begin by looking at some actual data. The *factor1* data file (*factor1.dta*) contains a set of variables from a study of adolescent attitudes and behaviors. Similar to the *esteem* data set used in earlier chapters, these variables are based on a series of questions designed to measure adolescent depressive symptoms and self-esteem. The labels indicate the types of questions that the respondents answered. Before beginning the analysis, it is important to point out that these variables are not technically appropriate for the type of analysis presented next since they are ordinal in nature (there are four categories for each). Before describing techniques that have been developed for analyzing categorical variables in a factor analytic context, we begin with more commonly used methods that assume continuous variables.

As an initial step, examine the descriptive statistics for the variables and a correlation matrix (exclude *id* from consideration and use only the variables *worthy* to *attitude*). The reason for this is to get a rough idea of how these variables relate to one another statistically; in other words, do they share variability? This also offers a beginning point for PCA.

From the descriptive statistics displayed in Example 10.1 notice that the variables range from 1 to 4, with higher numbers indicating greater agreement with the survey question. It is also interesting to compare the means from the variables. The first and last two variables have means close to 3, whereas the others have means closer to 1.

In the correlation matrix, the variable "person of worth" is positively correlated with "is capable," yet it is negatively correlated with "felt bothered." Hence, it appears that there are separate patterns of correlations, some positive, some negative, among these variables. If you look carefully enough, you should be able to determine which are thought to measure depressive symptoms and which are thought to measure self-esteem. However, in an exploratory analysis the typical situation is to have little idea which variables are related to the others or how they might be indicators of some latent variable. In other words, an initial first goal is to explore associations among the variables to determine if any reasonable patterns emerge.

Example 10.1

summarize worthy-attitude

Variable	Obs	Mean	Std. Dev.	Min	Max
worthy	739	3.151556	.862518	1	4
capable	738	2.409214	.8980737	1	4
bother	737	1.506106	.7483882	1	4
noeat	736	1.565217	.8594173	1	4
blues	737	1.466757	.7952265	1	4
depress	737	1.549525	.8553241	1	4
failure	737	1.221167	.5697895	1	4
lonely	736	1.637228	.9140922	1	4
crying	737	1.325645	.718397	1	4
sad	736	1.714674	.8914419	1	4
tired	737	1.613297	.8379498	1	4
goodqual	738	2.982385	.7872552	1	4
attitude	738	2.827913	.8584681	1	4

univar worthy-attitude   * note: univar *is a user-written command*

Variable	n	Mean	S.D.	Min	.25	Quantiles Mdn	.75	Max
worthy	739	3.15	0.86	1.00	3.00	3.00	4.00	4.00
capable	738	2.41	0.90	1.00	2.00	2.00	3.00	4.00
bother	737	1.51	0.75	1.00	1.00	1.00	2.00	4.00
noeat	736	1.57	0.86	1.00	1.00	1.00	2.00	4.00
blues	737	1.47	0.80	1.00	1.00	1.00	2.00	4.00
depress	737	1.55	0.86	1.00	1.00	1.00	2.00	4.00
failure	737	1.22	0.57	1.00	1.00	1.00	1.00	4.00
lonely	736	1.64	0.91	1.00	1.00	1.00	2.00	4.00
crying	737	1.33	0.72	1.00	1.00	1.00	1.00	4.00
sad	736	1.71	0.89	1.00	1.00	1.00	2.00	4.00
tired	737	1.61	0.84	1.00	1.00	1.00	2.00	4.00
goodqual	738	2.98	0.79	1.00	3.00	3.00	3.00	4.00
attitude	738	2.83	0.86	1.00	2.00	3.00	3.00	4.00

	worthy	capable	bother	noeat	blues	depress	failure
worthy	1.0000						
capable	0.4233	1.0000					
bother	-0.2362	-0.1672	1.0000				
noeat	-0.1069	-0.1669	0.3674	1.0000			
blues	-0.2014	-0.1770	0.4958	0.4144	1.0000		
depress	-0.2248	-0.1801	0.4417	0.3526	0.6681	1.0000	
failure	-0.2241	-0.1818	0.3157	0.2238	0.4546	0.5608	1.0000
lonely	-0.2430	-0.2269	0.4023	0.3261	0.5522	0.6371	0.5113
crying	-0.0953	-0.1375	0.3348	0.3187	0.4520	0.4604	0.4247
sad	-0.2000	-0.2040	0.4348	0.3500	0.5951	0.6649	0.5369
tired	-0.2062	-0.2490	0.3498	0.2993	0.4084	0.4816	0.3790
goodqual	0.4367	0.4116	-0.1433	-0.0547	-0.1027	-0.1797	-0.1806
attitude	0.3728	0.2831	-0.1850	-0.0386	-0.1442	-0.2592	-0.2385

PRINCIPAL COMPONENTS ANALYSIS

The aim of PCA is to reproduce the correlation matrix (a covariance matrix may also be used) by finding a set of underlying components (or constructs) (as shown later, these are similar to factors in an FA, although both fall under the label *latent variable*) that accounts for the variability in the data. It does this through an iterative approach in which it first finds the component that explains the largest proportion of variance among the variables, then the component, uncorrelated with the first, which explains the second largest proportion of variance, and so on. Chapter 3 of Kline's (1994) book on FA provides an excellent description of these steps. An interesting feature of PCA is that it leaves no variability unexplained. Thus, it is not, strictly speaking, a statistical technique since it assumes there is no random component to the data. Some analysts see this as a disadvantage because (a) randomness is ubiquitous in the social sciences and (b) tests of statistical significance play a limited role (if any) in PCA. Regardless of these limitations, it can still be a useful data reduction technique (Johnson, 1998). Moreover, in order to link PCA (and FA) to the regression models we examined in earlier chapters, it may be helpful (and it will become clearer) to think about the components as explanatory variables that predict the observed variables. Thus, in this situation, the observed variables are all outcome variables. Figures 10.1 and 10.2—shown later in the chapter—illustrate this idea.

It is now time to estimate a PCA in Stata using the aforementioned variables and then figure out what aspects of the output mean. The command `factor` estimates both a PCA and various types of FA. There is also a command called `pca` that directly estimates a principal components model. The key difference is that `pca` provides as many components as there are variables, whereas the `factor` command, using the subcommand pcf for *principal components factor* analysis, provides information on only a select number of factors that account for the most variability among the observed variables. Since we use the `factor` command extensively in this chapter, it is used here to estimate a PCA.

Example 10.2

factor worthy-attitude, pcf

Factor analysis/correlation   Number of obs   =   731
    Method: principal-component factors   Retained factors =   2
    Rotation: (unrotated)   Number of params =   25

Factor	Eigenvalue	Difference	Proportion	Cumulative
Factor1	5.08446	3.25174	0.3911	0.3911
Factor2	1.83272	0.90491	0.1410	0.5321
Factor3	0.92780	0.19189	0.0714	0.6035
Factor4	0.73591	0.07057	0.0566	0.6601
Factor5	0.66535	0.07073	0.0512	0.7112
Factor6	0.59461	0.01483	0.0457	0.7570
Factor7	0.57978	0.03212	0.0446	0.8016
Factor8	0.54766	0.02858	0.0421	0.8437
Factor9	0.51908	0.01874	0.0399	0.8836
Factor10	0.50034	0.07871	0.0385	0.9221
Factor11	0.42164	0.11431	0.0324	0.9546
Factor12	0.30732	0.02398	0.0236	0.9782
Factor13	0.28334	.	0.0218	1.0000

LR test: independent vs. saturated: chi2(78) = 3539.78 Prob>chi2 = 0.0000

Factor loadings (pattern matrix) and unique variances

Variable	Factor1	Factor2	Uniqueness
worthy	-0.4174	0.6337	0.4241
capable	-0.3964	0.5766	0.5103
bother	0.6222	0.0808	0.6063
noeat	0.5072	0.2120	0.6978
blues	0.7587	0.2366	0.3684
depress	0.8150	0.1568	0.3112
failure	0.6843	0.0640	0.5276
lonely	0.7829	0.1169	0.3734
crying	0.6440	0.2892	0.5017
sad	0.8153	0.2023	0.2944
tired	0.6604	0.0093	0.5637
goodqual	-0.3376	0.7107	0.3809
attitude	-0.3977	0.5648	0.5228

The first part of the output in Example 10.2 shows the *eigenvalues* and proportion of variance explained by the components. The eigenvalues may be thought of as measures of the total amount of the variance explained by each component (or factor, as shown later). Hence, the larger the eigenvalue, the more variance explained. The sum of the eigenvalues equals the number of variables in the model. Many programs, such as Stata and SPSS, assume that the analyst wants to retain only those components with eigenvalues ≥1. This is a common cut-off point in PCA and FA (it is also known as the *Guttman–Kaiser criterion*; Yeomans and Golder, 1982). An alternative is to use a *scree plot*. This plots the eigenvalues and helps show where the drop-off in their values occurs (type `screeplot` as a post-estimation command after estimating the model in Stata). There are other decision rules one might use to decide how many principal components to examine, but no clear guidance. It is best to consider these decision rules as general guidelines, but to rely most on the purposes of the research. What is the goal of the data reduction exercise?

In the initial analysis provided in Example 10.2, the first component explains about 39% of the shared variance among the set of variables, and the second explains about 14% of the variance. These measures of "variance explained" are computed by dividing the respective eigenvalues by the number of variables (e.g., component 1: eigenvalue/13: 5.085/13 = 0.39). The remaining components explain a much smaller proportion of the variance.[2] Among all the components, all of the variance in the set of variables is accounted for.

The next portion of the output provided by Stata is the *component matrix* or what Stata labels the `Factor loadings` (these should be called *component loadings* to distinguish PCA from FA). The loadings represent the correlations between the components and the variables (e.g., r(*worthy, component 1*) = 0.634). The sum of squares of each row represents the proportion of variance in each variable that is explained by the components. One way to think about this is to imagine that we have constructed our conceptual components and then used them in a linear regression model to predict each variable. The value of the R^2 from this model equals the sum of squares of the rows. This is also known as the *communality* or the reliability of the observed variable.

For example, suppose we constructed two components, labeled x_1 and x_2, from this analysis. We then estimated the following linear regression model:

Person of worth (*worthy*) = $\alpha + \beta_1 x_1 + \beta_2 x_2 + \varepsilon$.

We would expect the R^2 from this model to be $(-0.417)^2 + (0.634)^2 = 0.576$ (assuming that x_1 and x_2, which represent the components, are uncorrelated, as they are in a PCA). This can also be computed easily by subtracting the *uniqueness* from 1.0. In other words, the uniqueness is the proportion of the variability in the variable that is not accounted for by the components (e.g., $1 - 0.576 = 0.424$). As suggested earlier, another way to think about the component matrix is that it measures the strength (and direction) of the relationship between the hypothetical components and the manifest variables.

The results of PCA, as well as FA, are often shown with a diagram. Figure 10.1 provides a diagram of part of the results found in Example 10.2.

The diagram demonstrates that the two components are assumed to account for part of the variability in the person-of-worth variable. However, there is also error that remains even after the two components and their association with the variable are estimated. This is the uniqueness that is shown in the output (but the PCA model assumes that it can be explained

---

2. Recall that there are always as many components as there are variables in a PCA; the key is figuring out which components are most meaningful.

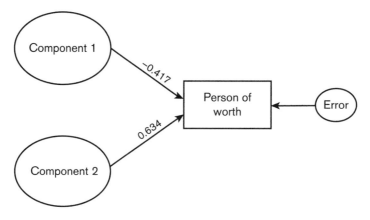

FIGURE 10.1

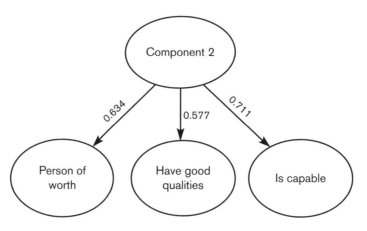

FIGURE 10.2

by other components). We could produce similar diagrams for the other observed variables in the model. Note that there is no line or arrow linking the two components. This is because they have a correlation of zero.

A more frequently used diagram is to show the association between each component and the observed variables. This provides a way to quickly determine the component loadings (correlations) for each constituent variable. Figure 10.2 shows component 2 and three of the observed variables.

PCA is almost always used solely for exploratory analysis and data reduction. For instance, we now see that the variables of interest appear to load well on two components, one of which seems to underlie feeling bothered, blue, or depressed, whereas the other seems to underlie feeling worthy, capable, and having good qualities. We may now determine more clearly which variables assess depressive symptoms and which assess self-esteem. However, we should first explore further how these constructs relate to one another.

FACTOR ANALYSIS

Recall that PCA assumes that all of the variance in the observed variables is accounted for by the model. This can be considered a limitation since there is likely variance that is—or should be—left unexplained by a set of components. In general, errors in statistics are the rule rather than the exception. Thus, most analysts are not content with PCA; they prefer FA as an alternative. This often provides similar results as PCA, especially for data sets that contain a large number of observations, but it offers alternative estimation routines and several post-estimation options. For example, we no longer have to assume that the constructs—or *factors* as we call them in this section—are uncorrelated; we may now test the hypothesis that they are correlated. In the model we consider here, it seems unwise to assume that the latent variables that seem to underlie depressive symptoms and self-esteem have a correlation of zero. Instead, a reasonable hypothesis derived from the research literature is that they are negatively associated (Sowislo and Orth, 2013).

The first FA model used is called *principal factor analysis* (it is sometimes called *principal axis factor analysis*, although Stata calls it simply *principal factor* or pf). In Stata, rather than asking for pcf with factor, we ask for pf or list nothing since it is the default option.[3] Principal factor analysis also assumes that the factors are uncorrelated, but it uses a different initial estimate of the communalities. Instead of using a 1.0 (as PCA uses), it uses the R^2 from a regression model with the other variables predicting the variable of interest. For instance, the R^2 from a linear regression model predicting the variable *worthy* using the other 12 variables is 0.317 (determine this for yourself).

The first part of the output shown in Example 10.3 is similar to the output for the PCA in Example 10.2, with eigenvalues and the proportion of variance explained by the particular factors. The first two constructs, factors here, explain a similar proportion of the variance among the observed variables as in the PCA. Moreover, the component matrix has been replaced by the factor matrix (although they are labeled the same in both outputs); yet the interpretation of these numbers is virtually identical. The factor loadings may be interpreted as the correlations between the factors and the variables. Hence, by squaring them, we may determine the proportion of variance of that particular variable that is explained by the factor. A rule of thumb is to look for loadings that are greater than (in absolute value terms) 0.4 (some analysts are stricter and claim that the threshold should be 0.5). Thus, if a factor loading is 0.4, the factor accounts for 16% of the variance in the particular variable. Similarly, a factor loading of 0.5 indicates that the factor accounts for 25% of the variance of the particular variable.

Stata provides information on only 5 of the 13 factors that it extracted. This is because factors 6–13 have negative eigenvalues, which does not make much sense given that

---

3. There are several estimation routines used in factor analysis, including maximum likelihood (ML). One advantage of ML is the availability of fit statistics that are useful for comparing models and determining statistical significance. For instance, we might compare a two-factor model with a three-factor model to see which one fits the data best. Since ML estimation is used, a likelihood ratio test is available. As with most ML models, this general approach works best with large samples.

eigenvalues are measures of variance explained.[4] Thus, only factors with positive eigenvalues are presented in the factor loadings portion of the output. We can also simplify this by asking Stata to present only the factors with eigenvalues of 1.0 or greater (using the `fac-tors` subcommand), but it is best in EFA to first allow the program to extract as many factors as its default estimation routine recommends.

Example 10.3

```
factor worthy-attitude, pf
```

```
Factor analysis/correlation                     Number of obs    =       731
    Method: principal factors                   Retained factors =         5
    Rotation: (unrotated)                        Number of params =        55
```

Factor	Eigenvalue	Difference	Proportion	Cumulative
Factor1	4.54869	3.39318	0.8728	0.8728
Factor2	1.15551	0.89801	0.2217	1.0945
Factor3	0.25750	0.13887	0.0494	1.1439
Factor4	0.11863	0.11677	0.0228	1.1667
Factor5	0.00186	0.02634	0.0004	1.1671
Factor6	-0.02447	0.01589	-0.0047	1.1624
Factor7	-0.04036	0.02715	-0.0077	1.1546
Factor8	-0.06752	0.03420	-0.0130	1.1417
Factor9	-0.10172	0.01227	-0.0195	1.1222
Factor10	-0.11399	0.03516	-0.0219	1.1003
Factor11	-0.14915	0.02713	-0.0286	1.0717
Factor12	-0.17628	0.02090	-0.0338	1.0378
Factor13	-0.19718	.	-0.0378	1.0000

```
LR test: independent vs. saturated:  chi2(78) = 3539.78 Prob>chi2 = 0.0000
```

Factor loadings (pattern matrix) and unique variances

Variable	Factor1	Factor2	Factor3	Factor4	Factor5	Uniqueness
worthy	-0.3772	0.5142	-0.0637	0.0287	0.0110	0.5883
capable	-0.3548	0.4557	-0.1059	-0.1135	0.0134	0.6422
bother	0.5708	0.0422	0.2299	-0.0606	0.0120	0.6158
noeat	0.4572	0.1385	0.2878	0.0584	0.0018	0.6856
blues	0.7328	0.1966	0.1495	-0.1334	-0.0112	0.3841
depress	0.7998	0.1298	-0.0810	-0.1617	0.0015	0.3108
failure	0.6396	0.0346	-0.1935	-0.0299	-0.0107	0.5513
lonely	0.7548	0.0857	-0.1137	0.0032	-0.0006	0.4099
crying	0.6014	0.2215	-0.0441	0.1744	-0.0059	0.5569
sad	0.7998	0.1733	-0.1074	0.0946	-0.0049	0.3097
tired	0.6083	-0.0138	-0.0037	0.1023	0.0275	0.6186
goodqual	-0.3036	0.5708	0.0105	0.0019	0.0013	0.5819
attitude	-0.3562	0.4397	0.1168	0.0562	-0.0189	0.6627

---

4. Note, however, that, in linear algebra applications, eigenvalues can be negative, such as when a matrix is negative definite. But in a factor analytic context, we typically do not consider the factors associated with negative eigenvalues to be meaningful. They can, though, indicate missing data issues that should be explored (Schwertman and Allen, 1979).

Consider that we are still assuming, in a multidimensional sense, that our factors are positioning themselves so as to explain the maximum amount of variance in the set of variables. But this is not actually the case. We may take an additional step and "rotate" factors so they explain more of the variance (see Kline, 1994, 57–59, for some diagrams). In Stata the rotate post-estimation command is used. It is useful to determine some of the options for this command (type help rotate).

There are two general types of rotation used in FA: *orthogonal* and *oblique*. The former results in factors that explain the most variance, yet are uncorrelated with each other. In the social and behavioral sciences, however, researchers often allow factors to be correlated, so oblique rotation is a frequent choice. Stata includes several orthogonal rotation schemes, including the most commonly used (and the Stata default), *varimax*. It also offers several oblique rotation schemes, including perhaps the most common, *oblimin*, but also *promax* and *oblimax*. To keep things simple, we use only varimax and oblimin in this section.

The output in Example 10.3 (cont.) shows the result of a varimax rotation. Our first rotated solution is derived by typing rotate after the last model. However, the last command resulted in five factors, which is more than we would likely wish to use in this analysis. Therefore, it was changed to factor worthy-attitude, pf factors(2) before the rotate command so that the analysis retains only the first two factors, both of which have eigenvalues greater than 1.0. The subcommand blanks is used so that only loadings greater than 0.4 are shown. The first panel of the output shows the variance and proportion explained by each factor. The next panel displays the factor loadings after rotation. The last panel (Factor rotation matrix) indicates how far the factors were rotated.

The results demonstrate more clearly that the first and last two variables load on the second factor, whereas the remaining nine variables load on the first factor. This shows well that there are two factors that we might now tentatively label as depressive symptoms and self-esteem.

Example 10.3 (Continued)

```
factor worthy-attitude, pf factors(2)          * note the addition
rotate, blanks(0.4)
```

```
Factor analysis/correlation              Number of obs    =      731
    Method: principal factors            Retained factors =        2
    Rotation: orthogonal varimax (Kaiser off)   Number of params =       25
```

Factor	Variance	Difference	Proportion	Cumulative
Factor1	4.14711	2.59001	0.7958	0.7958
Factor2	1.55710	.	0.2988	1.0945

```
LR test: independent vs. saturated:  chi2(78) = 3539.78 Prob>chi2 = 0.0000
```

Rotated factor loadings (pattern matrix) and unique variances

Variable	Factor1	Factor2	Uniqueness
worthy		0.6126	0.5933
capable		0.5499	0.6665
bother	0.5504		0.6725
noeat	0.4769		0.7718
blues	0.7557		0.4244
depress	0.7956		0.3435
failure	0.6124		0.5897
lonely	0.7382		0.4229
crying	0.6408		0.5893
sad	0.8106		0.3302
tired	0.5664		0.6298
goodqual		0.6404	0.5820
attitude		0.5354	0.6798

(blanks represent abs(loading)<.4)

Factor rotation matrix

	Factor1	Factor2
Factor1	0.9390	-0.3440
Factor2	0.3440	0.9390

The final exploratory tool in FA is to rotate the factors "obliquely." As mentioned earlier, this means that the factors are allowed to correlate (i.e., they are no longer orthogonal). Some statisticians recommend against oblique factors since they are hard to interpret (Johnson, 1998). However, most social and behavioral scientists use oblique rotation since they argue that many of the most interesting and important conceptual measures that define the factors are bound to be associated in the "real world." As noted earlier, we rely on oblimin as our oblique rotation scheme.

Since we already have an initial extraction using principal axis factoring, we simply ask Stata for an oblique rotation using oblimin. Once again, we use the subcommand blanks.

The factor loadings are not dramatically different when using an oblique rotation scheme. However, the obliquely rotated factors explain a greater proportion of the variability among the observed indicators than the orthogonal factors. This is often the case when two proposed constructs are related to one another.

In this case, the correlation between the two factors is −0.38 (the estat common postestimation command provides this). In other words, those reporting higher values on the self-esteem questions tend to report lower values on the depressive symptoms questions. One of the issues to consider when using oblique rotation, however, is how to treat an observed variable that loads on more than one correlated factor. This issue does not arise in our model, but what should the analyst do when it does? Suppose, for example, that the *attitude* variable loaded on both factors with loadings in excess of 0.5 or 0.6. What would we do? Perhaps this is realistic: an observed indicator is part of two relatively distinct concepts. Nonetheless, the solution to this dilemma is not always straightforward.

Example 10.3 (Continued)

```
rotate, oblique oblimin blanks(0.4)
```

```
Factor analysis/correlation                    Number of obs      =      731
    Method: principal factors                  Retained factors =        2
    Rotation: oblique oblimin (Kaiser off)     Number of params =       25
```

Factor	Variance	Proportion	Rotated factors are correlated
Factor1	4.42829	0.8497	
Factor2	2.17167	0.4167	

```
LR test: independent vs. saturated:  chi2(78) = 3539.78 Prob>chi2 = 0.0000
```

Rotated factor loadings (pattern matrix) and unique variances

Variable	Factor1	Factor2	Uniqueness
worthy		0.6227	0.5933
capable		0.5560	0.6665
bother	0.5414		0.6725
noeat	0.4955		0.7718
blues	0.7793		0.4244
depress	0.8004		0.3435
failure	0.5991		0.5897
lonely	0.7336		0.4229
crying	0.6751		0.5893
sad	0.8262		0.3302
tired	0.5422		0.6298
goodqual		0.6678	0.5820
attitude		0.5393	0.6798

(blanks represent abs(loading)<.4)

Factor rotation matrix

	Factor1	Factor2
Factor1	0.9821	-0.5472
Factor2	0.1884	0.8370

CREATING LATENT VARIABLES

Once we are satisfied that we have identified meaningful constructs, the next step is often to take a confirmatory approach and consider how to construct latent variables that, in this case, measure depressive symptoms and self-esteem (or whatever latent variables the analyst finds in the FA). A first approach that might come to mind is to use the information from the FA to construct these scores. Stata has a couple of ways to do this: regression and Bartlett's. The regression approach uses an ordinary least squares (OLS) type equation with the variables as is and the coefficients from a weighting matrix. This estimates predicted scores (latent variables) for each person in the data set. Bartlett's is a weighted least squares approach. Another method is to simply take the mean of the items to create new (latent) variables. Here are some of these options in practice.

Example 10.3 (Continued)

```
predict depress1 selfesteem1          * regression-based predicted scores
predict depress2 selfesteem2, bartlett
                             * predicted scores based on Bartlett's WLS procedure
egen depress3 = rowmean(bother-tired)
egen selfesteem3 = rowmean(worthy capable goodqual
   attitude)
```

The main difference among these commands is that the first two use all 13 variables to estimate the predicted variables (based on the results of the FA) and the next two use the specific variables that the factor analyses suggest gauge depressive symptoms and self-esteem. The latter approach seems more appropriate because it clearly specifies what variables underlie the two concepts. However, the correlations among these new self-esteem constructs exceed 0.98 and the correlations among these new depressive symptoms constructs exceed 0.99, so they are very similar measures. Moreover, consider that the correlations among the self-esteem and depressive symptoms constructs are larger when all the variables are used to compute them. This should not be surprising.

Some researchers argue that taking the mean of a set of manifest variables is inappropriate since it does not attenuate measurement error. It simply combines several variables with error into a composite variable that still includes error. Thus, to validly adjust for measurement error, other confirmatory factor analytic techniques, such as the SEMs discussed later in the chapter, should be used.

FACTOR ANALYSIS FOR CATEGORICAL VARIABLES

PCA and FA utilize a Pearson's correlation or covariance matrix among a set of observed variables. Thus, they are most appropriate for continuous variables. Yet, as we have learned time and time again, many (perhaps most) variables in the social sciences are categorical.

This was clearly the case for the manifest indicators of depressive symptoms and self-esteem. So what options are available when the analyst has only categorical or a mix of categorical and continuous variables?

A common approach is to estimate correlations that are appropriate for categorical variables and use these in the FA. For example, when variables are ordinal, one may use *polychoric* correlations (Holgado-Tello et al., 2010). *Polyserial* correlations are used when one variable is continuous (interval to be exact) and the other is categorical, whereas *tetrachoric* correlations are used when both variables are binary (help tetrachoric). An important point about these types of correlations is that they assume that a continuous (unobserved or latent) variable underlies the categorical variable and thus that a bivariate normal distribution underlies the correlation (recall that a similar assumption forms the basis of the probit model; see Chapter 3). In any event, these correlations may be used in place of the Pearson's correlations that are normally used in PCA or FA.

In Stata, one may employ a user-written command, polychoric, to compute correlations for ordinal variables. For our analysis, it provides polychoric correlations since it recognizes that the variables are ordinal (but it can also estimate polyserial correlations). These should exhibit less bias than the correlations computed using the corr command.

Example 10.4

polychoric worthy-attitude   * the output is edited

Polychoric correlation matrix

	worthy	capable	bother	noeat	blues
worthy	1				
capable	.49423249	1			
bother	-.31197986	-.21642641	1		
noeat	-.1436164	-.22012591	.48012016	1	
blues	-.29386803	-.24720255	.63106712	.54170922	1

Compared to the Pearson's correlations estimated earlier, the polychoric correlations shown in Example 10.4 are a bit farther away from zero.

Since Stata can use raw data or a correlation matrix to estimate an FA model (using the factormat command), we may save the polychoric correlation matrix and then use it in an FA. The commands in Example 10.4 (cont.) demonstrate how this is done. Note that we could also use these steps with tetrachoric correlations.[5]

The results are similar to those found with the earlier FA, although the factor loadings for the self-esteem items on Factor 1 are larger in this model than in the model shown in Example 10.3. However, rotating the factors reduces this tendency by a substantial margin (try rotate, oblique oblimin).

---

5. For more information on this approach, see http://www.philender.com/courses/multivariate /notes2/morefa.html.

Example 10.4 (Continued)

```
matrix R = r(R)                        * saves the previous correlation matrix
factormat R, n(731) pf factors(2)
                   * requests a principal axis factor analysis with a sample size of 731
```

```
Factor analysis/correlation                 Number of obs    =      731
  Method: principal factors                  Retained factors =        2
  Rotation: (unrotated)                      Number of params =       25
```

Factor	Eigenvalue	Difference	Proportion	Cumulative
Factor1	5.93303	4.59707	0.8343	0.8343
Factor2	1.33596	0.99175	0.1879	1.0222
Factor3	0.34421	0.16906	0.0484	1.0706
Factor4	0.17515	0.13604	0.0246	1.0952
Factor5	0.03911	0.01862	0.0055	1.1007
Factor6	0.02049	0.04002	0.0029	1.1036
Factor7	-0.01953	0.02775	-0.0027	1.1009
Factor8	-0.04728	0.02349	-0.0066	1.0942
Factor9	-0.07077	0.04464	-0.0100	1.0843
Factor10	-0.11541	0.01104	-0.0162	1.0680
Factor11	-0.12645	0.04282	-0.0178	1.0503
Factor12	-0.16927	0.01892	-0.0238	1.0265
Factor13	-0.18819	.	-0.0265	1.0000

```
LR test: independent vs. saturated:  chi2(78) = 5784.06 Prob>chi2 = 0.0000
```

Factor loadings (pattern matrix) and unique variances

Variable	Factor1	Factor2	Uniqueness
worthy	-0.4400	0.5531	0.5005
capable	-0.4016	0.4710	0.6169
bother	0.6708	0.0662	0.5457
noeat	0.5403	0.1852	0.6738
blues	0.8165	0.1921	0.2964
depress	0.8730	0.1307	0.2208
failure	0.8116	-0.0229	0.3407
lonely	0.8268	0.1021	0.3059
crying	0.7364	0.2473	0.3965
sad	0.8659	0.2016	0.2096
tired	0.6921	0.0122	0.5209
goodqual	-0.3632	0.6161	0.4885
attitude	-0.4026	0.4722	0.6149

Other options when faced with categorical variables are to utilize a user-written Stata program called `polychoricpca` (*polychoric principal components analysis*), use Stata's `gsem` command (shown later), or employ specialized software designed for SEM (e.g., MPlus, LISREL). An example of a polychoric PCA in Stata is found in Example 10.5.

`polychoricpca worthy-attitude, score(f) nscore(2)`

    *   output not shown*

This estimates a PCA using the polychoric correlations and then creates two new variables, with the prefix *f* (*f1* and *f2*), that are similar to the factor-based variables created in Example 10.3. In fact, the first polychoric component (*f1*) is correlated with the first depressive symptoms construct (*depress1*) at 0.96 and the second polychoric component (*f2*) is correlated with the first self-esteem factor (*selfesteem1*) at 0.81. Slightly larger correlations exist between these components and the factor variables predicted with the `factormat` command used earlier.

Of course, it is up to the analyst to decide which manifest variables should be part of which factors or components. As suggested by these various latent variable analyses and as shown when we created latent variables using the `egen` command, it may be better—especially since we have entered the domain of confirmatory analysis—to estimate separate models for the self-esteem and depressive symptom items and create latent variables from each. The `polychoricpca` commands that take these steps are illustrated in Example 10.6.

`polychoricpca worthy capable goodqual attitude,`
    `score(esteem) nscore(1)`

Principal component analysis

k	Eigenvalues	Proportion explained	Cum. explained
1	2.372585	0.593146	0.593146
2	0.687887	0.171972	0.765118
3	0.479057	0.119764	0.884882
4	0.460471	0.115118	1.000000

`polychoricpca bother-tired, score(depress) nscore(1)`

Principal component analysis

k	Eigenvalues	Proportion explained	Cum. explained
1	5.783241	0.642582	0.642582
2	0.801506	0.089056	0.731639
3	0.539937	0.059993	0.791632
4	0.486604	0.054067	0.845699
5	0.452355	0.050262	0.895960
6	0.291519	0.032391	0.928351
7	0.283725	0.031525	0.959876
8	0.184719	0.020524	0.980401
9	0.176394	0.019599	1.000000

There are now two models, both of which have a single eigenvalue that are much larger than the others. The new components (latent variables) computed from each, *esteem1* and *depress1*, have a Pearson's correlation of −0.343 (see Example 10.7).

Example 10.7

```
sum esteem1 depress1
```

Variable	Obs	Mean	Std. Dev.	Min	Max
esteem1	736	-.0150939	1.34964	-3.780652	2.60021
depress1	734	-.0465292	1.726182	-1.70851	5.925948

```
corr esteem1 depress1
```

	esteem1 depress1
esteem1	1.0000
depress1	-0.3432 1.0000

CONFIRMATORY FACTOR ANALYSIS USING STRUCTURAL EQUATION MODELING (SEM)

Stata has recently introduced a suite of commands for SEM, which is the most common approach for estimating CFA models. In fact, it should be reiterated that CFA is arguably the preferred approach for identifying latent variables. Some observers maintain that, since it has—or ought to have—theoretical underpinnings and is hypothesis driven, CFA should be used and EFA should be avoided. In any event, SEMs have become the main analytical approach for estimating CFA models. One advantage of a CFA model estimated in an SEM framework is that it provides a host of model fit statistics with which to judge the adequacy of one's hypothesized models. A second advantage is that SEMs allow researchers to explicitly model the error structure and thus efficiently adjust for measurement error. These issues are beyond the scope of this brief introduction.

Example 10.8 illustrates how we may use the `sem` command to create the two latent variables (see the Stata help menu for a complete set of instructions regarding `sem`). We will assume that we have already set up hypotheses about the existence of these variables and now wish to use a CFA model to examine them statistically. The syntax involves naming latent variables using upper case letters at the beginning of the word (e.g., *Esteem* rather than *esteem*).

Example 10.8

```
sem (Esteem -> worthy capable goodqual attitude)
    (Depress -> bother-tired), standardized
```

Note that after giving the latent variables names we then place an arrow toward the observed variables that we presume are indicators of each latent variable. Recall that, in an FA, the latent variables are designed to explain the shared variability among the observed variables, so the direction of the arrows makes sense. We also include the subcommand `standard-`

`ized` so that Stata provides standardized regression coefficients (factor loadings), which is a standard practice in CFA.

The Stata output is rather lengthy since it provides results for each observed variable (factor loadings and intercepts), error variances for each observed variable, and the covariance (correlation if `standardized` is used) between the two factors. Thus, the output provided in Example 10.8 is edited.

The results indicate, among other things, that the factor loading for *worthy* and *Esteem* is 0.68, whereas the factor loading for *bother* and *Depress* is 0.56. The estimated error variance for the variable *worthy* is 0.54. This is the proportion of variability in *worthy* that is left unexplained once *Esteem* is estimated. Thus, 54% of its variability is not accounted for by the model, whereas 46% is accounted for by the model. Similarly, the estimated error variance for the variable *bother* is 0.69. Hence, 31% of its variance is accounted for by the model (or $\{factor\ loading\}^2 \rightarrow 0.56^2 = 0.31$).

```
Structural equation model                     Number of obs      =        731
Estimation method  = ml
Log likelihood     = -9812.7622

 ( 1)   [worthy]Esteem = 1
 ( 2)   [bother]Depress = 1
```

Standardized	Coef.	OIM Std. Err.	z	P>|z|	[95% Conf. Interval]	
Measurement						
worthy <-						
Esteem	.6760602	.0300701	22.48	0.000	.6171239	.7349966
_cons	3.653141	.102451	35.66	0.000	3.452341	3.853941
capable <-						
Esteem	.6012726	.0319553	18.82	0.000	.5386413	.6639038
_cons	2.685707	.079383	33.83	0.000	2.530119	2.841295
goodqual <-						
Esteem	.6618867	.0304983	21.70	0.000	.602111	.7216623
_cons	3.821149	.1065604	35.86	0.000	3.612295	4.030004
attitude <-						
Esteem	.5641029	.0332669	16.96	0.000	.498901	.6293048
_cons	3.28548	.0935483	35.12	0.000	3.102129	3.468831
bother <-						
Depress	.5587546	.0274771	20.34	0.000	.5049005	.6126087
_cons	2.014941	.0643817	31.30	0.000	1.888755	2.141126
noeat <-						
Depress	.4570924	.0312388	14.63	0.000	.3958655	.5183194
_cons	1.821661	.0603141	30.20	0.000	1.703448	1.939875
blues <-						
Depress	.7521273	.0184229	40.83	0.000	.7160191	.7882355
_cons	1.845592	.0608097	30.35	0.000	1.726407	1.964777

var(e.worthy)	.5429426	.0406584			.4688257	.6287766
var(e.capable)	.6384713	.0384277			.5674271	.7184105
var(e.goodqual)	.561906	.0403729			.4880957	.6468781
var(e.attitude)	.6817879	.0375319			.6120562	.7594642
var(e.bother)	.6877933	.0307059			.6301687	.7506872
var(e.noeat)	.7910665	.0285581			.737028	.849067
var(e.blues)	.4343046	.0277127			.3832479	.4921631
var(e.depress)	.3243828	.0240476			.2805145	.3751113
var(e.failure)	.5804866	.0304904			.5236997	.6434311
var(e.lonely)	.4056674	.0266987			.3565733	.4615209
var(e.crying)	.610928	.030838			.5533803	.6744602
var(e.sad)	.3189318	.0238673			.2754216	.3693156
var(e.tired)	.6425207	.0308354			.5848397	.7058907
var(Esteem)	1	.			.	.
var(Depress)	1	.			.	.
cov(Esteem,Depress)	-.4001682	.0402175	-9.95	0.000	-.4789931	-.3213433

LR test of model vs. saturated: chi2(64) = 228.65, Prob > chi2 = 0.0000

(the output is edited)

The correlation (labeled *cov* in the output) between *Esteem* and *Depress* is −0.40. This is slightly larger than the correlation we found earlier using FA with oblique rotation (−0.38) (see Example 10.3). The difference is likely due to the different estimation technique used in the SEM approach (ML), as well as the fact that specific variables were used to define the factors in the SEM.

The sem command includes several post-estimation options, many of which are regularly reported by researchers who use this approach to analyze empirical models. For instance, fit statistics, including χ^2 tests, information indices (e.g., Bayesian information criterion [BIC]), and root mean squared errors (RMSEs) are often included in presentations of SEMs. Following the sem command, we may type estat gof, stats(all) to obtain these fit statistics (try it). The RMSEA (the last letter stands for approximation) for the model in Example 10.8, for instance, is 0.059.[6] We may also examine predicted values and residuals. In addition, there are several different estimation routines available for SEMs. The default is ML, but there are options for missing data and distribution free estimation. Books on SEMs provide a detailed discussion of these issues (Kaplan, 2009).

GENERALIZED SEM

Since the observed variables are categorical, we may underestimate the effects in the CFA using Stata's sem command. An alternative is to use polychoric correlations as the input

---

6. Some researchers suggest that a decent model fit is gauged by an RMSEA below 0.07, with a good fit below 0.05. However, there is considerable debate regarding these thresholds, as well as other goodness-of-fit issues involving SEMs (e.g., Chen et al., 2008; Fan and Sivo, 2007).

data. However, we may also use Stata's gsem (*generalized structural equation model*) suite of commands since, like the glm command, they allow various distributional assumptions regarding the observed variables. Example 10.9 provides the syntax to estimate the CFA of self-esteem and depression.

```
gsem (Esteem -> worthy capable goodqual attitude, ologit)
     (Depress -> bother-tired, ologit)
```

Note that the syntax is almost identical to what we used in Example 10.8, except we now specify the presumed distribution of the observed variables. Here, we identify them as ordinal variables and request that Stata use an ordinal logistic approach (ologit) to estimate the model.

The results are rather lengthy since Stata provides the coefficients and the cut-points for each of the observed variables (see Chapter 4). The generalized sem command does not provide standardized results (although other programs, such as MPlus, do). However, we can estimate the correlation between *Esteem* and *Depress* by using their variances and covariance in the following well-known formula:

$$\hat{\rho}_{xy} = \frac{\text{cov}(x, y)}{\hat{\sigma}_x \times \hat{\sigma}_y}.$$

Thus, the estimated correlation between *Esteem* and *Depress* is

$$\hat{\rho}_{xy} = \frac{-1.371}{\sqrt{3.868} \times \sqrt{2.668}} = -0.427.$$

This is slightly larger in magnitude than the estimate from the sem command (−0.40). Assuming we have specified the distribution of the observed variables correctly, this correlation is likely to be less biased than the earlier one.

Example 10.9 (Continued)

```
Generalized structural equation model
Log likelihood = -8190.7654

( 1)  [worthy]Esteem = 1
( 2)  [bother]Depress = 1
```

	Coef.	Std. Err.	z	P>|z|	[95% Conf. Interval]	
worthy <-						
Esteem	1	(constrained)				
capable <-						
Esteem	.7450775	.0987965	7.54	0.000	.5514399	.9387151
goodqual <-						
Esteem	.9893068	.1421666	6.96	0.000	.7106653	1.267948
attitude <-						
Esteem	.6928403	.0955177	7.25	0.000	.5056291	.8800515
bother <-						
Depress	1	(constrained)				
noeat <-						
Depress	.6957695	.084208	8.26	0.000	.5307249	.8608142
blues <-						
Depress	1.665931	.1789589	9.31	0.000	1.315178	2.016684
depress <-						
Depress	2.17293	.2438267	8.91	0.000	1.695039	2.650822
failure <-						
Depress	1.531565	.1850369	8.28	0.000	1.168899	1.894231
lonely <-						
Depress	1.702209	.1825876	9.32	0.000	1.344344	2.060074
crying <-						
Depress	1.299623	.150422	8.64	0.000	1.004802	1.594445
sad <-						
Depress	2.099021	.2315445	9.07	0.000	1.645202	2.55284
tired <-						
Depress	1.01778	.1119579	9.09	0.000	.7983468	1.237214

[the cut-points section of the output is omitted]

var(Esteem)	3.867672	.7589972			2.632744	5.68186
var(Depress)	2.668328	.4506412			1.91639	3.715306
cov(Depress,Esteem)	-1.370684	.2294717	-5.97	0.000	-1.82044	-.9209276

A BRIEF NOTE ON REGRESSION ANALYSES USING STRUCTURAL EQUATION MODELS

Now that we have examined various regression models and data reduction techniques, it is a relatively straightforward matter to combine them into a full SEM. Stata's sem and gsem commands, as well as specialty programs such as MPlus, AMOS, EQS, and others, are well designed for this task. In this context, the data reduction portion is referred to as the *measurement model* and the regression portion is referred to as the *structural model*.

For example, consider the *nlsy_delinq* data set (*nlsy_delinq.dta*). It includes variables that measure a number of self-reported delinquent activities among a sample of young people (*propdamage – stole49*). It also has some items designed to measure their general behaviors based on parents' reports (*mood – getsalong*). Suppose we hypothesize that parents' views of general behaviors predict young people's self-reported involvement in delinquent behaviors. We may test this hypothesis by combining a measurement model (a CFA that distinguishes general behaviors from delinquent behaviors) and a structural model (a regression model that predicts delinquent behaviors with these general behaviors) in an SEM. Although the observed variables are categorical, we estimate it with a linear model. However, you may wish to consider a generalized SEM (gsem) to determine whether its results differ.

Example 10.10

```
sem (General -> mood argues tense complains fearful
    getsalong) (Delinq -> propdamage shoplift stole50 stole49)
    (General -> Delinq) (male -> Delinq), standardized
```

```
Structural equation model              Number of obs    =       612
Estimation method  = ml
Log likelihood     = -3894.1956

 ( 1)   [propdamage]Delinq = 1
 ( 2)   [mood]General = 1
```

		OIM				
Standardized	Coef.	Std. Err.	z	P>|z|	[95% Conf. Interval]	
Structural						
Delinq <-						
male	.165038	.0464044	3.56	0.000	.074087	.2559889
General	-.121971	.0540125	-2.26	0.024	-.2278335	-.0161085

```
Measurement
   mood <-
       General     .6039277    .0336929    17.92   0.000    .5378908    .6699645
         _cons    3.063474    .0964414    31.77   0.000    2.874452    3.252496

   argues <-
       General     .5801782    .0346758    16.73   0.000    .5122149    .6481416
         _cons    2.936495    .0931587    31.52   0.000    2.753907    3.119083

   tense <-
       General      .597845    .0342996    17.43   0.000     .530619     .665071
         _cons    4.297035    .1293018    33.23   0.000    4.043608    4.550462

   complains <-
       General     .6661731    .0315102    21.14   0.000    .6044142    .7279319
         _cons    4.400819    .1321225    33.31   0.000    4.141864    4.659775
```

(the output is edited)

```
        var(e.mood)     .6352714    .0406962                 .5603127    .7202581
      var(e.argues)     .6633932    .0402363                 .5890385    .7471337
       var(e.tense)     .6425813    .0410117                 .5670241    .7282067
   var(e.complains)     .5562134    .0419825                 .4797263    .6448956
     var(e.fearful)     .7071313    .0393686                 .6340311    .7886595
    var(e.getsalong)    .7938067    .0353186                 .7275159    .8661378
   var(e.propdamage)    .6287273    .0452858                 .5459491    .7240566
     var(e.shoplift)    .5921431    .0463552                 .5079153    .6903383
      var(e.stole50)    .5621301    .0479187                 .4756377    .6643507
      var(e.stole49)    .6891045    .0429198                 .6099151    .7785756
       var(e.Delinq)    .9589376     .019798                 .9209088    .9985367
       var(General)            1           .                        .           .

 cov(male,General)      .0261316    .0465193    0.56   0.574   -.0650445    .1173077

LR test of model vs. saturated: chi2(42)  =    94.34, Prob > chi2 = 0.0000
```

The model includes the variable *male* since many studies indicate that boys tend to be involved in juvenile delinquency more than girls (e.g., Daigle et al., 2007). The results of the SEM shown in Example 10.10 indicate a modest association ($\beta = -0.122$, $p = 0.024$) between parents' observed behaviors and self-reported delinquent behaviors. In addition, males tend to be involved in delinquent behaviors more than females ($\beta = 0.165$). The model fits the data fairly well, with an RMSEA of 0.045 (95% CIs: 0.033, 0.057), although other fit statistics and diagnostics should be considered (Kline, 2010).

As shown earlier, FA models are often depicted in a visual way. Similarly, SEMs are frequently shown as diagrams. Figure 10.3 provides one way that the SEM estimated here might be presented.

Observed variables, such as *male*, are usually shown in rectangles, whereas latent variables are placed in ovals. The standardized regression coefficients are on the arrows that run from *male* and *general behaviors* to *delinquent behaviors*. Moreover, the error term is provided. A more complete diagram would include the observed variables that are explained by the latent vari-

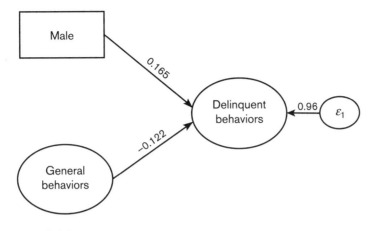

ables and their estimated errors, but they are omitted here because including them all creates a cluttered graph. The presentation of SEMs typically includes various fit statistics—which may include the Akaike's information criterion and BIC—that function to tell the analyst whether the data fit the model well. A description of these, as well as other important issues, may be found in books that specialize in SEMs (e.g., Kaplan, 2009; Raykov and Marcoulides, 2012).[7]

If we compare these results to a standard measurement approach wherein we sum the variables or take their means (e.g., `egen delinq = rowmean (propdamage-stole49)`) and then estimate an OLS regression model, we find, as research on measurement error suggests, that the standardized coefficient (beta weight) from the OLS model ($\beta = -0.094$) is closer to zero than the standardized coefficient from the SEM (-0.122).

These examples provide only a limited perspective on capabilities of SEMs. Most SEM software also includes a host of tools for estimating and testing various types of CFA and other measurement and regression models, as well as options for testing the assumptions of these models (Acock, 2013).

FINAL WORDS

This chapter provides a brief overview of a set of techniques designed to explore latent variables. Although there are several other approaches, CFA remains the most common technique. As shown in this chapter, it may be used with continuous and categorical data and lends itself to structural models designed to predict some outcome based on a set of explanatory variables. The examples provide merely a glimpse of these issues. But, of course, as

---

7. Although SEMs are often utilized in the manner shown here, it is important to point out that they may also be used for most of the regression models discussed in this book. In fact, some researchers claim that linear and other types of regression models are simply special cases of SEMs. For instance, the glm model shown in Example 2.3 is easy to estimate with Stata's sem command:

```
sem (sei <- female nonwhite educate pasei).
```

emphasized throughout this book, there are many additional steps that should be taken before we conclude with any reasonable confidence that the variables examined in these examples are related in a meaningful way. Some of these steps address assumptions of the regression models, but others are beyond the scope of this presentation and concern the essential roles of theory and measurement, as well as our ability to make causal inferences in the social sciences.

EXERCISES FOR CHAPTER 10

1. This exercise uses the personality data set (*personality.dta*). These data include information from several hundred employees who work for a large retail company. As the name of the data set implies, the project that collected the data was designed to measure personality attributes of these employees. The variables provide a brief description of their personality attributes, such as relaxed, tense, or imaginative, with higher values on each variable denoting that the person is described well by this attribute (e.g., someone who scores a 5 on PERS01 is a very talkative person). Our goal is to determine whether a set of latent variables underlies these observed personality characteristics.
   a. Estimate a factor analysis model with all of the personality variables in the data set. Use a principal axis factor model. Retain only those factors that have an eigenvalue greater than 1.
   b. What are the correlations between factor 1 and the following variables: *talkative, full of energy*, and *disorganized*?
   c. What are the communalities for the following variables: *finds fault, tense*, and *reliable*?

2. Use varimax rotation on the factor analysis model estimated in exercise 1. What proportion of the variability is accounted for by each of the factors (use the same eigenvalue threshold as in 1a)?
   a. Based on the varimax rotated factor model, what are the correlations between factor 1 and the following variables: *talkative, full of energy*, and *disorganized*? What is the correlation between factor 2 and factor 3?
   b. Use oblimin rotation on the factor analysis model estimated in exercise 1. What is the correlation between factor 1 and factor 2? How would you describe a person who had a high value on factor 3 (do not simply list her/his attributes)?

3. Estimate a factor analysis model using polychoric correlations[8] and principal axis factoring with the following observed variables: *PERS01, PERS03, PERS05, PERS07, PERS13*, and *PERS16*.
   a. How many factors should you extract from this model?
   b. How would you characterize the factor(s)?

4. Estimate the factor analysis model you concluded with in exercise 2 using a generalized structural equation model (in Stata, use the `gsem` command). If using Stata, you first need to rename the variables since capital letters identify latent variables. Even though the model coefficients are on different scales, how do the results of the generalized SEM model compare in general to the results of the model in exercise 3?

_____

8. If you have not yet done so, you will need to download the Stata user-written routine titled `polychoric` (type `findit polychoric`) or use a similar program in another brand of statistical software (e.g., package *polycor* in R).

SAS, SPSS, and R Code for Examples in Chapters

The following code is designed to help interested users replicate most of the analyses found in the examples in the chapters (only code for selected figures is provided). Please note that specific commands are provided, but the code to read and manage data, and to install requisite packages in R, are not consistently listed. You should also be aware that statistical software packages typically include unique options and output, so that output based on the three programs might differ. Moreover, some of the examples below may provide only partial replication. All three software packages are updated routinely, so users are encouraged to check the most recent software documentation to determine if conventions have changed, new features added, and so forth. In some cases, the producers may have updated their software so that it now allows full replication of the examples. For example, R packages are routinely updated by their authors (see the help menu for `update.packages`) and SAS and SPSS users regularly produce new code and macros for statistical procedures.

There are also now relatively simple ways to call one program into another. For example, if SPSS does not include a command for estimating a particular model, the user may set up SPSS to call R to estimate the model, which can then be output to SPSS (see http://www.ibm.com/developerworks/library/ba-call-r-spss). There are also ways to call R from Stata (Newson, 2014) and SAS (http://support.sas.com/rnd/app/studio/Rinterface2.html).

SECTION 1: SAS CODE

CHAPTER 1

Figure 1.2

```
proc sgplot data=gss;
  density sei / type=kernel;
run;
```

Example 1.1

```
proc reg data=gss;
  model sei=female / stb;
run;
```

Example 1.2

```
proc reg data=gss;
  model sei=female educate nonwhite pasei/ stb;
run;
```

Figure 1.4

```
proc reg data=gss;
  model sei = female educate nonwhite pasei;
  plot predicted.*residual.;
run;
```

Example 1.3

```
proc reg data=gss;
  model sei=female educate nonwhite pasei;
  output out=gss student=rstudent predicted=pred;
run;

data zgss;
set gss;
zpred = pred;
run;

proc standard data=gss mean=0 std=1 out=zgss;
  var zpred ;
run;

ods graphics on;
proc loess data=zgss;
  model rstudent=zpred;
run;
ods graphics off;
```

Example 1.4

```
* Breusch-Pagan Test ;

proc model data=gss;
  parms b0 b1 b2 b3 b4;
  sei = b0 + b1*female + b2 * educate + b3*nonwhite + b4 *pasei;
  fit sei / white breusch=(1 female educate nonwhite pasei);
run;
```

```
* Simple coding for White's Test was not available in SAS at the
  time of publication ;
```

Example 1.5

```
data gssabs;
set gss;
absresid=abs(rstudent);
run;

proc reg data=gssabs;
  model absresid=female educate nonwhite pasei;
run;
```

Example 1.6

```
proc robustreg data=gss method=m;
  model sei = female educate nonwhite pasei;
run;
```

Figure 1.5

```
proc sgplot data=gss;
  density rstudent / type=kernel;
run;
```

Figure 1.6

```
proc univariate data=gss;
  qqplot rstudent/ normal(mu=0 sigma=1);
run;
```

Figure 1.7

```
proc univariate data=CH13;
  ppplot rstudent/ normal(mu=0 sigma=1);
run;
```

Figure 1.8

```
proc reg data=gss;
  model sei = female educate nonwhite pasei;
  output out=gss residual=res;
run;

data gsscpr;
set gss;
b2 = 3.72279;
cpr = (b2 * educate) + res;
run;
```

```
proc sgplot data=gsscpr;
  scatter y = cpr x = educate;
  loess y = cpr x = educate/ smooth = .5;
run;
```

Example 1.7

```
* Link Test ;

data gsssq;
set gss;
pred2=pred**2;
run;

proc reg data= gsssq;
  model sei = pred pred2;
run;

* Ramsey RESET Test ;

proc autoreg data=gss;
  model sei = female educate nonwhite pasei/ RESET;
run;
```

Figure 1.10

```
proc reg data=gss;
  model sei=female educate nonwhite pasei;
  output out=gss residual=resid H=lev cookd=cookd;
run;

proc standard data=gss mean=0 std= 1 out=gss;
  var resid;
run;

data gss;
set gss;
resid2=resid**2;
run;

proc means data=gss;
  var resid2 lev;
run;

proc sgplot data=gss;
  scatter y= lev x= resid2;
```

```
  refline 0.0023/ axis=Y;
  refline 0.999/axis= X;
run;
```

Figure 1.11

```
proc sgplot data=gss;
  scatter y= cookd x= lev;
  refline 0.0018/ axis=Y;
  refline 0.01/axis= X;
run;
```

Example 1.8

```
proc quantreg data=gss;
  model sei = female educate nonwhite pasei / quantile=.5;
run;
```

Example 1.9

```
data gss;
set gss;
logsei = log(sei);
run;

proc reg data=gss;
  model logsei=female educate nonwhite pasei/ stb;
run;
```

Example 1.10

```
data gss;
set gss;
edu_pa=educate*pasei;
run;

proc reg data=gss;
  model sei = female educate nonwhite pasei edu_pa/vif;
run;
```

Example 1.11

```
proc reg data=gss;
  model sei=female educate nonwhite pasei;
  output out=zgss2 student= rstudent predicted=pred;
run;

proc standard data=zgss2 mean=0 std= 1 out=zgss2;
```

```
    var educate pasei;
run;

data zgss2;
set zgss2;
zedu_zpa=educate*pasei;
run;

proc reg data=zgss2;
  model sei = female educate nonwhite pasei zedu_zpa/vif;
run;

* The remaining examples and figures in Chapter 1 may be
  replicated by adapting the code already provided ;
```

CHAPTER 2

Example 2.1

```
proc freq data=gss;
  tables polviews;
run;
```

Example 2.2

```
proc freq data=gss;
  tables polviews*female / chisq;
run;
```

Figure 2.1

```
proc univariate data=gss;
  where childs>0 AND marital^= 5;
  var childs;
  histogram/endpoints=0 to 8 by 1;
run;
```

Example 2.3

```
proc mixed data= gss method=ml mmeq;
  model sei = female nonwhite educate pasei/s;
run;
```

Example 2.4

```
data gsssub;
set gss;
if female ^= . and nonwhite ^= . and educate ^= . and pasei ^= .;
```

```
run;

proc mixed data= gsssub method=ml mmeq;
  model sei = female nonwhite /s;
run;

proc mixed data= gsssub2 method=ml mmeq;
  model sei= /s;
run;
```

CHAPTER 3

Example 3.1

```
proc reg data=depress;
  model satlife = age;
  plot residual.*predicted.;
  output out=depress student=rstudent;
run;
```

Figure 3.1

```
* See Example 3.1 output for plot ;
```

Figure 3.2

```
proc univariate data=depress;
  qqplot rstudent/ normal(mu=0 sigma=1);
run;
```

Example 3.2

```
proc freq data=depress;
  tables male*satlife;
run;
```

Example 3.3

```
proc logistic data=depress descending;
  model satlife = male;
  contrast 'female' intercept 1 male 0 1 / estimate=prob;
  contrast 'male' intercept 1 male 1 0 / estimate=prob;
run;
```

Example 3.4

```
proc logistic data=depress descending;
  class male / param=ref;
  model satlife = male age iq;
```

```
  contrast 'female' intercept 1 male 0 1 age 37.4776271 iq 91.7909091/
    estimate=prob;
  contrast 'male' intercept 1 male 1 0 age 37.4776271 iq 91.7909091/
    estimate=prob;
run;
```

Example 3.5

```
* See the Example 3.4 code output ;
```

Example 3.6

```
proc logistic data=depress descending;
  model satlife = male age iq;
   test1: test intercept + 1.1793*male + 0*age + 0*iq;
run;
```

Example 3.7

```
proc logistic data=depress descending;
  model satlife = male age iq;
  output out=depress resdev=deviance H=lev predicted=pred
    reschi=resid;
run;

proc sgplot data=depress;
  density deviance / type=kernel;
run;

proc sgplot data=depress;
  scatter y= deviance x=pred;
run;

proc sgplot data=depress;
  scatter y = deviance x = age;
  loess y = deviance x = age;
run;

proc sgplot data=depress;
  scatter y= resid x= lev;
  refline .1284/ axis=X;
run;


* Delta-deviance was not an option to save in SAS at the time of
  publication. They could be computed, however, using some
  simple programming steps. ;
```

Figure 3.5

```
proc sgplot data=depress;
  scatter y=detad x=lev/ jitter;
  refline .1284/ axis=x;
  refline 4/ axis=y;
run;

proc means data=depress;
run;
```

Example 3.8

```
* Delta-deviance was not an option to save in SAS at the time
  of publication. They could be computed, however, using some
  simple programming steps. ;
```

Example 3.9

```
proc logistic data = depress descending;
  model satlife = male / link = probit;
  contrast 'female' intercept 1 male 0 1 / estimate=prob;
  contrast 'male' intercept 1 male 1 0 / estimate=prob;
run;
```

Example 3.10

```
proc logistic data=depress descending;
  model satlife = male sleep1 age iq/ link=probit;
  contrast "male good" intercept 1 male 1 0 sleep1 1 0 age 37.4576271
    iq 91.7909091/estimate=prob;
  contrast "female good" intercept 1 male 0 1 sleep1 1 0 age 37.4576271
    iq 91.7909091/estimate=prob;
  contrast "male poor" intercept 1 male 1 0 sleep1 0 1 age 37.4576271 iq
    91.7909091/estimate=prob;
  contrast "female poor" intercept 1 male 0 1 sleep1 0 1 age 37.4576271
    iq 91.7909091/estimate=prob;
run;

proc logistic data=depress descending;
  model satlife = male sleep1 age iq/ link=probit;
  contrast "female poor 30" intercept 1 male 0 1 sleep1 0 1 age 30 iq
    91.7909091/estimate=prob;
  contrast "female poor 40" intercept 1 male 0 1 sleep1 0 1 age 40 iq
    91.7909091/estimate=prob;
run;
```

Example 3.11

```
* There was no dprobit option in SAS at the time of publication. ;
```

Example 3.12

```
* There was no dprobit option in SAS at the time of publication. ;
```

Figure 3.7

```
proc logistic data=depress descending;
  model satlife = male sleep1 age iq/ link=probit;
  output out=depress resdev=deviance predicted=pred;
run;

proc sgplot data=depress;
  scatter y= deviance x=pred/jitter;
run;
```

Example 3.13

```
proc logistic data=depress descending;
  model satlife = male / firth;
  contrast 'female' intercept 1 male 0 1 / estimate=prob;
  contrast 'male' intercept 1 male 1 0 / estimate=prob;
run;
```

Example 3.14

```
proc genmod data = depress descending;
  model satlife = male / dist = binomial link = log;
run;
```

Example 3.15

```
proc genmod data = depress descending;
  class id;
  model satlife = male/ dist = poisson link = log;
  repeated subject = id;
run;
```

CHAPTER 4

Example 4.1

```
proc freq data=gss;
tables spanking;
run;
```

Figure 4.1

```
proc sgplot data=gss;
  reg y=spanking x=educate;
run;
```

Example 4.2

```
proc reg data=gss;
  model spanking=educate;
  output out=gss student=rstudent;
run;
```

Figure 4.2

```
proc sgplot data=gss;
  density rstudent/ type=kernel;
run;
```

Figure 4.3

```
proc univariate data=gss;
  qqplot rstudent/ normal(mu=0 sigma=1);
run;
```

Example 4.3

```
proc freq dat=gss;
  tables female*spanking;
run;
```

Example 4.4

```
data gss1;
set gss;
if spanking='strongly agree' then spanking=1;
if spanking='agree' then spanking=2;
if spanking='disagree' then spanking=3;
if spanking='strongly disagree' then spanking=4;
run;

proc logistic data = gss1 descending order=internal;
  class spanking(ref='strongly disagree') female (ref='male') /
    param=reference;
  model spanking = female;
run;
```

Example 4.5

```
* There was no omodel or Brant test option available in SAS at
  the time of publication ;
```

Example 4.6

```
proc logistic data = gss1 descending order=internal;
  class spanking(ref='strongly agree') female (ref='male') /
    param=reference;
  model spanking = female/ link=probit;
run;

data compgss;
set gss1;
if (female^=.) AND (spanking^=.);
run;

proc freq data=compgss;
  tables female spanking/ missing;
run;

proc logistic data = compgss order=internal;
  class spanking(ref='strongly agree') female (ref='male') /
    param=reference;
  model spanking = female/ link=probit;
  output out=compgss predprobs=c;
run;

proc means data=comgss;
  class female;
  var CP_strongly_agree CP_agree CP_disagree CP_
    strongly_disagree;
run;

proc logistic data = compgss order=internal;
  class spanking(ref='strongly agree') female (ref='male') /
    param=reference;
  model spanking = female/ link=probit;
  output out=compgss predprobs=i;
run;

proc means data=compgss;
  class female;
  var IP_strongly_agree IP_agree IP_disagree IP_
    strongly_disagree;
run;
```

Example 4.7

```
proc logistic data = gss1 descending order=internal;
  class spanking(ref='strongly disagree') female (ref='male') /
    param=reference;
  model spanking = female educate polviews;
run;
```

Example 4.8

```
proc logistic data = gss1 order=internal;
  class spanking(ref='strongly agree') female (ref='male') /
    param=reference;
  model spanking = female educate polviews/ link=probit;
  output out=gssmult predprobs=i;
run;

data submult;
set gssmult;
if (female^=.) AND (spanking^=.);
run;

proc means data=submult;
  class female;
  var IP_strongly_agree IP_agree IP_disagree IP_
    strongly_disagree;
run;

proc means data=submult;
  where educate = 10 and female = 0;
  var IP_strongly_agree IP_agree IP_disagree IP_
    strongly_disagree;
run;

proc means data=submult;
  where educate = 16 and female = 0;
  var IP_strongly_agree IP_agree IP_disagree IP_
    strongly_disagree;
run;
```

Figure 4.5

```
proc gchart data=submult;
  vbar female / discrete type=mean mean sumvar=IP_strongly_
    agree;
run;
```

```
proc gchart data=submult;
  vbar female / discrete type=mean mean sumvar=IP_
    strongly_disagree;
run;
```

Example 4.9

```
data gsslog;
set submult;
if spanking = "1" then newspank= "1"
if spanking = "2" then newspank= "0"
if spanking = "3" then newspank = "0"
if spanking = "4" then newspank= "0"
run;

proc logistic data=gsslog descending;
  model newspank = female educate polviews;
  output out=gsslog resdev=deviance H=lev predicted=pred
    reschi=resid;
run;

* Delta-deviance was not an option to save in SAS at the time of
  publication. They could be computed, however, using some
  simple programming steps. ;
```

Figure 4.6

```
proc sgplot data=gsslog;
  scatter y=deviance x=pred;
run;
```

Example 4.10

```
* There was no gologit2 option available in SAS at the time of
  publication ;
```

Example 4.11

```
* See Examples 1.3 and 4.7 for code that may be adapted for this
  example ;
```

CHAPTER 5

Example 5.1

```
proc freq data = gss;
  tables polview1;
run;
```

Example 5.2

```
proc freq data = gss;
  tables nonwhite*polview1/ nocol;
run;
```

Example 5.3

```
proc logistic data = gss;
  class polview1 (ref = "2")/ param = ref;
  model polview1 = nonwhite / link = glogit;
  contrast "white" intercept 1 nonwhite 0 /estimate=prob;
  contrast "nonwhite" intercept 1 nonwhite 1 /estimate=prob;
run;

proc logistic data = gss;
  class polview1 (ref = "3")/ param = ref;
  model polview1 = nonwhite / link = glogit;
  contrast "white" intercept 1 nonwhite 0 /estimate=prob;
  contrast "nonwhite" intercept 1 nonwhite 1 /estimate=prob;
run;
```

Example 5.4

```
proc logistic data = gss;
  class polview1 (ref = "2") / param = ref;
  model polview1 = female age educate / link = glogit;
run;
```

Example 5.5

```
proc logistic data = gss;
  class polview1 (ref = "3") / param = ref;
  model polview1 = female age educate / link = glogit;
  output out=gss predicted=pred;
run;

proc logistic data = gss;
  class polview1 (ref = "2") / param = ref;
  model polview1 = female age educate/ link = glogit;
  contrast "male 12" intercept 1 female 0 educate 12 age 44.695573/
    estimate=prob;
  contrast "male 16" intercept 1 female 0 educate 16 age 44.695573/
    estimate=prob;
run;
```

Example 5.6

```
* The format of this data set is not compatible for running a
  multinomial probit regression model in SAS. Proc MDC can be
  used for some multinomial probit regression models; more
  information can be found on the SAS support website ;
```

Example 5.7

```
data gss_lib;
set gss;
if (polview1=1) then libmod=1;
if (polview1=2) then libmod=0;
if (polview=3) then libmod=.;
run;

proc logistic data=gss_lib descending;
  model libmod = female age educate;
  output out=gss_lib resdev=deviance H=lev predicted=pred
    reschi=resid;
run;


* Delta-deviance was not available as a save option in SAS at
  the time of publication. They could be computed, however, with
  a few simple programming steps. ;
```

Figure 5.3

```
* Delta-deviance was not available as a save option in SAS at
  the time of publication. They could be computed, however, with
  a few simple programming steps. ;
```

Example 5.8

```
* Delta-deviance was not available as a save option in SAS at
  the time of publication. They could be computed, however, with
  a few simple programming steps. ;
```

Example 5.9

```
* There was no mlogtest option in SAS at the time of publication ;
```

CHAPTER 6

Example 6.1

```
proc freq data=gss;
  tables volteer;
run;
```

Example 6.2

```
proc means data=gss;
  var volteer;
run;
```

Figure 6.3

```
proc univariate data=gss;
  var volteer;
  histogram/normal(mu=0 sigma=1);
run;
```

Example 6.3

```
proc means data=gss;
  class female;
  var volteer;
run;
```

Example 6.4

```
proc genmod data = gss;
  model volteer = female / dist=poisson;
  output out=gss pred=predict;
run;

proc means data=gss;
  class female;
  var predict;
run;
```

Example 6.5

```
proc genmod data = gss;
  model volteer = female nonwhite educate income / dist=poisson;
  output out=gss pred=predict_m resdev=deviance leverage=lev
    cooksd=cookd;
  store p1;
run;

ods output ParameterEstimates = est;
proc plm source = p1;
  show parameters;
run;

data est_exp;
set est;
```

```
irr = exp(estimate);
run;

proc print data = est_exp;
run;

data predictevents;
set gss;
  do white = 0;
  nonwhite=white;
  output;
  end;
  do twelve= 12;
  educate=twelve;
  output;
  end;
  do inc = 5;
  income=inc;
  output;
  end;
run;

proc plm source=p1;
  score data = predictevents out=pred /ilink;
run;

proc means data = pred mean;
  class female white twelve inc;
  var predicted;
run;

proc means data=gss;
  class female nonwhite;
  var predict_m;
run;
```

Figure 6.5
```
ods graphics on;
proc loess data=gss;
  model predict_m=educate;
  run;
ods graphics off;
```

Example 6.6

```
* SAS uses the genmod command for Poisson regression; see
  Example 6.5 for additional saved statistics. ;
```

Figure 6.6

```
proc sgplot data=gss;
  scatter y= deviance x= lev;
  refline 4/ axis=Y;
  refline 0.0077/axis= X;
run;
```

Figure 6.7

```
proc sgplot data=gss;
  density adjdev / type=kernel;
run;
```

Example 6.7

```
proc genmod data = gss;
  model volteer = female nonwhite educate income/ dist=poisson
    scale=deviance;
run;
```

Example 6.8

```
proc genmod data = gss;
  model volteer = female nonwhite educate income /dist=negbin;
  output out=gss pred=predict_nb resdev=deviance_nb
    leverage=lev_nb cooksd=cookd_nb;
  store p1;
run;

ods output ParameterEstimates = est;
proc plm source = p1;
  show parameters;
run;

data est_exp;
set est;
irr = exp(estimate);
run;

proc print data = est_exp;
run;
```

```
data predictevents_nb;
set gss;
do white = 0;
nonwhite=white;
output;
end;
do twelve= 12;
educate=twelve;
output;
end;
do inc = 5;
income=inc;
output;
end;
run;

proc plm source=p1;
  score data = predictevents_nb out=pred_nb /ilink;
run;

proc means data = pred_nb mean;
  class female white twelve inc;
  var predicted;
run;

proc means data=gss;
  class female nonwhite;
  var predict_nb;
run;
```

Example 6.9

```
* SAS uses the genmod command for negative binomial regression.
  See Example 6.8 for additional saved statistics. ;

data gss;
set gss;
adjdev_nb = deviance_nb + (1/(6*sqrt(predict_nb)));
run;
```

Figure 6.9

```
proc sgplot data=gss;
  scatter y= deviance_nb x= lev_nb;
  refline 4/ axis=Y;
```

```
    refline 0.0077/axis= X;
run;
```

Figure 6.10

```
data gss_vol;
set gss;
if volteer ^=0;
run;

proc univariate data=gss_vol;
  var volteer;
  histogram/normal(mu=0 sigma=1)endpoints=1 to 10 by 1;
run;
```

Example 6.10

```
proc genmod data = gss;
  model volteer = female educate nonwhite income/dist=zip;
  zeromodel educate / link = logit;
  estimate "Example 6.10 Male" intercept 1 female 0 nonwhite 0
    educate 12 income 5;
  estimate "Example 6.10 Female" intercept 1 female 1 nonwhite 0
    educate 12 income 5;
  store p1;
run;

ods output ParameterEstimates = est;
proc plm source = p1;
  show parameters;
run;

data est_exp;
set est;
irr = exp(estimate);
run;
proc print data = est_exp;
run;
```

```
* Although there was no Vuong test option in SAS at the time of
  publication, there is a %VUONG macro available on the SAS
  support website (support.sas.com) that can be installed to run
  this test ;
```

Example 6.11

```
proc genmod data = gss;
  model volteer = female educate nonwhite income/dist=zinb;
```

```
  zeromodel educate / link = logit;
  store p1;
run;

ods output ParameterEstimates = est_zinb;
proc plm source = p1;
  show parameters;
run;
data est_exp_zinb;
set est_zinb;
irr = exp(estimate);
run;
proc print data = est_exp_zinb;
run;
```

* Although there was no Vuong test option in SAS at the time of
  publication, there is a %VUONG macro available on the SAS
  support website (support.sas.com) that can be installed to run
  this test ;

Example 6.12

* In SAS, estimates are calculated in the same code as the
  original model. See the example 6.10 output ;

Example 6.13

```
proc genmod data = gss;
  model volteer = female educate nonwhite income/dist=zip;
  zeromodel educate / link = logit;
  output out=gss pred=predict resraw=resid;
run;

proc standard data=gss mean=0 std=1 out=gss;
  var resid;
run;
```

Figure 6.12

```
proc sgplot data=gss;
  where volteer>0;
  scatter y= resid x= predict/ jitter;
run;
```

Figure 6.13

```
data gss;
set gss;
```

```
anscombe = (1.5*((volteer)**(2/3))-(predict**(2/3)))/(predict**(1/6));
run;

proc sgplot data=gss;
  where volteer>0;
  scatter y= anscombe x= predict/ jitter;
run;
```

Figure 6.14

```
proc sgplot data=gss;
  where volteer>0;
  density anscombe / type=kernel;
run;
```

CHAPTER 7

Example 7.1

```
proc print data=event1;
run;
```

Example 7.2

```
* Time variables in SAS are set with each individual survival
  analysis. See Example 7.3 ;
```

Figure 7.2

```
* Survival plots in SAS are provided by default with each
  individual survival analysis. See the output from Example 7.3 ;
```

Example 7.3

```
proc lifetest data=event1 atrisk;
  time marriage*marriage(0);
run;
```

Example 7.4

```
* Mean and median survival output are provided by default with
  PROC LIFETEST. See the output from Example 7.3 ;
```

Figure 7.3

```
proc lifetest data=event1 atrisk;
  strata cohab;
  time marriage*marriage(0);
run;
```

Example 7.5

```
* Log-rank and Wilcoxon tests are provided by default with PROC
  LIFETEST. See the output from figure 7.3 ;
```

Example 7.6

```
* Time variables in SAS are set with each individual survival
  analysis. See Example 7.7 ;
```

Example 7.7

```
proc lifereg data=event2;
  model marriage*evermarr(0)= educate age1sex attend14 cohab
    race/dist=lnormal;
run;

data estimates7_7;
bo=exp(2.7760);
b1=exp(0.1230);
b2=exp(-0.0734);
b3=exp(1.0892);
b4=exp(0.3829);
b5=exp(2.3255);
run;

proc print data=estimates7_7;
run;
```

Example 7.8

```
proc lifereg data=event2;
  model marriage*evermarr(0)= educate age1sex attend14 cohab
    race/dist=exponential;
run;

data estimates7_8;
bo=exp(-6.7878);
b1=exp(-0.0099);
b2=exp(0.0586);
b3=exp(-0.6869);
b4=exp(0.2791);
b5=exp(-1.0460);
run;

proc print data=estimates7_8;
run;
```

Example 7.9

```
proc lifereg data=event2;
  model marriage*evermarr(0)= educate age1sex attend14 cohab
    race/dist=weibull;
run;

data estimates7_9;
bo=exp(-5.8129*(1/3.2512));
b1=exp(-0.0673*(1/3.2512));
b2=exp(0.1143*(1/3.2512));
b3=exp(-1.3647*(1/3.2512));
b4=exp(0.2504*(1/3.2512));
b5=exp(-2.5386*(1/3.2512));
run;

proc print data=estimates7_9;
run;
```

Example 7.10

```
proc lifereg data=event2;
  model marriage*evermarr(0)= educate age1sex attend14 cohab
    race/dist=lnormal;
  output out=event2 CRES=cs7_10;
run;

proc lifetest data=event2;
  time cs7_10*evermarr(0);
run;

* There was no option to save KM estimates in SAS's PROC
  LIFETEST and PROC LIFEREG at the time of publication ;
```

Figure 7.4

```
* There was no option to save KM estimates in PROC LIFETEST and
  PROC LIFEREG in SAS at the time of publication ;
```

Figure 7.6

```
proc reliability data=event2;
  distribution lnorm;
  model marriage*evermarr(0)= educate age1sex attend14 cohab race
    /obstats (DRESID) ;
  ods output ModObstats = figure7_6;
run;
```

```
proc sgplot data=figure7_6;
  scatter y = dresid x = marriage;
run;
```

Example 7.11

```
proc phreg data = event2;
  model marriage*evermarr(0) = educate age1sex attend14 cohab
    race/ ties=efron;
  proportionality_test: test educate, age1sex, attend14, cohab,
    race;
run;
```

Example 7.12

```
data event2;
set event2;
cohab_marr=cohab*marriage;
run;

proc phreg data = event2;
  model marriage*evermarr(0) = educate age1sex attend14 cohab
    cohab_marr/ ties=efron;
run;
```

Figure 7.7

```
ods graphics on;
proc lifetest data=event2 plots=lls;
  time marriage*evermarr(0);
  strata cohab;
run;
ods graphics off;
```

Example 7.13

```
proc phreg data=event2;
  model marriage*evermarr(0) = age1sex attend14 race educatet
    cohabt/ ties=efron;
  educatet = educate*log(marriage);
  cohabt = cohab*log(marriage);
run;
```

Example 7.14

```
proc print data=event3;
run;
```

Example 7.15

```
proc logistic data=event3 descending;
  class year (ref='1') / param = ref;
  model delinq1 = stress cohes year;
run;
```

Example 7.16

```
proc phreg data=event3;
  model (year_s year)*delinq1(0) = stress cohes/ ties=efron;
  id newid;
run;
```

Example 7.17

```
proc logistic data=event4 descending;
  class year (ref='1') / param=ref;
  model delinq = stress cohes year;
run;
```

Example 7.18

```
proc phreg data=event4;
  model (year_s year)*delinq(0) = stress cohes/ ties=exact;
  id newid;
run;
```

CHAPTER 8

Example 8.1

```
proc print event4;
run;
```

Example 8.2

```
proc panel data=esteem;
  model esteem= stress cohes male nonwhite/ fittwo;
  id newid year;
run;
```

Example 8.3

```
proc panel data=esteem;
  model esteem= stress cohes male nonwhite/ rantwo;
  id newid year;
run;
```

Example 8.5

```
proc genmod data=event4 descending;
  class newid;
  model delinq = stress cohes/ dist=binomial;
  repeated subjects=newid/ type=ind corrw;
  estimate "Stress O.R." stress 1 / exp;
  estimate "Choes O.R." cohes 1 / exp;
  estimate "Intercept O.R." intercept 1/ exp;
run;
```

Example 8.6

```
proc genmod data=event4 descending;
  class newid;
  model delinq = stress cohes/ dist=binomial;
  repeated subjects=newid/ type=exch corrw;
  estimate "Stress O.R." stress 1 / exp;
  estimate "Choes O.R." cohes 1 / exp;
  estimate "Intercept O.R." intercept 1/ exp;
run;
```

Example 8.7

```
proc genmod data=event4 descending;
  class newid;
  model delinq = stress cohes/ dist=binomial;
  repeated subjects=newid/ type=ar(1) corrw;
  estimate "Stress O.R." stress 1 / exp;
  estimate "Choes O.R." cohes 1 / exp;
  estimate "Intercept O.R." intercept 1/ exp;
run;
```

Example 8.8

```
proc genmod data=event4 descending;
  class newid;
  model delinq = stress cohes/ dist=binomial;
  repeated subjects=newid/ type=un corrw;
  estimate "Stress O.R." stress 1 / exp;
  estimate "Choes O.R." cohes 1 / exp;
  estimate "Intercept O.R." intercept 1/ exp;
run;
```

Example 8.9

```
proc genmod data=esteem descending;
  class newid;
  model bi_depress = male/ dist=binomial;
  repeated subjects=newid/ type=exch corrw;
  estimate "Male O.R." male 1 / exp;
  estimate "Intercept O.R." intercept 1 / exp;
run;
```

Example 8.10

```
* There was no exact equivalent of Stata's xtlogit command in
  SAS at the time of publication. However, there are multiple
  procedures that examine random effects in logistic models,
  such as PROC GENMOD or PROC GLIMMIX ;
```

Example 8.11

```
* See the comment that accompanies Example 8.10. ;
```

CHAPTER 9

Example 9.1

```
proc reg data=multilevel_data12;
  model income=male married;
run;
```

Example 9.2

```
proc genmod data=multilevel_data12;
  class idcomm;
  model income=married male;
  repeated subject=idcomm;
run;
```

Example 9.3

```
proc means data=multilevel_data12;
  var income;
  by idcomm;
run;
```

Example 9.4

```
proc mixed data = multilevel_data12 method=ML;
  class idcomm;
  model income= / solution;
  random intercept / subject = idcomm;
run;
```

Figure 9.1

```
proc sgplot data=multilevel_data12;
  where idcomm=700 OR idcomm=1500 OR idcomm=2075 OR idcomm=4480;
  reg x=male y=income/group=idcomm;
run;
```

Example 9.5

```
proc mixed data = multilevel_data12 method=ML;
  class idcomm;
  model income= male / solution;
  random intercept / subject = idcomm;
run;
```

Example 9.6

```
proc mixed data=multilevel_data12;
  class idcomm;
  model income = male / s;
  random Int / type=un sub=idcomm s;
run;
```

```
* Note that the random effects are represented as variances
  rather than standard deviations ;
```

Example 9.7

```
proc mixed data = multilevel_data12 method=ML;
  class idcomm;
  model income= male / solution residual outp=example8_7;
  random intercept male / subject = idcomm solution;
  ODS output SolutionR = Town;
run;
```

```
* See the comment that accompanies Example 9.6 ;
```

Figure 9.2

```
proc standard data=example8_7 mean=0 std=1 out=zexample8_7;
  var resid;
run;
```

```
proc sgplot data=ezxample8_7;
  density resid / type=kernel;
run;
```

Figure 9.3

```
data rintercept rslope;
set Town;
if (Effect= 'Intercept') then output rintercept;
if (Effect= 'male') then output rslope;
run;

data rintercept;
set rintercept;
Intercept=Estimate;
drop Estimate StdErrPred DF Effect tvalue Probt;
run;

data rslope;
set rslope;
Slope=Estimate;
drop Estimate StdErrPred DF Effect tvalue Probt;
run;

data random;
merge rintercept rslope;
by idcomm;
run;

proc sgplot data=random;
scatter x=Intercept y=Slope / datalabel=idcomm;
run;
```

Example 9.8

```
proc mixed data = multilevel_data12 method=ML;
  class idcomm;
  model income= male pop2000 disadvantage/ solution;
  random intercept male / subject = idcomm;
run;
```

Example 9.9

```
proc mixed data = multilevel_data12 method=ML;
  class idcomm;
  model income= age disadvantage age*disadvantage/ solution
    outp=figure9_4;
  random intercept age / subject = idcomm;
run;
```

Figure 9.4

```
data figure9_4;
set figure9_4;
dis_cat=.;
if (disadvantage <- 0.57) then dis_cat=1;
if (disadvantage>0.57) then dos_cat=2;
run;

proc sgplot data=figure9_4;
  reg x=age y=pred/group=dis_cat;
run;
```

Example 9.10

```
proc glimmix data = multilevel_data12;
  class idcomm;
  model trust= age educate/ solution dist=binomial;
  random intercept age /subject=idcomm;
  nloptions tech=nrridg;
  estimate "Age Exp" age 1 / exp;
  estimate "Educate Exp" educate 1/ exp;
  estimate "Intercept Exp" intercept 1 / exp;
run;
```

Figure 9.6

```
* Anscombe residuals were not available in SAS PROC GLIMMIX at
  the time of publication. However, they could be computed with
  a few programming steps. ;
```

Figure 9.7

```
proc sgplot data=esteem;
  where newid=14 OR newid=23 OR newid=34 OR newid=53;
  reg x=age y=esteem/group=newid;
run;
```

Example 9.11

```
proc mixed data = esteem method=ML;
  class newid;
  model esteem= / solution;
  random intercept / subject = newid;
run;
```

Example 9.12

```
proc mixed data = esteem method=ML;
  class newid;
  model esteem= age / solution;
  random intercept / subject = newid;
run;

proc mixed data = esteem method=ML;
  class newid;
  model esteem= age / solution;
  random intercept age/ subject = newid;
run;

proc mixed data = esteem;
  class newid;
  model esteem= age;
  random intercept age/ subject = newid type=UN;
run;
```

Example 9.13

```
proc mixed data = esteem method=ML;
  class newid;
  model esteem= age stress cohes male/ solution;
  random intercept age stress cohes/ subject = newid;
run;
```

Example 9.14

```
proc mixed data = esteem method=ML;
  class newid;
  model esteem = age stress cohes male age*male / solution
    outp=figure9_8;
  random intercept age stress cohes/ subject = newid;
run;
```

Figure 9.8

```
proc sgplot data=figure9_8;
  reg x=age y=pred/group=male;
run;
```

Example 9.15

```
proc mixed data = esteem method=ML;
  class newid;
```

```
    model esteem= age stress cohes male nonwhite/ solution corrb
      residual outp=example9_15;
    random intercept age stress cohes / subject = newid type=UN
      gcorr;
run;
```

Figure 9.9

```
proc standard data=example9_15 mean=0 std=1 out=zexample8_15;
  var resid;
run;

proc sgplot data=zexample9_15;
  density resid / type=kernel;
run;
```

Figure 9.10

```
proc sgplot data=zexample9_15;
  scatter=pred y=resid;
run;
```

Example 9.16

```
proc mixed data = esteem method=ML;
  class newid;
  model esteem= age stress cohes male nonwhite/ solution;
  random intercept age stress cohes/ subject = newid solution;
  ODS output SolutionR = Town;
run;
```

Example 9.17

```
proc glimmix data = event4 method=MMPL;
  class newid;
  model stress= age/ solution dist=poisson;
  random intercept/subject=newid;
  estimate "Age IRR" age 1 / exp;
  estimate "Intercept IRR" intercept 1 / exp;
run;
```

CHAPTER 10

Example 10.1

```
proc means data=factor1;
  var worthy capable bother noeat blues depress failure lonely
    crying sad tired goodqual attitude;
run;
```

```
proc corr data=factor1;
  var worthy capable bother noeat blues depress failure lonely
    crying sad tired goodqual attitude;
  with worthy capable bother noeat blues depress failure lonely
    crying sad tired goodqual attitude;
run;
```

Example 10.2

```
proc factor data=factor1 method=prin rotate=none mineigen=1 round
    print;
  var worthy capable bother noeat blues depress failure lonely
    crying sad tired goodqual attitude;
run;
```

Example 10.3

```
proc factor data=factor1 method=prin priors=smc rotate=none
    nfactors=5 round print;
  var worthy capable bother noeat blues depress failure lonely
    crying sad tired goodqual attitude;
  run;
```

```
proc factor data=factor1 method=prin priors=smc rotate=varimax
  nfactors=2 flag=40 round print;
  var worthy capable bother noeat blues depress failure lonely
    crying sad tired goodqual attitude;
run;
```

```
proc factor data=factor1 method=prin priors=smc rotate=oblimin
    nfactors=2 flag=40 round print out=example9_3;
  var worthy capable bother noeat blues depress failure lonely
    crying sad tired goodqual attitude;
run;
```

```
proc means data=example9_3;
  var Factor1 Factor2;
run;
```

Example 10.4

```
proc freq data = factor1;
  tables (worthy capable bother noeat blues depress failure
    lonely crying sad tired goodqual attitude)*(worthy capable
    bother noeat blues depress failure lonely crying sad tired
    goodqual attitude) /plcorr;
```

```
ods output measures=example10_4 (where=(statistic="Polychoric
    Correlation")keep = statistic table value);
run;

proc print data = example10_4;
run;

proc corr data=factor1 outplc=R;
  var worthy capable bother noeat blues depress failure lonely
    crying sad tired goodqual attitude;
  with worthy capable bother noeat blues depress failure lonely
    crying sad tired goodqual attitude;
run;

proc factor data=R method=prin priors=smc rotate=none nfactors=2
  round print;
run;
```

Example 10.5

```
proc factor data=R method=prin rotate=none nfactors=2 round
  print;
  run;
```

Example 10.6

```
proc factor data=R method=prin rotate=none nfactors=1 round
  print;
  var worthy capable goodqual attitude;
run;

proc factor data=R method=prin rotate=none nfactors=2 round
  print;
  var bother noeat blues depress failure lonely crying sad tired;
run;
```

Example 10.7

```
* Because only a correlation matrix for polychoric factor
  analysis could be used in SAS at the time of publication,
  factors could not be saved as variables and analyzed ;
```

Example 10.8

```
proc calis data=factor1;
  path Esteem ---> worthy, Esteem ---> capable, Esteem --->
    goodqual, Esteem ---> attitude,
```

```
  Depress1 ---> bother, Depress1 ---> noeat, Depress1 ---> blues,
    Depress1 ---> depress,
  Depress1 ---> failure, Depress1 ---> lonely, Depress1 --->
    crying, Depress1 ---> sad,
  Depress1 ---> tired;
  pvar Esteem = 1, Depress1 = 1;
run;
```

Example 10.9

```
* PROC CALIS did not have an option for ordered logistic SEM at
  the time of publication ;
```

Example 10.10

```
proc calis data=nlsy_delinq;
  path
  General---> mood, General---> argues, General---> tense,
    General---> complains,
  General---> fearful, General---> gestating,
  Delinq---> propdamage, Delinq---> shoplift, Delinq---> stole50,
    Delinq---> stole49,
  General---> Delinq,
  Male---> Delinq;
run;
```

CHAPTER 1

Figure 1.2

```
GGRAPH
  /GRAPHDATASET NAME ="graphdataset" VARIABLES =sei MISSING
    =LISTWISE REPORTMISSING =NO
  /GRAPHSPEC SOURCE =INLINE.
BEGIN GPL
SOURCE: s=userSource (id("graphdataset"))
DATA: sei=col(source(s), name("sei"))
GUIDE: axis(dim(1), label("Respondent's Socioeconomic
  Index"))
GUIDE: axis(dim(2), label("Frequency"))
SCALE: linear(dim(1), min(15), max(100))
SCALE: linear(dim(2), include(0))
ELEMENT: interval(position(summary.count(bin.rect(sei))), shape.
  interior(shape.square))
ELEMENT: line(position(density.kernel.gaussian(sei)))
END GPL.
```

Example 1.1

```
REGRESSION
/MISSING LISTWISE
/STATISTICS COEFF OUTS R ANOVA
/CRITERIA=PIN(.05) POUT(.10)
/NOORIGIN
/DEPENDENT sei
/METHOD=ENTER female.
```

Example 1.2

```
REGRESSION
/STATISTICS COEFF OUTS R ANOVA
/DEPENDENT sei
/METHOD=ENTER female educate nonwhite pasei.
```

Figure 1.4

```
REGRESSION
  /DEPENDENT sei
  /METHOD=ENTER female educate nonwhite pasei
  /SCATTERPLOT(*resid *pred).
```

Example 1.3

```
REGRESSION
  /DEPENDENT SEI
  /METHOD=ENTER female educate nonwhite pasei
  /SAVE zpred(ZPR_1) sresid(SRE_1).

GGRAPH
  /GRAPHDATASET NAME="graphdataset" VARIABLES=ZPR_1 SRE_1
    MISSING=LISTWISE REPORTMISSING=NO
  /GRAPHSPEC SOURCE=INLINE.
BEGIN GPL
SOURCE: s=userSource(id("graphdataset"))
DATA: ZPR_1=col (source(s), name("ZPR_1"))
DATA: SRE_1=col (source(s), name("SRE_1"))
GUIDE: axis(dim(1), label ("Standardized Predicted Value"))
GUIDE axis(dim(2), label ("Studentized Residual"))
ELEMENT: point(position(ZPR_1*SRE_1))
ELEMENT: line(position(smooth.loess.gaussian(ZPR_1*SRE_1)))
END GPL.
```

Example 1.4

```
* There was no SPSS procedure for the Breusch-Pagan or White
  test available at the time of publication; however, see www.
  spsstools.net for a user-written macro.
```

Example 1.5

```
COMPUTE absresid=abs(SRE_1).
EXECUTE.

REGRESSION
  /DEPENDENT absresid
  /METHOD=ENTER female educate nonwhite pasei.
```

Example 1.6

```
GENLIN sei WITH female educate nonwhite pasei
  /MODEL female educate nonwhite pasei INTERCEPT=YES
  DISTRIBUTION=NORMAL LINK=IDENTITY
/CRITERIA SCALE=MLE COVB=ROBUST PCONVERGE=1E-006(ABSOLUTE)
  SINGULAR=1E-012 ANALYSISTYPE=3(WALD) CILEVEL=95 CITYPE=WALD
  LIKELIHOOD=FULL
  /MISSING CLASSMISSING=EXCLUDE
  /PRINT MODELINFO FIT SUMMARY SOLUTION.
```

Figure 1.5

```
GGRAPH
  /GRAPHDATASET NAME="graphdataset" VARIABLES=SRE_1
    MISSING=LISTWISE REPORTMISSING=NO
  /GRAPHSPEC SOURCE=INLINE.
BEGIN GPL
  SOURCE: s=userSource(id("graphdataset"))
DATA: SRE_1=col(source(s), name("SRE_1"))
TRANS: rug = eval(-60)
GUIDE: axis(dim(1), label("SRE_1"))
GUIDE: axis(dim(2), label("Frequency"))
SCALE: linear(dim(2), min(-30))
ELEMENT: point(position(SRE_1*rug), transparency.
  exterior(transparency."0.8"))
ELEMENT: line(position(density.kernel.gaussian(SRE_1*1)))
END GPL.
```

Figure 1.6

```
PPLOT
  /VARIABLES=SRE_1
  /NOLOG
  /NOSTANDARDIZE
  /TYPE=Q-Q
  /FRACTION=BLOM
  /TIES=MEAN
  /DIST=NORMAL.
```

Figure 1.7

```
PPLOT
  /VARIABLES=SRE_1
  /NOLOG
  /NOSTANDARDIZE
  /TYPE=P-P
  /FRACTION=BLOM
  /TIES=MEAN
  /DIST=NORMAL.
```

Figure 1.8

```
REGRESSION
  /DEPENDENT sei
  /METHOD=ENTER female educate nonwhite pasei
  /PARTIALPLOT ALL.
```

Example 1.7

```
* Link test.

REGRESSION
  /DEPENDENT sei
  /METHOD=ENTER female educate nonwhite pasei
  /SAVE PRED(pred).

COMPUTE pred2=pred*pred.
EXECUTE.

REGRESSION
  /DEPENDENT sei
  /METHOD=ENTER pred pred2.
```

* There was no Ramsey RESET test in SPSS at the time of publication; however, it is straightforward to estimate since the model behind it is simply $yi = \alpha + \beta_1 x_1 + \ldots + \beta_k x_k + \beta y^{*2} + \varepsilon$, where y^{*2} is the predicted values from the OLS regression model squared. Some versions of the test include higher-order versions of the predicted values (e.g., y^{*3}), so an F-test is then used to determine whether the coefficients associated with these higher-order terms are statistically significant.

Figure 1.10

```
REGRESSION
/STATISTICS COEFF OUTS R ANOVA
/DEPENDENT sei
/METHOD=ENTER female educate nonwhite pasei
/SAVE ZRESID(ZRE_1) COOK(COO_1) LEVER(LEV_1).

COMPUTE ZRE2=ZRE_1*ZRE_1.
EXECUTE.

GGRAPH
  /GRAPHDATASET NAME="graphdataset" VARIABLES=ZRE2 LEV_1
    MISSING=LISTWISE REPORTMISSING=NO
  /GRAPHSPEC SOURCE=INLINE.
BEGIN GPL
SOURCE: s = userSource(id("graphdataset"))
DATA: ZRE2=col(source(s), name("ZRE2"))
DATA: LEV_1=col(source(s), name("LEV_1"))
GUIDE: axis(dim(1), label("Normal Residuals Squared"))
GUIDE: axis(dim(2), label("Centered Leverage Value"))
```

```
GUIDE: form.line(position(*, .0018))
GUIDE: form.line(position(.9978,*))
ELEMENT: point(position(ZRE2*LEV_1))
END GPL.
```

Figure 1.11

```
GGRAPH
  /GRAPHDATASET NAME="graphdataset" VARIABLES= LEV_1 COO_1
    MISSING=LISTWISE REPORTMISSING=NO
  /GRAPHSPEC SOURCE=INLINE.
BEGIN GPL
SOURCE: s=userSource(id("graphdataset"))
DATA: LEV_1=col(source(s), name("LEV_1"))
DATA: COO_1=col(source(s), name("COO_1"))
GUIDE: axis(dim(1), label("Leverage"))
GUIDE: axis(dim(2), label("Cook's D"))
GUIDE: form.line(position(*, .0018), color(color.red))
GUIDE: form.line(position(.01,*), color(color.red))
SCALE: linear(dim(2), min(-.0005))
ELEMENT: point(position(LEV_1*COO_1))
END GPL.
```

Example 1.8

```
* No quantile regression option was available in SPSS at the
  time of publication.
```

Example 1.9

```
COMPUTE Ln_SEI=LN(sei).
EXECUTE.


REGRESSION
  /DEPENDENT Ln_SEI
  /METHOD=ENTER female educate nonwhite pasei.
```

Example 1.10

```
COMPUTE edu_pa=educate*pasei.
EXECUTE.


REGRESSION
  /STATISTICS COEFF OUTS R ANOVA TOL
  /DEPENDENT sei
  /METHOD=ENTER female educate nonwhite pasei edu_pa.
```

Example 1.11

```
DESCRIPTIVES VARIABLES = pasei educate
  /SAVE.
COMPUTE zedu_zpa=Zeducate*Zpasei.
EXECUTE.


REGRESSION
  /STATISTICS COEFF OUTS R ANOVA TOL
  /DEPENDENT sei
  /METHOD=ENTER female Zeducate nonwhite Zpasei zedu_zpa.
```

* The remaining examples and figures in Chapter 1 may be
  replicated by adapting the code already provided.

CHAPTER 2

Example 2.1

```
FREQUENCIES VARIABLES=polviews
  /ORDER=ANALYSIS.
```

Example 2.2

```
CROSSTABS
  /TABLES=polviews BY female
  /FORMAT=AVALUE TABLES
  /STATISTICS=CHISQ
  /CELLS=COUNT ROW
  /COUNT ROUND CELL.
```

Figure 2.1

```
USE ALL.
COMPUTE filter_$=(marital ~= 5 AND childs > 0).
VARIABLE LABELS filter_$ 'marital ~= 5 AND childs > 0 (FILTER)'.
VALUE LABELS filter_$ 0 'Not Selected' 1 'Selected'.
FORMATS filter_$ (f1.0).
FILTER BY filter_$.
EXECUTE.


GGRAPH
  /GRAPHDATASET NAME="graphdataset" VARIABLES=childs COUNT()
    [name="COUNT"]
  MISSING=LISTWISE REPORTMISSING=NO
  /GRAPHSPEC SOURCE=INLINE.
BEGIN GPL
```

```
SOURCE: s = userSource(id("graphdataset"))
DATA: childs=col(source(s), name("childs"), unit.
  category())
DATA: COUNT=col(source(s), name("COUNT"))
GUIDE: axis(dim(1), label("number of children"))
GUIDE: axis(dim(2), label("Count"))
SCALE: cat(dim(1), include("8"))
SCALE: linear(dim(2), include(0))
ELEMENT: interval(position(childs*COUNT), shape.interior(shape.
    square))
END GPL.
```

Example 2.3

```
GENLIN sei WITH pasei female nonwhite educate
  /MODEL pasei female nonwhite educate INTERCEPT=YES
DISTRIBUTION=NORMAL LINK=IDENTITY
  /CRITERIA SCALE=MLE COVB=MODEL PCONVERGE=1E-006(ABSOLUTE)
    SINGULAR=1E-012 ANALYSISTYPE=3(WALD) CILEVEL=95 CITYPE=WALD
    LIKELIHOOD=FULL
  /MISSING CLASSMISSING=EXCLUDE
  /PRINT MODELINFO FIT SUMMARY SOLUTION.
```

Example 2.4

```
USE ALL.
COMPUTE filter_$=(MISSING(pasei) ~= 1 AND MISSING(female) ~= 1 AND
  MISSING(nonwhite) ~= 1 AND MISSING(educate) ~= 1 AND MISSING(sei)
  ~= 1).
VARIABLE LABELS filter_$ 'MISSING(pasei) ~= 1 AND MISSING(female)
  ~= 1 AND MISSING(nonwhite) ~= 1 AND MISSING(educate) ~= 1 AND
  MISSING(sei) ~= 1 (FILTER)'.
VALUE LABELS filter_$ 0 'Not Selected' 1 'Selected'.
FORMATS filter_$ (f1.0).
FILTER BY filter_$.
EXECUTE.

* Constrained Model.
GENLIN sei WITH female nonwhite
  /MODEL female nonwhite INTERCEPT=YES
DISTRIBUTION=NORMAL LINK=IDENTITY
  /CRITERIA SCALE=MLE COVB=MODEL PCONVERGE=1E-006(ABSOLUTE)
    SINGULAR=1E-012
  ANALYSISTYPE=3(WALD) CILEVEL=95 CITYPE=WALD LIKELIHOOD=FULL
```

```
/MISSING CLASSMISSING=EXCLUDE
/PRINT MODELINFO FIT SUMMARY SOLUTION.

* Null Model.
GENLIN sei
  /MODEL INTERCEPT=YES
  DISTRIBUTION=NORMAL LINK=IDENTITY
  /CRITERIA SCALE=MLE COVB=MODEL PCONVERGE=1E-006(ABSOLUTE)
    SINGULAR=1E-012
  ANALYSISTYPE=3(WALD) CILEVEL=95 CITYPE=WALD LIKELIHOOD=FULL
  /MISSING CLASSMISSING=EXCLUDE
  /PRINT MODELINFO FIT SUMMARY SOLUTION.
```

CHAPTER 3

Example 3.1

```
REGRESSION
  /DEPENDENT satlife
  /METHOD=ENTER age
  /SCATTERPLOT(*resid *pred)
  /SAVE SRESID(SRE).
```

Figure 3.1

```
* See the Example 3.1 output for this plot.
```

Figure 3.2

```
PPLOT
  /VARIABLES=SRE
  /NOLOG
  /NOSTANDARDIZE
  /TYPE=Q-Q
  /FRACTION=BLOM
  /TIES=MEAN
  /DIST=NORMAL.
```

Example 3.2

```
CROSSTABS
  /TABLES=male BY satlife
  /FORMAT=AVALUE TABLES
  /CELLS=COUNT ROW
  /COUNT ROUND CELL.
```

Example 3.3

```
LOGISTIC REGRESSION VARIABLES satlife
  /METHOD=ENTER male
  /CONTRAST (male)=Indicator(1)
  /SAVE=PRED(PRE)
  /CRITERIA=PIN(0.05) POUT(0.10) ITERATE(20) CUT(0.5).


SUMMARIZE
  /TABLES=PRE BY sex
  /FORMAT=NOLIST TOTAL
  /TITLE='Case Summaries'
  /MISSING=VARIABLE
  /CELLS=MEAN.
```

Example 3.4

```
LOGISTIC REGRESSION VARIABLES satlife
  /METHOD=ENTER male age iq
  /CONTRAST (male)=Indicator(1)
  /CRITERIA=PIN(0.05) POUT(0.10) ITERATE(20) CUT(0.5).
```

Example 3.5

```
LOGISTIC REGRESSION VARIABLES satlife
  /METHOD=ENTER male age iq
  /CONTRAST (male)=Indicator
  /SAVE=PRED(PRE_2)
  /CRITERIA=PIN(0.05) POUT(0.10) ITERATE(20) CUT(0.5).


SUMMARIZE
  /TABLES=PRE_2 BY male
  /FORMAT=NOLIST TOTAL
  /TITLE='Case Summaries'
  /MISSING=VARIABLE
  /CELLS=MEAN.
```

Example 3.6

```
LOGISTIC REGRESSION VARIABLES satlife
  /METHOD=ENTER male
  /METHOD=BSTEP(WALD) age iq
  /CONTRAST (male)=Indicator(1)
  /CRITERIA=PIN(0.05) POUT(0.10) ITERATE(20) CUT(0.5).
```

Example 3.7

```
LOGISTIC REGRESSION VARIABLES satlife
  /METHOD=ENTER male
  /CONTRAST (male)=Indicator(1)
  /SAVE=PRED (PRE) LEVER (LEV) ZRESID (ZRE) DEV (DEV)
  /CRITERIA=PIN(0.05) POUT(0.10) ITERATE(20) CUT(0.5).

GGRAPH
/GRAPHDATASET NAME="graphdataset" VARIABLES=DEV MISSING=LISTWISE
  REPORTMISSING=NO
/GRAPHSPEC SOURCE=INLINE.
BEGIN GPL
SOURCE: s=userSource(id("graphdataset"))
DATA: DEV_1=col(source(s), name("DEV"))
TRANS: rug = eval(-60)
GUIDE: axis(dim(1), label("DEV"))
GUIDE: axis(dim(2), label("Frequency"))
SCALE: linear(dim(2), min(-5))
SCALE: linear(dim(1), min(-3), max(3))
ELEMENT: point(position(DEV*rug), transparency.
  exterior(transparency."0.8"))
ELEMENT: line(position(density.kernel.gaussian(DEV*1)))
END GPL.

GGRAPH
  /GRAPHDATASET NAME="graphdataset" VARIABLES=PRE DEV
    MISSING=LISTWISE REPORTMISSING=NO
  /GRAPHSPEC SOURCE=INLINE.
BEGIN GPL
SOURCE: s = userSource(id("graphdataset"))
DATA: PRE=col(source(s), name("PRE"))
DATA: DEV=col(source(s), name("DEV"))
GUIDE: axis(dim(1), label("Predicted probability"))
GUIDE: axis(dim(2), label("Deviance value"))
SCALE: linear(dim(1), min(0), max(1))
SCALE: linear(dim(2), min(-3), max(3))
ELEMENT: point.jitter(position(PRE*DEV))
END GPL.

GGRAPH
  /GRAPHDATASET NAME="graphdataset" VARIABLES=educate DEV
    MISSING=LISTWISE REPORTMISSING=NO
  /GRAPHSPEC SOURCE=INLINE.
```

```
BEGIN GPL
SOURCE: s = userSource(id("graphdataset"))
DATA: educate=col(source(s), name("educate"))
DATA: DEV=col(source(s), name("DEV"))
GUIDE: axis(dim(1), label("Years of Formal Education"))
GUIDE: axis(dim(2), label("Deviance value"))
SCALE: linear(dim(1), min(0), max(1))
SCALE: linear(dim(2), min(-3), max(3))
ELEMENT: point.jitter(position(educate*DEV))
END GPL.

GGRAPH
  /GRAPHDATASET NAME="graphdataset" VARIABLES=LEV ZRE
    MISSING=LISTWISE REPORTMISSING=NO
  /GRAPHSPEC SOURCE=INLINE.
BEGIN GPL
SOURCE: s = userSource(id("graphdataset"))
DATA: LEV=col(source(s), name("LEV"))
DATA: ZRE=col(source(s), name("ZRE"))
GUIDE: axis(dim(1), label("Leverage value"))
GUIDE: axis(dim(2), label("Normalized residual"))
GUIDE: form.line(position(.051,*))
ELEMENT: point.jitter(position(LEV*ZRE))
END GPL.
```
* Delta-deviance was not an option to save in SPSS at the time
  of publication; they could be computed, however, using some
  simple programming steps.

Figure 3.5
```
GGRAPH
  /GRAPHDATASET NAME="graphdataset" VARIABLES=LEV ddev
    MISSING=LISTWISE REPORTMISSING=NO
  /GRAPHSPEC SOURCE=INLINE.
BEGIN GPL
SOURCE: s = userSource(id("graphdataset"))
DATA: LEV_=col(source(s), name("LEV"))
DATA: ddev=col(source(s), name("ddev"))
GUIDE: axis(dim(1), label("Leverage Value"))
GUIDE: axis(dim(2), label("Delta Deviance Value"))
GUIDE: form.line(position(*,4))
GUIDE: form.line(position(.1098,*))
SCALE: linear(dim(2), max(4.5))
```

```
ELEMENT: point.jitter(position(LEV*ddev))
END GPL.
```

Example 3.8

```
* Delta-deviance was not an option to save in SPSS at the time
  of publication; they could be computed, however, using some
  simple programming steps.
```

Example 3.9

```
GENLIN satlife (REFERENCE=FIRST) WITH male
  /MODEL male INTERCEPT=YES
DISTRIBUTION= BINOMIAL LINK=PROBIT
  /CRITERIA METHOD=FISHER(1) SCALE=1 COVB=MODEL MAXITERATIONS=100
    MAXSTEPHALVING=5
  PCONVERGE=1E-006(ABSOLUTE) SINGULAR=1E-012 ANALYSISTYPE=3(WALD)
    CILEVEL=95 CITYPE=WALD
  LIKELIHOOD=KERNEL
  /MISSING CLASSMISSING=EXCLUDE
  /PRINT CPS DESCRIPTIVES MODELINFO FIT SUMMARY SOLUTION
  /SAVE MEANPRED (MeanPred).

SUMMARIZE
  /TABLES=MeanPred BY male
  /FORMAT=NOLIST TOTAL
  /TITLE='Case Summaries'
  /MISSING=VARIABLE
  /CELLS=MEAN.
```

Example 3.10

```
GENLIN satlife (REFERENCE=FIRST) WITH male sleep1 age iq
  /MODEL male sleep1 age iq INTERCEPT=YES
DISTRIBUTION=BINOMIAL LINK=PROBIT
  /CRITERIA METHOD=FISHER(1) SCALE=1 COVB=MODEL MAXITERATIONS=100
    MAXSTEPHALVING=5
  PCONVERGE=1E-006(ABSOLUTE) SINGULAR=1E-012 ANALYSISTYPE=3(WALD)
    CILEVEL=95 CITYPE=WALD
  LIKELIHOOD=KERNEL
  /MISSING CLASSMISSING=EXCLUDE
  /PRINT CPS DESCRIPTIVES MODELINFO FIT SUMMARY SOLUTION
  /SAVE MEANPRED (MeanPred_2) XBPRED DEVIANCERESID.
SUMMARIZE
  /TABLES=MeanPred_2 BY male BY sleep1
  /FORMAT=NOLIST TOTAL
```

```
/TITLE='Case Summaries'
/MISSING=VARIABLE
/CELLS=MEAN.

USE ALL.
COMPUTE filter_$=(Age_Cat = 1).
VARIABLE LABELS filter_$ 'Age_Cat = 1 (FILTER)'.
VALUE LABELS filter_$ 0 'Not Selected' 1 'Selected'.
FORMATS filter_$ (f1.0).
FILTER BY filter_$.
EXECUTE.

SUMMARIZE
  /TABLES=MeanPred_2 BY male BY sleep1 BY age
  /FORMAT=NOLIST TOTAL
  /TITLE='Case Summaries'
  /MISSING=VARIABLE
  /CELLS=MEAN.
```

Example 3.11

```
* There was no dprobit option in SPSS at the time of
  publication.
```

Example 3.12

```
* There was no dprobit option in SPSS at the time of
  publication.
```

Figure 3.7

```
GGRAPH
  /GRAPHDATASET NAME="graphdataset" VARIABLES=MeanPred_2
    DevianceResidual MISSING=LISTWISE
  REPORTMISSING=NO
  /GRAPHSPEC SOURCE=INLINE.
BEGIN GPL
  SOURCE: s = userSource(id("graphdataset"))
DATA: MeanPred_2=col(source(s), name("MeanPred_2"))
DATA: DevianceResidual=col(source(s), name("DevianceResidual"))
GUIDE: axis(dim(1), label("Predicted Value of Mean of
    Response"))
GUIDE: axis(dim(2), label("Deviance Residual"))
ELEMENT: point(position(MeanPred_2*DevianceResidual))
END GPL.
```

Example 3.13

* Although there was no built-in command at the time of writing, a user-written program called STATS_FIRTHLOG may be used to estimate this model.

Example 3.14

```
GENLIN satlife (REFERENCE=FIRST) WITH male
  /MODEL male INTERCEPT=YES
DISTRIBUTION=BINOMIAL LINK=LOG
  /CRITERIA METHOD=FISHER(1) SCALE=1 COVB=MODEL MAXITERATIONS=100
    MAXSTEPHALVING=5
  PCONVERGE=1E-006(ABSOLUTE) SINGULAR=1E-012 ANALYSISTYPE=3(WALD)
    CILEVEL=95 CITYPE=WALD
  LIKELIHOOD=KERNEL
  /MISSING CLASSMISSING=EXCLUDE
  /PRINT CPS DESCRIPTIVES MODELINFO FIT SUMMARY SOLUTION
    (EXPONENTIATED).
```

Example 3.15

```
GENLIN satlife WITH male
  /MODEL male INTERCEPT=YES
DISTRIBUTION=POISSON LINK=LOG
  /CRITERIA METHOD=FISHER(1) SCALE=1 COVB=ROBUST MAXITERATIONS=100
    MAXSTEPHALVING=5
  PCONVERGE=1E-006(ABSOLUTE) SINGULAR=1E-012 ANALYSISTYPE=3(WALD)
    CILEVEL=95 CITYPE=WALD
  LIKELIHOOD=KERNEL
  /MISSING CLASSMISSING=EXCLUDE
  /PRINT CPS DESCRIPTIVES MODELINFO FIT SUMMARY SOLUTION
    (EXPONENTIATED).
```

CHAPTER 4

Example 4.1

```
FREQUENCIES VARIABLES=spanking
  /ORDER=ANALYSIS.
```

Figure 4.1

```
GGRAPH
  /GRAPHDATASET NAME="graphdataset" VARIABLES=spanking educate
  /GRAPHSPEC SOURCE=INLINE.
BEGIN GPL
SOURCE: s = userSource(id("graphdataset"))
DATA: educate=col(source(s), name("educate"))
```

```
DATA: spanking=col(source(s), name("spanking"))
GUIDE: axis(dim(2), label("Agree to Spanking"))
GUIDE: axis(dim(1), label("Years of Formal Education"))
SCALE: linear(dim(2), min(0))
ELEMENT: line(position(smooth.linear(educate*spanking)))
ELEMENT: point(position(educate*spanking))
END GPL.
```

Example 4.2

```
REGRESSION
  /MISSING LISTWISE
  /STATISTICS COEFF OUTS R ANOVA
  /CRITERIA=PIN(.05) POUT(.10)
  /NOORIGIN
  /DEPENDENT spanking
  /METHOD=ENTER educate
  /SAVE SRESID.
```

Figure 4.2

```
GGRAPH
  /GRAPHDATASET NAME="graphdataset" VARIABLES=SRE_1
    MISSING=LISTWISE REPORTMISSING=NO
  /GRAPHSPEC SOURCE=INLINE.
BEGIN GPL
SOURCE: s = userSource(id("graphdataset"))
DATA: SRE_1=col(source(s), name("SRE_1"))
TRANS: rug = eval(-60)
GUIDE: axis(dim(1), label("Studentized Residuals"))
GUIDE: axis(dim(2), label("Frequency"))
SCALE: linear(dim(2), min(-30))
ELEMENT: point(position(SRE_1*rug), transparency.
    exterior(transparency."0.8"))
ELEMENT: line(position(density.kernel.gaussian(SRE_1*1)))
END GPL.
```

Figure 4.3

```
PPLOT
  /VARIABLES=SRE_1
  /NOLOG
  /NOSTANDARDIZE
  /TYPE=Q-Q
  /FRACTION=BLOM
```

```
/TIES=MEAN
/DIST=NORMAL.
```

Example 4.3

```
CROSSTABS
  /TABLES=female BY spanking
  /FORMAT=AVALUE TABLES
  /CELLS=COUNT ROW
  /COUNT ROUND CELL.
```

Example 4.4

```
GENLIN spanking (ORDER=ASCENDING) WITH female
  /MODEL female
DISTRIBUTION=MULTINOMIAL LINK=CUMLOGIT
  /CRITERIA METHOD=FISHER(1) SCALE=1 COVB=MODEL MAXITERATIONS=100
    MAXSTEPHALVING=5
  PCONVERGE=1E-006(ABSOLUTE) SINGULAR=1E-012 ANALYSISTYPE=3(WALD)
    CILEVEL=95 CITYPE=WALD
  LIKELIHOOD=KERNEL
  /MISSING CLASSMISSING=EXCLUDE
  /PRINT CPS DESCRIPTIVES MODELINFO FIT SUMMARY SOLUTION
    (EXPONENTIATED).
```

Example 4.5

```
PLUM spanking WITH female
  /CRITERIA=CIN(95) DELTA(0) LCONVERGE(0) MXITER(100) MXSTEP(5)
    PCONVERGE(1.0E-6)
  SINGULAR(1.0E-8)
  /LINK=LOGIT
  /PRINT=TPARALLEL.
```

Example 4.6

```
GENLIN spanking (ORDER=ASCENDING) WITH female
  /MODEL female
DISTRIBUTION=MULTINOMIAL LINK=CUMPROBIT
  /CRITERIA METHOD=FISHER(1) SCALE=1 COVB=MODEL MAXITERATIONS=100
    MAXSTEPHALVING=5
  PCONVERGE=1E-006(ABSOLUTE) SINGULAR=1E-012 ANALYSISTYPE=3(WALD)
    CILEVEL=95 CITYPE=WALD
  LIKELIHOOD=KERNEL
  /MISSING CLASSMISSING=EXCLUDE
  /PRINT CPS DESCRIPTIVES MODELINFO FIT SUMMARY SOLUTION
    (EXPONENTIATED)
```

```
  /SAVE MEANPRED(:25).

CROSSTABS
  /TABLES=spanking BY female
  /FORMAT=AVALUE TABLES
  /CELLS=COUNT COLUMN
  /COUNT ROUND CELL.

SUMMARIZE
  /TABLES=CumMeanPredicted_1 CumMeanPredicted_2
    CumMeanPredicted_3 BY female
  /FORMAT=NOLIST TOTAL
  /TITLE='Case Summaries'
  /MISSING=VARIABLE
  /CELLS=MEAN.

PLUM spanking WITH female
  /CRITERIA=CIN(95) DELTA(0) LCONVERGE(0) MXITER(100) MXSTEP(5)
    PCONVERGE(1.0E-6)
  SINGULAR(1.0E-8)
  /LINK=PROBIT
  /SAVE=ESTPROB.

SUMMARIZE
  /TABLES=EST1_1 EST2_1 EST3_1 EST4_1 BY female
  /FORMAT=NOLIST TOTAL
  /TITLE='Case Summaries'
  /MISSING=VARIABLE
  /CELLS=MEAN.
```

Example 4.7

```
GENLIN spanking (ORDER=ASCENDING) WITH female polviews educate
  /MODEL female polviews educate
DISTRIBUTION=MULTINOMIAL LINK=CUMLOGIT
  /CRITERIA METHOD=FISHER(1) SCALE=1 COVB=MODEL MAXITERATIONS=100
    MAXSTEPHALVING=5
  PCONVERGE=1E-006(ABSOLUTE) SINGULAR=1E-012 ANALYSISTYPE=3(WALD)
    CILEVEL=95 CITYPE=WALD
  LIKELIHOOD=KERNEL
  /MISSING CLASSMISSING=EXCLUDE
  /PRINT CPS DESCRIPTIVES MODELINFO FIT SUMMARY SOLUTION
    (EXPONENTIATED).
```

Example 4.8

```
GENLIN spanking (ORDER=ASCENDING) WITH female polviews educate
  /MODEL female polviews educate
DISTRIBUTION=MULTINOMIAL LINK=CUMPROBIT
  /CRITERIA METHOD=FISHER(1) SCALE=1 COVB=MODEL MAXITERATIONS=100
    MAXSTEPHALVING=5
  PCONVERGE=1E-006(ABSOLUTE) SINGULAR=1E-012 ANALYSISTYPE=3(WALD)
    CILEVEL=95 CITYPE=WALD
  LIKELIHOOD=KERNEL
  /MISSING CLASSMISSING=EXCLUDE
  /PRINT CPS DESCRIPTIVES MODELINFO FIT SUMMARY SOLUTION
    (EXPONENTIATED).

PLUM spanking WITH female educate polviews
  /CRITERIA=CIN(95) DELTA(0) LCONVERGE(0) MXITER(100) MXSTEP(5)
    PCONVERGE(1.0E-6)
  SINGULAR(1.0E-8)
  /LINK=PROBIT
  /PRINT=FIT PARAMETER SUMMARY
  /SAVE=ESTPROB.

SUMMARIZE
  /TABLES=EST1_2 EST2_2 EST3_2 EST4_2 BY female
  /FORMAT=NOLIST TOTAL
  /TITLE='Case Summaries'
  /MISSING=VARIABLE
  /CELLS=MEAN.

USE ALL.
  COMPUTE filter_$=(educate = 10 AND female = 0).
  VARIABLE LABELS filter_$ 'educate = 10 AND female = 0 (FILTER)'.
  VALUE LABELS filter_$ 0 'Not Selected' 1 'Selected'.
  FORMATS filter_$ (f1.0).
  FILTER BY filter_$.
  EXECUTE.

DESCRIPTIVES VARIABLES=EST1_2 EST2_2 EST3_2 EST4_2
  /STATISTICS=MEAN.

USE ALL.
COMPUTE filter_$=(educate = 16 AND female = 0).
VARIABLE LABELS filter_$ 'educate = 16 AND female = 0 (FILTER)'.
VALUE LABELS filter_$ 0 'Not Selected' 1 'Selected'.
```

```
FORMATS filter_$ (f1.0).
FILTER BY filter_$.
EXECUTE.


DESCRIPTIVES VARIABLES=EST1_2 EST2_2 EST3_2 EST4_2
  /STATISTICS=MEAN.


PLUM spanking WITH female educate polviews
  /CRITERIA=CIN(95) DELTA(0) LCONVERGE(0) MXITER(100) MXSTEP(5)
    PCONVERGE(1.0E-6)
  SINGULAR(1.0E-8)
  /LINK=PROBIT
  /PRINT=TPARALLEL.
```

Figure 4.5

```
FILTER OFF.
  USE ALL.
  EXECUTE.


GGRAPH
  /GRAPHDATASET NAME="graphdataset" VARIABLES=female
    MEAN(EST1_2) MEAN(EST4_2)
  MISSING=LISTWISE REPORTMISSING=NO
TRANSFORM=VARSTOCASES(SUMMARY="#SUMMARY" INDEX="#INDEX")
  /GRAPHSPEC SOURCE=INLINE.
BEGIN GPL
SOURCE: s = userSource(id("graphdataset"))
DATA: female=col(source(s), name("female"), unit.category())
DATA: SUMMARY=col(source(s), name("#SUMMARY"))
DATA: INDEX=col(source(s), name("#INDEX"), unit.category())
COORD: rect(dim(1,2), cluster(3,0))
GUIDE: axis(dim(3), label("sex of respondent"))
GUIDE: axis(dim(2), label("Mean"))
GUIDE: legend(aesthetic(aesthetic.color.interior), label(""))
SCALE: cat(dim(3), include("0", "1"))
SCALE: linear(dim(2), include(0))
SCALE: cat(aesthetic(aesthetic.color.interior), include("0", "1"))
SCALE: cat(dim(1), include("0", "1"))
ELEMENT: interval(position(INDEX*SUMMARY*female), color.
  interior(INDEX), shape.interior(shape.square))
END GPL.
```

Example 4.9

```
RECODE spanking (4=1) (3=0) (2=0) (1=0) INTO newspank.
VARIABLE LABELS newspank 'newspank'.
EXECUTE.

GENLIN newspank (REFERENCE=FIRST) WITH female educate polviews
  /MODEL female educate polviews INTERCEPT=YES
DISTRIBUTION=BINOMIAL LINK=LOGIT
  /CRITERIA METHOD=FISHER(1) SCALE=1 COVB=MODEL MAXITERATIONS=100
    MAXSTEPHALVING=5
  PCONVERGE=1E-006(ABSOLUTE) SINGULAR=1E-012 ANALYSISTYPE=3(WALD)
    CILEVEL=95 CITYPE=WALD
  LIKELIHOOD=KERNEL
  /MISSING CLASSMISSING=EXCLUDE
  /PRINT MODELINFO FIT SUMMARY SOLUTION (EXPONENTIATED)
  /SAVE MEANPRED LEVERAGE DEVIANCERESID.

* Delta-deviance was not an option to save in SPSS at the time
  of publication; they could be computed, however, using some
  simple programming steps.
```

Figure 4.6

```
GGRAPH
  /GRAPHDATASET NAME="graphdataset" VARIABLES=MeanPredicted
    DevianceResidual
  MISSING=LISTWISE REPORTMISSING=NO
  /GRAPHSPEC SOURCE=INLINE.
BEGIN GPL
  SOURCE: s = userSource(id("graphdataset"))
DATA: MeanPredicted=col(source(s), name("MeanPredicted"))
DATA: DevianceResidual=col(source(s), name("DevianceResidual"))
GUIDE: axis(dim(1), label("Predicted Value of Mean of Response"))
GUIDE: axis(dim(2), label("Deviance Residual"))
ELEMENT: point(position(MeanPredicted*DevianceResidual))
END GPL.
```

Example 4.10

```
* There no gologit2 option available in SPSS at the time of
  publication.
```

Example 4.11

```
* See Examples 1.3 and 4.7 for code that may be adapted for this
  example.
```

CHAPTER 5

Example 5.1

```
FREQUENCIES VARIABLES=polview1
  /ORDER=ANALYSIS.
```

Example 5.2

```
CROSSTABS
  /TABLES=nonwhite BY polview1
  /FORMAT=AVALUE TABLES
  /CELLS=COUNT ROW
  /COUNT ROUND CELL.
```

Example 5.3

```
NOMREG polview1 (BASE=2) WITH nonwhite
  /CRITERIA CIN(95) DELTA(0) MXITER(100) MXSTEP(5) CHKSEP(20)
    LCONVERGE(0)
  PCONVERGE(0.000001) SINGULAR(0.00000001)
  /MODEL
  /INTERCEPT=INCLUDE
  /PRINT= PARAMETER SUMMARY LRT MFI IC CELLPROB
  /SAVE ESTPROB.


SUMMARIZE
  /TABLES=EST1_1 EST2_1 EST3_1 BY nonwhite
  /FORMAT=NOLIST TOTAL
  /TITLE='Case Summaries'
  /MISSING=VARIABLE
  /CELLS=MEAN.
```

Example 5.4

```
NOMREG polview1 (BASE=2 ORDER=ASCENDING) WITH female age educate
  /CRITERIA CIN(95) DELTA(0) MXITER(100) MXSTEP(5) CHKSEP(20)
    LCONVERGE(0)
  PCONVERGE(0.000001) SINGULAR(0.00000001)
  /MODEL
  /STEPWISE=PIN(.05) POUT(0.1) MINEFFECT(0) RULE(SINGLE)
    ENTRYMETHOD(LR) REMOVALMETHOD(LR)
  /INTERCEPT=INCLUDE
  /PRINT= PARAMETER SUMMARY LRT MFI IC
  /SAVE ESTPROB.
```

Example 5.5

```
NOMREG polview1 (BASE=3 ORDER=ASCENDING) WITH female age
    educate
  /CRITERIA CIN(95) DELTA(0) MXITER(100) MXSTEP(5) CHKSEP(20)
    LCONVERGE(0)
  PCONVERGE(0.000001) SINGULAR(0.00000001)
  /MODEL
  /STEPWISE=PIN(.05) POUT(0.1) MINEFFECT(0) RULE(SINGLE)
    ENTRYMETHOD(LR) REMOVALMETHOD(LR)
  /INTERCEPT=INCLUDE
  /PRINT= PARAMETER SUMMARY LRT MFI IC
  /SAVE ESTPROB.

USE ALL.
COMPUTE filter_$=(female = 0 AND educate =12 OR educate=16).
VARIABLE LABELS filter_$ 'female = 0 AND educate =12 OR educate=16
  (FILTER)'.
VALUE LABELS filter_$ 0 'Not Selected' 1 'Selected'.
FORMATS filter_$ (f1.0).
FILTER BY filter_$.
EXECUTE.

SUMMARIZE
  /TABLES=EST1_1 EST2_1 EST3_1 BY educate BY female
  /FORMAT=NOLIST TOTAL
  /TITLE='Case Summaries'
  /MISSING=VARIABLE
  /CELLS=MEAN.

USE ALL.
COMPUTE filter_$=(female = 0 AND educate =10 OR educate=12 OR
  educate=14 OR educate=16 AND
  age >29 AND age<50).
VARIABLE LABELS filter_$ 'female = 0 AND educate =10 OR educate=12
    OR educate=14 OR educate=16
  AND age >29 AND age<50 (FILTER)'.
VALUE LABELS filter_$ 0 'Not Selected' 1 'Selected'.
FORMATS filter_$ (f1.0).
FILTER BY filter_$.
EXECUTE.

SUMMARIZE
  /TABLES=EST1_3 EST2_3 EST3_3 BY educate
  /FORMAT=NOLIST TOTAL
```

```
/TITLE='Case Summaries'
/MISSING=VARIABLE
/CELLS=MEAN.
```

Example 5.6

```
* There was no multinomial probit regression option in SPSS at
  the time of publication.
```

Example 5.7

```
RECODE polview1 (1=1) (2=0) INTO libmod.
VARIABLE LABELS libmod 'libmod'.
EXECUTE.

LOGISTIC REGRESSION VARIABLES libmod
  /METHOD=ENTER female age educate
  /CONTRAST (female)=Indicator(1)
  /SAVE=PRED (PRE) COOK (COO) LEVER (LEV) SRESID (SRE) ZRESID
    (ZRE) DEV (DEV)
  /CRITERIA=PIN(0.05) POUT(0.10) ITERATE(20) CUT(0.5).

* Delta-deviance was not available as a save option in SPSS at
  the time of publication; they could be computed, however, with
  a few simple programming steps.
```

Figure 5.3

```
* Delta-deviance was not available as a save option in SPSS at
  the time of publication; they could be computed, however, with
  a few simple programming steps.
```

Example 5.8

```
* Delta-deviance was not available as a save option in SPSS at
  the time of publication; they could be computed, however, with
  a few simple programming steps.
```

Example 5.9

```
* There was no mlogtest option in SPSS at the time of
  publication.
```

CHAPTER 6

Example 6.1

```
FREQUENCIES VARIABLES=volteer
  /ORDER=ANALYSIS.
```

Example 6.2

```
DESCRIPTIVES VARIABLES=volteer
  /STATISTICS=MEAN STDDEV MIN MAX.
```

Figure 6.3

```
GGRAPH
  /GRAPHDATASET NAME="graphdataset" VARIABLES=volteer COUNT()
    [name="COUNT"]
  MISSING=LISTWISE REPORTMISSING=NO
  /GRAPHSPEC SOURCE=INLINE.
BEGIN GPL
  SOURCE: s=userSource(id("graphdataset"))
  DATA: volteer=col(source(s), name("volteer"), unit
    .category())
  DATA: COUNT=col(source(s), name("COUNT"))
  GUIDE: axis(dim(1), label("number of volunteer activities in
    past year"))
  GUIDE: axis(dim(2), label("Frequency"))
  SCALE: linear(dim(2), include(0))
  ELEMENT: interval(position(volteer*COUNT), shape.interior(shape.
    square))
END GPL.
```

Example 6.3

```
SUMMARIZE
  /TABLES=volteer BY female
  /FORMAT=NOLIST TOTAL
  /TITLE='Case Summaries'
  /MISSING=VARIABLE
  /CELLS=MEAN MIN MAX STDDEV COUNT.
```

Example 6.4

```
GENLIN volteer WITH female
  /MODEL female INTERCEPT=YES
DISTRIBUTION=POISSON LINK=LOG
  /CRITERIA METHOD=FISHER(1) SCALE=1 COVB=MODEL MAXITERATIONS=100
    MAXSTEPHALVING=5
  PCONVERGE=1E-006(ABSOLUTE) SINGULAR=1E-012 ANALYSISTYPE=3(WALD)
    CILEVEL=95 CITYPE=WALD
  LIKELIHOOD=FULL
  /MISSING CLASSMISSING=EXCLUDE
  /PRINT MODELINFO FIT SUMMARY SOLUTION
  /SAVE MEANPRED (MeanPred).
```

```
SUMMARIZE
  /TABLES=MeanPred BY female
  /FORMAT=NOLIST TOTAL
  /TITLE='Case Summaries'
  /MISSING=VARIABLE
  /CELLS=MEAN.
```

Example 6.5

```
GENLIN volteer WITH female nonwhite educate income
  /MODEL female nonwhite educate income INTERCEPT=YES
DISTRIBUTION=POISSON LINK=LOG
  /CRITERIA METHOD=FISHER(1) SCALE=1 COVB=MODEL MAXITERATIONS=100
    MAXSTEPHALVING=5
  PCONVERGE=1E-006(ABSOLUTE) SINGULAR=1E-012 ANALYSISTYPE=3(WALD)
    CILEVEL=95 CITYPE=WALD
  LIKELIHOOD=FULL
  /MISSING CLASSMISSING=EXCLUDE
  /PRINT MODELINFO FIT SUMMARY SOLUTION (EXPONENTIATED)
  /SAVE MEANPRED(MeanPred_2) COOK(COOK) LEVERAGE(LEV)
    DEVIANCERESID(DRESID).

USE ALL.
COMPUTE filter_$=(nonwhite = 0 AND educate = 12 AND income = 5).
VARIABLE LABELS filter_$ 'nonwhite = 0 AND educate = 12 AND income
  = 5 (FILTER)'.
VALUE LABELS filter_$ 0 'Not Selected' 1 'Selected'.
FORMATS filter_$ (f1.0).
FILTER BY filter_$.
EXECUTE.

SUMMARIZE
  /TABLES=MeanPred_2 BY female
  /FORMAT=NOLIST TOTAL
  /TITLE='Case Summaries'
  /MISSING=VARIABLE
  /CELLS=MEAN.
```

Figure 6.5

```
GGRAPH
  /GRAPHDATASET NAME="graphdataset" VARIABLES=educate
    MeanPred_2 MISSING=LISTWISE
  REPORTMISSING=NO
  /GRAPHSPEC SOURCE=INLINE.
```

```
BEGIN GPL
SOURCE: s = userSource(id("graphdataset"))
DATA: educate=col(source(s), name("educate"))
DATA: MeanPred_2=col(source(s), name("MeanPred_2"))
GUIDE: axis(dim(1), label("years of formal education"))
GUIDE: axis(dim(2), label("Predicted Value of Mean of Response"))
ELEMENT: line(position(smooth.loess.gaussian(educate*MeanPred_2)))
END GPL.
```

Example 6.6

```
* SPSS uses the GENLIN command for Poisson regression; see
  Example 6.5 for additional statistics that may be saved.

COMPUTE adjdev = DevianceResidual + (1/(6*sqrt(MeanPredicted))).
EXECUTE.
```

Figure 6.6

```
GGRAPH
  /GRAPHDATASET NAME="graphdataset" VARIABLES=LEV DRESID
    MISSING=LISTWISE REPORTMISSING=NO
  /GRAPHSPEC SOURCE=INLINE.
BEGIN GPL
SOURCE: s = userSource(id("graphdataset"))
DATA: LEV=col(source(s), name("LEV"))
DATA: DRESID=col(source(s), name("DRESID"))
GUIDE: axis(dim(1), label("Leverage Value"))
GUIDE: axis(dim(2), label("Deviance Residual"))
GUIDE: form.line(position(*,4))
GUIDE: form.line(position(0.0077,*))
ELEMENT: point(position(LEV*DRESID))
END GPL.
```

Figure 6.7

```
GGRAPH
  /GRAPHDATASET NAME="graphdataset" VARIABLES=adjdev
    MISSING=LISTWISE REPORTMISSING=NO
  /GRAPHSPEC SOURCE=INLINE.
BEGIN GPL
SOURCE: s = userSource(id("graphdataset"))
DATA: adjdev=col(source(s), name("adjdev"))
TRANS: rug = eval(-60)
GUIDE: axis(dim(1), label("adjdev"))
GUIDE: axis(dim(2), label("Frequency"))
```

```
SCALE: linear(dim(2), min(-30))
ELEMENT: point(position(adjdev*rug), transparency.
    exterior(transparency."0.8"))
ELEMENT: line(position(density.kernel.gaussian(adjdev*1)))
END GPL.
```

Example 6.7

```
GENLIN volteer WITH female nonwhite educate income
  /MODEL female nonwhite educate income INTERCEPT=YES
DISTRIBUTION=POISSON LINK=LOG
  /CRITERIA METHOD=FISHER(1) SCALE=DEVIANCE COVB=MODEL
    MAXITERATIONS=100 MAXSTEPHALVING=5
  PCONVERGE=1E-006(ABSOLUTE) SINGULAR=1E-012 ANALYSISTYPE=3(WALD)
    CILEVEL=95 CITYPE=WALD
  LIKELIHOOD=FULL
  /MISSING CLASSMISSING=EXCLUDE
  /PRINT MODELINFO FIT SUMMARY SOLUTION.
```

Example 6.8

```
GENLIN volteer WITH female nonwhite educate income
  /MODEL female nonwhite educate income INTERCEPT=YES
DISTRIBUTION=NEGBIN(1) LINK=LOG
  /CRITERIA METHOD=FISHER(1) SCALE=PEARSON COVB=MODEL
    MAXITERATIONS=100 MAXSTEPHALVING=5
  PCONVERGE=1E-006(ABSOLUTE) SINGULAR=1E-012 ANALYSISTYPE=3(WALD)
    CILEVEL=95 CITYPE=WALD
  LIKELIHOOD=FULL
  /MISSING CLASSMISSING=EXCLUDE
  /PRINT MODELINFO FIT SUMMARY SOLUTION (EXPONENTIATED)
  /SAVE MEANPRED(MeanPred_2) COOK(COO_1) LEVERAGE(LEV_2)
    DEVIANCERESID(DRESID_2).

USE ALL.
COMPUTE filter_$=(nonwhite = 0 AND educate = 12 AND income = 5).
VARIABLE LABELS filter_$ 'nonwhite = 0 AND educate = 12 AND income
  = 5 (FILTER)'.
VALUE LABELS filter_$ 0 'Not Selected' 1 'Selected'.
FORMATS filter_$ (f1.0).
FILTER BY filter_$.
EXECUTE.

SUMMARIZE
```

```
 /TABLES=MeanPred_2 BY female
 /FORMAT=NOLIST TOTAL
 /TITLE='Case Summaries'
 /MISSING=VARIABLE
 /CELLS=MEAN.
```

Example 6.9

```
* SPSS uses the GENLIN command for negative binomial regression;
  see Example 6.8 for additional statistics that may be saved.

COMPUTE adjdev1 = DRESID_1 + (1/(6*sqrt(MeanPred_2))).
EXECUTE.
```

Figure 6.9

```
GGRAPH
  /GRAPHDATASET NAME="graphdataset" VARIABLES=LEV_2 DRESID_2
    MISSING=LISTWISE
  REPORTMISSING=NO
  /GRAPHSPEC SOURCE=INLINE.
BEGIN GPL
SOURCE: s = userSource(id("graphdataset"))
DATA: LEV_2=col(source(s), name("LEV _ 2"))
DATA: DRESID_2=col(source(s), name("DRESID_2"))
GUIDE: axis(dim(1), label("Leverage Value"))
GUIDE: axis(dim(2), label("Deviance Residual"))
GUIDE: form.line(position(*,4))
GUIDE: form.line(position(0.0077,*))
ELEMENT: point(position(LEV_2*DRESID_2))
END GPL.
```

Figure 6.10

```
USE ALL.
COMPUTE filter_$=(volteer > 0).
VARIABLE LABELS filter_$ 'volteer > 0 (FILTER)'.
VALUE LABELS filter_$ 0 'Not Selected' 1 'Selected'.
FORMATS filter_$ (f1.0).
FILTER BY filter_$.
EXECUTE.

GGRAPH
  /GRAPHDATASET NAME="graphdataset" VARIABLES=volteer COUNT()
    [name="COUNT"]
  MISSING=LISTWISE REPORTMISSING=NO
```

```
   /GRAPHSPEC SOURCE=INLINE.
BEGIN GPL
SOURCE: s = userSource(id("graphdataset"))
DATA: volteer=col(source(s), name("volteer"), unit.category())
DATA: COUNT=col(source(s), name("COUNT"))
GUIDE: axis(dim(1), label("number of volunteer activities in past
   year"))
GUIDE: axis(dim(2), label("Count"))
SCALE: linear(dim(2), include(0))
ELEMENT: interval(position(volteer*COUNT), shape.interior(shape.
   square))
END GPL.
```

Example 6.10

* There were no built-in zero inflated model options in SPSS at
  the time of publication; however, there is an extension
  command called STATS ZEROINFL that may be used in conjunction
  with an R plugin available for SPSS.

Example 6.11

* There were no built-in zero inflated model options in SPSS at
  the time of publication; however, there is an extension
  command called STATS ZEROINFL that may be used in conjunction
  with an R plugin available for SPSS.

Example 6.12

* There were no built-in zero inflated model options in SPSS at
  the time of publication; however, there is an extension
  command called STATS ZEROINFL that may be used in conjunction
  with an R plugin available for SPSS.

Example 6.13

* There were no built-in zero inflated model options in SPSS at
  the time of publication; however, there is an extension
  command called STATS ZEROINFL that may be used in conjunction
  with an R plugin available for SPSS.

Figure 6.12

* There were no built-in zero inflated model options in SPSS at
  the time of publication; however, there is an extension
  command called STATS ZEROINFL that may be used in conjunction
  with an R plugin available for SPSS.

Figure 6.13

```
* There were no built-in zero inflated model options in SPSS at
  the time of publication; however, there is an extension
  command called STATS ZEROINFL that may be used in conjunction
  with an R plugin available for SPSS.
```

Figure 6.14

```
* There were no built-in zero inflated model options in SPSS at
  the time of publication; however, there is an extension
  command called STATS ZEROINFL that may be used in conjunction
  with an R plugin available for SPSS.
```

CHAPTER 7

Example 7.1

```
GET
FILE =' \\tsclient\home\Downloads\event1.sav'.
DATASET NAME DataSet1 WINDOW=FRONT.
```

Example 7.2

```
* The time variable in SPSS is set for each individual analysis;
  for instance, see Example 7.3.
```

Figure 7.2

```
COMPUTE time=marriage.
  EXECUTE.

KM time
  /STATUS=marriage(1 THRU 1071)
  /PRINT NONE
  /PLOT SURVIVAL.
```

Example 7.3

```
KM time
  /STATUS=marriage(1 THRU 1071)
  /PRINT TABLE.

SURVIVAL TABLE=time
  /INTERVAL=THRU 1071 BY 1
  /STATUS=marriage(0 1071)
  /PRINT=TABLE.
```

Example 7.4

```
KM time
```

```
/STATUS=marriage(1 THRU 1071)
/PRINT MEAN.
```

Figure 7.3

```
KM time BY cohab
  /STATUS=marriage(1 THRU 1071)
  /PRINT NONE
  /PLOT SURVIVAL.
```

Example 7.5

```
KM time BY cohab
  /STATUS=marriage(1 THRU 1071)
  /PRINT NONE
  /TEST BRESLOW
  /COMPARE OVERALL POOLED.
```

Example 7.6

```
* The time variable in SPSS is set for each individual analysis;
  for instance, see Example 7.11.
```

Example 7.7

```
* There was no survival analysis using a lognormal distribution
  option in SPSS at the time of publication.
```

Example 7.8

```
* There was no survival analysis using the exponential
  distribution option in SPSS at the time of publication.
```

Example 7.9

```
* There was no survival analysis using a Weibull distribution
  option in SPSS at the time of publication.
```

Example 7.10

```
* There was no survival analysis using the lognormal
  distribution option in SPSS at the time of publication.
```

Figure 7.4

```
* There was no survival analysis using the lognormal
  distribution option in SPSS at the time of publication.
```

Figure 7.6

```
* There was no survival analysis using the lognormal
  distribution option in SPSS at the time of publication.
```

Example 7.11

```
DATASET ACTIVATE DataSet2.
  COXREG marriage
    /STATUS=evermarr(1)
    /METHOD=ENTER educate age1sex attend14 cohab race
    /PRINT=CI(95)
    /CRITERIA=PIN(.05) POUT(.10) ITERATE(20).
```

* There was no test of the proportional-hazards rank assumption
  in SPSS at the time of publication.

Example 7.12

```
COMPUTE cohab_marriage=cohab*marriage.
EXECUTE.
COXREG marriage
  /STATUS=evermarr(1)
  /METHOD=ENTER educate age1sex attend14
    cohab race cohab_marriage
  /PRINT=CI(95)
  /CRITERIA=PIN(.05) POUT(.10) ITERATE(20).
```

* There was no test of the proportional-hazards rank assumption
  in SPSS at the time of publication.

Figure 7.7

* There was no log-cumulative hazard function plot option in
  SPSS at the time of publication.

Example 7.13

* The default in SPSS is to include only one time-dependent
  covariate per model, so the example provides two separate
  models, one with *educate* as the time-dependent covariate and
  the other with *cohab* as the time-dependent covariate.

```
TIME PROGRAM.
COMPUTE T_COV_ = T_ * educate.
COXREG marriage
  /STATUS=evermarr(1)
  /METHOD=ENTER age1sex attend14 race cohab T_COV_
  /PRINT=CI(95)
  /CRITERIA=PIN(.05) POUT(.10) ITERATE(20).

TIME PROGRAM.
```

```
COMPUTE T_COV_ = T_ * cohab.
COXREG marriage
  /STATUS=evermarr(1)
  /METHOD=ENTER age1sex attend14 race T_COV_ educate
  /PRINT=CI(95)
  /CRITERIA=PIN(.05) POUT(.10) ITERATE(20).
```

Example 7.14

```
GET
  FILE='\\tsclient\home\Downloads\event3.sav'.
DATASET NAME DataSet3 WINDOW=FRONT.
```

Example 7.15

```
LOGISTIC REGRESSION VARIABLES delinq1
  /METHOD=ENTER stress cohes year
  /CONTRAST (year)=Indicator(1)
  /PRINT=CI(95)
  /CRITERIA=PIN(0.05) POUT(0.10) ITERATE(20) CUT(0.5).
```

Example 7.16

```
COXREG year
  /STATUS=delinq1(1)
  /METHOD=ENTER stress cohes
  /PRINT=CI(95)
  /CRITERIA=PIN(.05) POUT(.10) ITERATE(20).
```

Example 7.17

```
LOGISTIC REGRESSION VARIABLES delinq
  /METHOD=ENTER stress cohes year
  /CONTRAST (year)=Indicator(1)
  /PRINT=CI(95)
  /CRITERIA=PIN(0.05) POUT(0.10) ITERATE(20) CUT(0.5).
```

Example 7.18

```
* There was no option to change the tie breaking method in SPSS
  at the time of publication.

COXREG year
  /STATUS=delinq(1)
  /METHOD=ENTER stress cohes
  /PRINT=CI(95)
  /CRITERIA=PIN(.05) POUT(.10) ITERATE(20).
```

CHAPTER 8

Example 8.1

```
GET
FILE='\\tsclient\home\Downloads\event4.sav'.
DATASET NAME DataSet4 WINDOW=FRONT.
```

Example 8.2

* There was no equivalent of the *xtreg, fe* Stata command in SPSS at the time of publication; here is an example of a fixed-effects model in SPSS, but which is similar to the *xtreg, mle* command in Stata.

```
MIXED esteem WITH stress cohes male nonwhite
   /CRITERIA=CIN(95) MXITER(100) MXSTEP(10) SCORING(1)
      SINGULAR(0.000000000001) HCONVERGE(0, ABSOLUTE) LCONVERGE(0,
      ABSOLUTE) PCONVERGE(0.000001, ABSOLUTE)
   /FIXED=stress cohes male nonwhite | SSTYPE(3)
   /METHOD=ML
   /PRINT=SOLUTION.
```

Example 8.3

* There was no equivalent of *xtreg, re* Stata command in SPSS at the time of publication; there are options in SPSS to add random effects to a regression model along with fixed effects using the MIXED procedure.

Example 8.5

```
GENLIN delinq (REFERENCE=FIRST) WITH stress cohes
  /MODEL stress cohes INTERCEPT=YES
DISTRIBUTION=BINOMIAL LINK=LOGIT
  /CRITERIA METHOD=FISHER(1) SCALE=1 MAXITERATIONS=100
     MAXSTEPHALVING=5
PCONVERGE=1E-006(ABSOLUTE) SINGULAR=1E-012 ANALYSISTYPE=3(WALD)
  CILEVEL=95
  LIKELIHOOD=FULL
  /REPEATED SUBJECT=newid WITHINSUBJECT=year SORT=YES
     CORRTYPE=INDEPENDENT ADJUSTCORR=YES COVB=MODEL
     MAXITERATIONS=100 PCONVERGE=1e-006(ABSOLUTE) UPDATECORR=1
  /MISSING CLASSMISSING=EXCLUDE
  /PRINT MODELINFO FIT SUMMARY SOLUTION (EXPONENTIATED)
     WORKINGCORR.
```

Example 8.6

```
GENLIN delinq (REFERENCE=FIRST) WITH stress cohes
  /MODEL stress cohes INTERCEPT=YES
DISTRIBUTION=BINOMIAL LINK=LOGIT
  /CRITERIA METHOD=FISHER(1) SCALE=1 MAXITERATIONS=100
    MAXSTEPHALVING=5
  PCONVERGE=1E-006(ABSOLUTE) SINGULAR=1E-012 ANALYSISTYPE=3(WALD)
    CILEVEL=95
  LIKELIHOOD=FULL
  /REPEATED SUBJECT=newid WITHINSUBJECT=year SORT=YES
    CORRTYPE=EXCHANGEABLE ADJUSTCORR=YES
  COVB=MODEL MAXITERATIONS=100 PCONVERGE=1e-006(ABSOLUTE)
    UPDATECORR=1
  /MISSING CLASSMISSING=EXCLUDE
  /PRINT MODELINFO FIT SUMMARY SOLUTION (EXPONENTIATED)
    WORKINGCORR.
```

Example 8.7

```
GENLIN delinq (REFERENCE=FIRST) WITH stress cohes
  /MODEL stress cohes INTERCEPT=YES
DISTRIBUTION=BINOMIAL LINK=LOGIT
  /CRITERIA METHOD=FISHER(1) SCALE=1 MAXITERATIONS=100
    MAXSTEPHALVING=5
PCONVERGE=1E-006(ABSOLUTE) SINGULAR=1E-012 ANALYSISTYPE=3(WALD)
  CILEVEL=95
  LIKELIHOOD=FULL
  /REPEATED SUBJECT=newid WITHINSUBJECT=year SORT=YES
    CORRTYPE=AR(1) ADJUSTCORR=YES
  COVB=MODEL MAXITERATIONS=100 PCONVERGE=1e-006(ABSOLUTE)
    UPDATECORR=1
  /MISSING CLASSMISSING=EXCLUDE
  /PRINT MODELINFO FIT SUMMARY SOLUTION (EXPONENTIATED)
    WORKINGCORR.
```

Example 8.8

```
GENLIN delinq (REFERENCE=FIRST) WITH stress cohes
  /MODEL stress cohes INTERCEPT=YES
DISTRIBUTION=BINOMIAL LINK=LOGIT
  /CRITERIA METHOD=FISHER(1) SCALE=1 MAXITERATIONS=100
    MAXSTEPHALVING=5
  PCONVERGE=1E-006(ABSOLUTE) SINGULAR=1E-012 ANALYSISTYPE=3(WALD)
    CILEVEL=95
  LIKELIHOOD=FULL
```

```
  /REPEATED SUBJECT=newid WITHINSUBJECT=year SORT=YES
     CORRTYPE=UNSTRUCTURED ADJUSTCORR=YES
  COVB=MODEL MAXITERATIONS=100 PCONVERGE=1e-006(ABSOLUTE)
     UPDATECORR=1
  /MISSING CLASSMISSING=EXCLUDE
  /PRINT MODELINFO FIT SUMMARY SOLUTION (EXPONENTIATED)
     WORKINGCORR.
```

Example 8.9

```
GENLIN bi_depress (REFERENCE=FIRST) WITH male
  /MODEL male INTERCEPT=YES
DISTRIBUTION=BINOMIAL LINK=LOGIT
  /CRITERIA METHOD=FISHER(1) SCALE=1 MAXITERATIONS=100
     MAXSTEPHALVING=5
  PCONVERGE=1E-006(ABSOLUTE) SINGULAR=1E-012 ANALYSISTYPE=3(WALD)
     CILEVEL=95
  LIKELIHOOD=FULL
  /REPEATED SUBJECT=newid WITHINSUBJECT=year SORT=YES
     CORRTYPE=EXCHANGEABLE ADJUSTCORR=YES
  COVB=ROBUST MAXITERATIONS=100 PCONVERGE=1e-006(ABSOLUTE)
     UPDATECORR=1
  /MISSING CLASSMISSING=EXCLUDE
  /PRINT MODELINFO FIT SUMMARY SOLUTION (EXPONENTIATED).
```

Example 8.10

```
* There was no precise equivalent of Stata's xtlogit command in
  SPSS at the time of publication; however, see the GENLIN model
  documentation for similar models using the repeated option;
  for example, consider the following

GENLIN bi_depress (reference=0) WITH male
  /MODEL male
  DISTRIBUTION=BINOMIAL
  /REPEATED SUBJECT=newid
CORRTYPE=INDEPENDENT COVB=MODEL.
```

Example 8.11

```
* See the comment accompanying Example 8.10.
```

CHAPTER 9

Example 9.1

```
DATASET NAME DataSet1 WINDOW=FRONT.
REGRESSION
```

```
/MISSING LISTWISE
/STATISTICS COEFF OUTS R ANOVA
/CRITERIA=PIN(.05) POUT(.10)
/NOORIGIN
/DEPENDENT income
/METHOD=ENTER male married.
```

Example 9.2

```
GENLIN income WITH male married
  MODEL male married INTERCEPT=YES
  DISTRIBUTION=NORMAL LINK=IDENTITY
  /CRITERIA SCALE=MLE PCONVERGE=1E-006(ABSOLUTE) SINGULAR=1E-012
    ANALYSISTYPE=3(WALD)
  CILEVEL=95 LIKELIHOOD=KERNEL
  /REPEATED SUBJECT=idcomm SORT=YES CORRTYPE=INDEPENDENT
    ADJUSTCORR=YES COVB=ROBUST
  /MISSING CLASSMISSING=EXCLUDE
  /PRINT MODELINFO FIT SUMMARY SOLUTION.
```

Example 9.3

```
SUMMARIZE
  /TABLES=income BY idcomm
  /FORMAT=NOLIST TOTAL
  /TITLE='Case Summaries'
  /MISSING=VARIABLE
  /CELLS=MEAN SEMEAN.
```

Example 9.4

```
MIXED income
  /CRITERIA=CIN(95) MXITER(100) MXSTEP(10) SCORING(1)
    SINGULAR(0.000000000001) HCONVERGE(0, ABSOLUTE) LCONVERGE(0,
    ABSOLUTE) PCONVERGE(0.000001, ABSOLUTE)
  /FIXED=| SSTYPE(3)
  /METHOD=ML
  /PRINT=SOLUTION
  /RANDOM=INTERCEPT | SUBJECT(idcomm) COVTYPE(VC).
```

Figure 9.1

```
USE ALL.
COMPUTE filter_$=(idcomm = 4480 OR idcomm=1500 OR idcomm=700 OR
  idcomm=2075).
VARIABLE LABELS filter_$ 'idcomm = 4480 OR idcomm=1500 OR
  idcomm=700 OR idcomm=2075 (FILTER)'.
```

```
VALUE LABELS filter_$ 0 'Not Selected' 1 'Selected'.
FORMATS filter_$ (f1.0).
FILTER BY filter_$.
EXECUTE.

GGRAPH
  /GRAPHDATASET NAME="GraphDataset" VARIABLES= male income
    idcomm
  /GRAPHSPEC SOURCE=INLINE.
  BEGIN GPL
SOURCE: s=userSource( id( "GraphDataset" ) )
DATA: male=col( source(s), name( "male" ) )
DATA: income=col( source(s), name( "income" ) )
DATA: idcomm = col(source(s), name("idcomm"), unit.category())
ELEMENT: point( position(male * income))
ELEMENT: line(position(smooth.linear(male * income)),
  shape(idcomm))
END GPL.
```

Example 9.5

```
FILTER OFF.
  USE ALL.
  EXECUTE.

MIXED income WITH male
  /CRITERIA=CIN(95) MXITER(100) MXSTEP(10) SCORING(1)
    SINGULAR(0.000000000001) HCONVERGE(0, ABSOLUTE) LCONVERGE(0,
    ABSOLUTE) PCONVERGE(0.000001, ABSOLUTE)
  /FIXED=male | SSTYPE(3)
  /METHOD=ML
  /PRINT=SOLUTION
  /RANDOM=INTERCEPT | SUBJECT(idcomm) COVTYPE(VC).
```

Example 9.6

```
* There was no equivalent of xtreg, re Stata command in SPSS at
  the time of publication; there are options in SPSS to add
  random effects to a regression model along with fixed effects
  using the MIXED procedure.
```

Example 9.7

```
MIXED income WITH male
  /CRITERIA=CIN(95) MXITER(100) MXSTEP(10) SCORING(1)
    SINGULAR(0.000000000001) HCONVERGE(0, ABSOLUTE) LCONVERGE(0,
    ABSOLUTE) PCONVERGE(0.000001, ABSOLUTE)
```

```
/FIXED=male | SSTYPE(3)
/METHOD=ML
/PRINT=SOLUTION
/RANDOM=INTERCEPT male | SUBJECT(idcomm) COVTYPE(VC)
/SAVE=PRED RESID.
```

Figure 9.2

```
DESCRIPTIVES VARIABLES=RESID_1
  /SAVE
  /STATISTICS=MEAN STDDEV MIN MAX.

GGRAPH
  /GRAPHDATASET NAME ="graphdataset" VARIABLES =ZRESID_1 MISSING
    =LISTWISE REPORTMISSING =NO
  /GRAPHSPEC SOURCE =INLINE.
BEGIN GPL
SOURCE: s=userSource (id("graphdataset"))
DATA: ZRESID_1=col(source(s), name("ZRESID_1"))
GUIDE: axis(dim(1), label("Residuals"))
GUIDE: axis(dim(2), label("Frequency"))
SCALE: linear(dim(1), min(-4), max(4))
SCALE: linear(dim(2), include(0))
ELEMENT:interval(position(summary.count(bin.rect(ZRESID_1))),
  shape.interior(shape.square))
ELEMENT: line(position(density.kernel.gaussian(ZRESID_1)))
END GPL.
```

Figure 9.3

* At the time of publication, SPSS did not include an option to
  save *BLUPs*.

Example 9.8

```
MIXED income WITH male pop2000 disadvantage
  /CRITERIA=CIN(95) MXITER(100) MXSTEP(10) SCORING(1)
    SINGULAR(0.000000000001) HCONVERGE(0, ABSOLUTE) LCONVERGE(0,
    ABSOLUTE) PCONVERGE(0.000001, ABSOLUTE)
  /FIXED=male pop2000 disadvantage | SSTYPE(3)
  /METHOD=ML
  /PRINT=SOLUTION
  /RANDOM=INTERCEPT male | SUBJECT(idcomm) COVTYPE(VC).
```

Example 9.9

```
COMPUTE age_dis=age * disadvantage.
EXECUTE.

MIXED income WITH age disadvantage age_dis
  /CRITERIA=CIN(95) MXITER(100) MXSTEP(10) SCORING(1)
    SINGULAR(0.000000000001) HCONVERGE(0, ABSOLUTE) LCONVERGE(0,
    ABSOLUTE) PCONVERGE(0.000001, ABSOLUTE)
  /FIXED=age disadvantage age_dis | SSTYPE(3)
  /METHOD=ML
  /PRINT=SOLUTION
  /RANDOM=INTERCEPT age | SUBJECT(idcomm) COVTYPE(VC)
  /SAVE PRED(Pred8_5).
```

Figure 9.4

```
RECODE disadvantage (Lowest thru -0.57=0) (0.57 thru Highest=1)
    (-0.57 thru 0.57=SYSMIS) INTO Dis_Cat.
  VARIABLE LABELS Dis_Cat 'Dis_Cat'.
  EXECUTE.

GGRAPH
  /GRAPHDATASET NAME="GraphDataset" VARIABLES= age Pred9_5
    Dis_Cat
  /GRAPHSPEC SOURCE=INLINE.
BEGIN GPL
SOURCE: s=userSource( id( "GraphDataset" ) )
DATA: age=col( source(s), name( "age" ) )
DATA: Pred8_5=col( source(s), name( "Pred9_5" ) )
DATA: Dis_Cat = col(source(s), name("Dis_Cat"), unit.category())
ELEMENT: point( position(age * Pred9_5))
ELEMENT: line(position(smooth.linear(age * Pred9_5)), shape
    (Dis_Cat))
END GPL.
```

Example 9.10

```
GENLINMIXED
  /DATA_STRUCTURE SUBJECTS=idcomm
  /FIELDS TARGET=trust TRIALS=NONE OFFSET=NONE
  /TARGET_OPTIONS DISTRIBUTION=BINOMIAL LINK=LOGIT
  /FIXED EFFECTS=educate age USE_INTERCEPT=TRUE
  /RANDOM EFFECTS=age USE_INTERCEPT= TRUE COVARIANCE_
    TYPE=VARIANCE_COMPONENTS
```

```
   /BUILD_OPTIONS TARGET_CATEGORY_ORDER=ASCENDING INPUTS_
     CATEGORY_ORDER=ASCENDING MAX_ITERATIONS=100 CONFIDENCE_
     LEVEL=95 DF_METHOD=RESIDUAL COVB=MODEL
  /EMMEANS_OPTIONS SCALE=TRANSFORMED PADJUST=LSD.
```

Figure 9.6

```
* Anscombe residuals were not available in SPSS GENLINMIXED at
  the time of publication; however, they could be computed with
  a few programming steps.
```

Figure 9.7

```
COMPUTE filter_$=(newid=14 OR newid=23 OR newid=34 OR newid=53).
VARIABLE LABELS filter_$ 'newid=14 OR newid=23 OR newid=34 OR
  newid=53 (FILTER)'.
VALUE LABELS filter_$ 0 'Not Selected' 1 'Selected'.
FORMATS filter_$ (f1.0).
FILTER BY filter_$.
EXECUTE.

GGRAPH
  /GRAPHDATASET NAME="GraphDataset" VARIABLES= age esteem newid
  /GRAPHSPEC SOURCE=INLINE.
BEGIN GPL
SOURCE: s=userSource( id( "GraphDataset" ) )
DATA: age=col( source(s), name( "age" ) )
DATA: esteem=col( source(s), name( "esteem" ) )
DATA: newid = col(source(s), name("newid"), unit.category())
ELEMENT: point( position(age * esteem))
ELEMENT: line(position(smooth.linear(age * esteem)), shape(newid))
END GPL.
```

Example 9.11

```
MIXED esteem
  /CRITERIA=CIN(95) MXITER(100) MXSTEP(10) SCORING(1)
    SINGULAR(0.000000000001) HCONVERGE(0, ABSOLUTE) LCONVERGE(0,
    ABSOLUTE) PCONVERGE(0.000001, ABSOLUTE)
  /FIXED=| SSTYPE(3)
  /METHOD=ML
  /PRINT=SOLUTION.
```

Example 9.12

```
GET
FILE='\\tsclient\home\Downloads\esteem.sav'.
```

```
DATASET NAME DataSet1 WINDOW=FRONT.
MIXED esteem WITH age
  /CRITERIA=CIN(95) MXITER(100) MXSTEP(10) SCORING(1)
    SINGULAR(0.000000000001) HCONVERGE(0, ABSOLUTE) LCONVERGE(0,
    ABSOLUTE) PCONVERGE(0.000001, ABSOLUTE)
  /FIXED=age | SSTYPE(3)
  /METHOD=ML
  /PRINT=SOLUTION
  /RANDOM=INTERCEPT | SUBJECT(newid) COVTYPE(VC).


MIXED esteem WITH age
  /CRITERIA=CIN(95) MXITER(100) MXSTEP(10) SCORING(1)
    SINGULAR(0.000000000001) HCONVERGE(0, ABSOLUTE) LCONVERGE(0,
    ABSOLUTE) PCONVERGE(0.000001, ABSOLUTE)
  /FIXED=age | SSTYPE(3)
  /METHOD=ML
  /PRINT=SOLUTION
  /RANDOM=INTERCEPT age | SUBJECT(newid) COVTYPE(VC).
```

* There was no likelihood-ratio test in SPSS MIXED at the time of publication.

Example 9.13

```
MIXED esteem WITH age stress cohes male
  /CRITERIA=CIN(95) MXITER(100) MXSTEP(10) SCORING(1)
    SINGULAR(0.000000000001) HCONVERGE(0, ABSOLUTE) LCONVERGE(0,
    ABSOLUTE) PCONVERGE(0.000001, ABSOLUTE)
  /FIXED=age stress cohes male | SSTYPE(3)
  /METHOD=ML
  /PRINT=SOLUTION
  /RANDOM=INTERCEPT age stress cohes | SUBJECT(newid) COVTYPE(VC).
```

Example 9.14

```
MIXED esteem WITH age stress cohes male age_male
  /CRITERIA=CIN(95) MXITER(100) MXSTEP(10) SCORING(1)
    SINGULAR(0.000000000001) HCONVERGE(0, ABSOLUTE) LCONVERGE(0,
    ABSOLUTE) PCONVERGE(0.000001, ABSOLUTE)
  /FIXED=age stress cohes male age_male | SSTYPE(3)
  /METHOD=ML
  /PRINT=SOLUTION
  /RANDOM=INTERCEPT age stress cohes | SUBJECT(newid) COVTYPE(VC)
  /SAVE=PRED (Pred8_14).
```

Figure 9.8

```
GGRAPH
  /GRAPHDATASET NAME="GraphDataset" VARIABLES= age Pred8_14 male
  /GRAPHSPEC SOURCE=INLINE.
BEGIN GPL
SOURCE: s=userSource( id( "GraphDataset" ) )
DATA: age=col( source(s), name( "age" ) )
DATA: Pred8_14=col( source(s), name( "Pred8_14" ) )
DATA: male = col(source(s), name("male"), unit.category())
ELEMENT: point( position(age * Pred8_14))
ELEMENT: line(position(smooth.linear(age * Pred8_14)), shape(male))
END GPL.
```

Example 9.15

```
MIXED esteem WITH age stress cohes male nonwhite
  /CRITERIA=CIN(95) MXITER(100) MXSTEP(10) SCORING(1)
    SINGULAR(0.000000000001) HCONVERGE(0, ABSOLUTE) LCONVERGE(0,
    ABSOLUTE) PCONVERGE(0.000001, ABSOLUTE)
  /FIXED=age stress cohes male nonwhite | SSTYPE(3)
  /METHOD=ML
  /PRINT=SOLUTION
  /RANDOM=INTERCEPT age stress cohes | SUBJECT(newid) COVTYPE(UN)
  /SAVE=FIXPRED PRED RESID.
```

Figure 9.9

```
DESCRIPTIVES VARIABLES=RESID_1
  /SAVE
  /STATISTICS=MEAN STDDEV MIN MAX.

GGRAPH
  /GRAPHDATASET NAME ="graphdataset" VARIABLES =ZRESID_1 MISSING
    =LISTWISE REPORTMISSING =NO
  /GRAPHSPEC SOURCE =INLINE.
BEGIN GPL
SOURCE: s=userSource (id("graphdataset"))
DATA: ZRESID_1=col(source(s), name("ZRESID_1"))
GUIDE: axis(dim(1), label("Residuals"))
GUIDE: axis(dim(2), label("Frequency"))
SCALE: linear(dim(1), min(-10), max(5))
SCALE: linear(dim(2), include(0))
ELEMENT: interval(position(summary.count(bin.rect(ZRESID_1))),
  shape.interior(shape.square))
```

```
ELEMENT: line(position(density.kernel.gaussian(ZRESID_1)))
END GPL
```

Figure 9.10

```
 GGRAPH
 /GRAPHDATASET NAME="graphdataset" VARIABLES=FXPRED_2 ZRESID_1
   MISSING=LISTWISE REPORTMISSING=NO
 /GRAPHSPEC SOURCE=INLINE.
BEGIN GPL
SOURCE: s = userSource(id("graphdataset"))
DATA: FXPRED_2=col(source(s), name("FXPRED_2"))
DATA: ZRESID_1=col(source(s), name("ZRESID_1"))
GUIDE: axis(dim(1), label("Fixed Predicted Values"))
GUIDE: axis(dim(2), label("Zscore: Residuals"))
ELEMENT: point(position(FXPRED_2 *Z RESID_1))
END GPL.
```

Example 9.16

```
MIXED esteem WITH age stress cohes male nonwhite
  /CRITERIA=CIN(95) MXITER(100) MXSTEP(10) SCORING(1)
    SINGULAR(0.000000000001) HCONVERGE(0, ABSOLUTE) LCONVERGE(0,
    ABSOLUTE) PCONVERGE(0.000001, ABSOLUTE)
  /FIXED=age stress cohes male nonwhite | SSTYPE(3)
  /METHOD=ML
  /PRINT=SOLUTION
  /RANDOM=INTERCEPT age stress cohes | SUBJECT(newid)
    COVTYPE(VC).
```

Example 9.17

```
GENLIN stress WITH age
  /MODEL age INTERCEPT=YES
  DISTRIBUTION=POISSON LINK=LOG
  /CRITERIA METHOD=FISHER(1) SCALE=1 MAXITERATIONS=100
    MAXSTEPHALVING=5 PCONVERGE = 1E-006(ABSOLUTE) SINGULAR=
    1E-012 ANALYSISTYPE=3(WALD) CILEVEL=95 LIKELIHOOD=
    FULL
  /REPEATED SUBJECT=newid SORT=YES CORRTYPE=AR(1) ADJUSTCORR=YES
    COVB=ROBUST MAXITERATIONS=100 PCONVERGE=1e-006(ABSOLUTE)
    UPDATECORR=1
  /MISSING CLASSMISSING=EXCLUDE
  /PRINT MODELINFO FIT SUMMARY SOLUTION (EXPONENTIATED).
```

CHAPTER 10

Example 10.1

```
DESCRIPTIVES VARIABLES=worthy capable bother noeat blues depress
    failure lonely crying sad tired goodqual attitude
  /STATISTICS=MEAN STDDEV MIN MAX.

CORRELATIONS
  /VARIABLES=worthy capable bother noeat blues depress failure
    lonely crying sad tired goodqual attitude
  /PRINT=TWOTAIL NOSIG
  /MISSING=PAIRWISE.
```

Example 10.2

```
FACTOR
  /VARIABLES worthy capable bother noeat blues depress failure
    lonely crying sad tired goodqual attitude
  /MISSING LISTWISE
  /ANALYSIS worthy capable bother noeat blues depress failure
    lonely crying sad tired goodqual attitude
  /PRINT INITIAL EXTRACTION
  /CRITERIA MINEIGEN(1) ITERATE(25)
  /EXTRACTION PC
  /ROTATION NOROTATE
  /METHOD=CORRELATION.
```

Example 10.3

```
FACTOR
  /VARIABLES worthy capable bother noeat blues depress failure
    lonely crying sad tired goodqual attitude
  /MISSING LISTWISE
  /ANALYSIS worthy capable bother noeat blues depress failure
    lonely crying sad tired goodqual attitude
  /PRINT INITIAL EXTRACTION
  /CRITERIA FACTORS(5) ITERATE(25)
  /EXTRACTION PAF
  /ROTATION NOROTATE
  /METHOD=CORRELATION.

FACTOR
  /VARIABLES worthy capable bother noeat blues depress failure
    lonely crying sad tired goodqual attitude
  /MISSING LISTWISE
```

```
  /ANALYSIS worthy capable bother noeat blues depress failure
    lonely crying sad tired goodqual attitude
  /PRINT INITIAL EXTRACTION ROTATION
  /FORMAT BLANK(.40)
  /CRITERIA MINEIGEN(1) ITERATE(25)
  /EXTRACTION PAF
  /CRITERIA ITERATE(25)
  /ROTATION VARIMAX
  /METHOD=CORRELATION.

FACTOR
  /VARIABLES worthy capable bother noeat blues depress failure
    lonely crying sad tired goodqual attitude
  /MISSING LISTWISE
  /ANALYSIS worthy capable bother noeat blues depress failure
    lonely crying sad tired goodqual attitude
  /PRINT INITIAL EXTRACTION ROTATION
  /FORMAT BLANK(.40)
  /CRITERIA MINEIGEN(1) ITERATE(25)
  /EXTRACTION PAF
  /CRITERIA ITERATE(25) DELTA(0)
  /ROTATION OBLIMIN
  /METHOD=CORRELATION.

FACTOR
  /VARIABLES worthy capable bother noeat blues depress failure
    lonely crying sad tired goodqual attitude
  /MISSING LISTWISE
  /ANALYSIS worthy capable bother noeat blues depress failure
    lonely crying sad tired goodqual attitude
  /PRINT INITIAL
  /FORMAT BLANK(.40)
  /CRITERIA MINEIGEN(1) ITERATE(25)
  /EXTRACTION PAF
  /CRITERIA ITERATE(25) DELTA(0)
  /ROTATION OBLIMIN
  /SAVE REG(ALL)
  /METHOD=CORRELATION.

FACTOR
  /VARIABLES worthy capable bother noeat blues depress failure
    lonely crying sad tired goodqual attitude
  /MISSING LISTWISE
```

```
/ANALYSIS worthy capable bother noeat blues depress failure
  lonely crying sad tired goodqual attitude
/FORMAT BLANK(.40)
/CRITERIA MINEIGEN(1) ITERATE(25)
/EXTRACTION PAF
/CRITERIA ITERATE(25) DELTA(0)
/ROTATION OBLIMIN
/SAVE REG(ALL)
/METHOD=CORRELATION.

FACTOR
/VARIABLES worthy capable bother noeat blues depress failure
  lonely crying sad tired goodqual attitude
/MISSING LISTWISE
/ANALYSIS worthy capable bother noeat blues depress failure
  lonely crying sad tired goodqual attitude
/FORMAT BLANK(.40)
/CRITERIA MINEIGEN(1) ITERATE(25)
/EXTRACTION PAF
/CRITERIA ITERATE(25) DELTA(0)
/ROTATION OBLIMIN
/SAVE BART(ALL)
/METHOD=CORRELATION.
```

Example 10.4

```
NONPAR CORR
/VARIABLES=worthy capable bother noeat blues depress failure
  lonely crying sad tired goodqual attitude
/PRINT=SPEARMAN TWOTAIL NOSIG
/MISSING=PAIRWISE.
```

* Factor analysis with polychoric correlations was not
  available in SPSS at the time of publication; however, as of
  SPSS version 19 the extension command SPSSINC HETCOR was
  available for use with the R plug-in to estimate polychoric
  correlations; these may be used as input in a factor analysis
  with SPSS.

Example 10.5

* Factor analysis with polychoric correlations was not available
  in SPSS at the time of publication; however, as of SPSS
  version 19 the extension command SPSSINC HETCOR was available
  for use with the R plug-in to estimate polychoric correlations;
  these may be used as input in a factor analysis with SPSS.

Example 10.6

* Factor analysis with polychoric correlations was not
  available in SPSS at the time of publication; however, as of
  SPSS version 19 the extension command SPSSINC HETCOR was
  available for use with the R plug-in to estimate polychoric
  correlations; these may be used as input in a factor analysis
  with SPSS.

Example 10.7

* Factor analysis with polychoric correlations was not available
  in SPSS at the time of publication; however, as of SPSS
  version 19 the extension command SPSSINC HETCOR was available
  for use with the R plug-in to estimate polychoric correlations;
  these may be used as input in a factor analysis with SPSS.

Example 10.8

* Structural equation modeling (SEM) was not available in SPSS
  at the time of publication; SPSS Amos® software may be used
  for SEMs.

Example 10.9

* Structural equation modeling (SEM) was not available in SPSS
  at the time of publication; SPSS Amos® software may be used
  for SEMs.

Example 10.10

* Structural equation modeling (SEM) was not available in SPSS
  at the time of publication; SPSS Amos® software may be used
  for SEMs.

CHAPTER 1

Figure 1.2

```
plot(density(na.omit(gss$sei)))
```

Example 1.1

```
example.1.1 <- lm(sei~female, data=gss)
summary(example.1.1)
anova(example.1.1)

lm(scale(sei) ~ scale(female), data=gss
```

Example 1.2

```
example.1.2 <- lm(sei ~ female + educate + nonwhite + pasei,
    data=gss)
summary(example.1.2)
anova(example.1.2)

lm(scale(sei) ~ scale(female) + scale(educate) + scale(nonwhite) +
    scale(pasei), data=gss)
```

Figure 1.4

```
plot(fitted(example.1.2), residuals(example.1.2))
```

Example 1.3

```
rstdnt <- rstudent(example.1.2)
pred <- fitted(example.1.2)
zpred <- scale(pred, center=TRUE, scale=TRUE)
plot (rstdnt, zpred)
abline(lm(zpred~rstdnt), col="red")
lines(lowess(rstdnt, zpred), col="blue")
```

Example 1.4

```
# Breusch-Pagan Test
install.packages("lmtest")
library(lmtest)
bptest(example.1.2, data=gss, studentize=FALSE)

# White Test
install.packages("het.test")
library(het.test)
install.packages("vars")
```

```
library("vars")
model.var <- VAR(na.omit(gss))
whites.htest(model.var)
```

Example 1.5

```
abs.1.5 <- abs(rstdnt)
example.1.5 <- lm(abs.1.5~female+educate+nonwhite+pasei, data=gss)
summary(example.1.5)
anova(example.1.5)
```

Example 1.6

```
install.packages("MASS")
library(MASS)
example.1.6 <- rlm(sei ~ female + educate + nonwhite + pasei,
    data=gss)
summary(example.1.6)
```

Figure 1.5

```
plot(density(na.omit(rstdnt)))
```

Figure 1.6

```
qqnorm(rstudent(example.1.7))
abline(0,1)
```

Figure 1.7

```
probDist <- pnorm(rstudent(example.1.7))
plot(ppoints(length(rstudent(example.1.7))), sort(probDist),
  main="PP Plot", xlab="Observed Probability", ylab="Expected
  Probability")
abline(0,1)
```

Figure 1.8

```
library(car)
crPlots(example.1.2, terms = "educate", layout = c(1,1),
  ask=FALSE, main="Education Component+Residual Plot",
  smoother=loessLine)
```

Example 1.7

```
# Link Test
install.packages("devtools")
library(devtools)
devtools::install_version("LDdiag","0.1")
library(LDdiag)
```

```
example.1.7 <- lm(sei~female+educate+nonwhite+pasei, data=gss)
pregibon(example.1.7)

# Ramsey RESET Test
library(LDdiag)
ramsey(example.1.7)
```

Figure 1.10

```
rstandard2 <- (rstandard(example.1.7))^2
lev <- hatvalues(example.1.7)
plot(rstandard2, lev, xlab="Normalized Residuals Squared",
  ylab="Leverage")
abline(v=mean(rstandard2))
abline(h=mean(lev))
```

Figure 1.11

```
cookd <- cooks.distance(example.1.7)
plot(lev, cookd, xlab="Leverage", ylab="Cook's D")
abline(v=0.01)
abline(h=0.0018)
```

Example 1.8

```
install.packages("quantreg")
library(quantreg)
quant.1.7 <- rq(example.1.7, tau=.5, data = gss, method="br", model
  = TRUE)
summary(quant.1.7)
```

Example 1.9

```
example.1.9 <- lm(I(log(sei))~female+educate+nonwhite+pasei,
    data=gss)
summary(example.1.9)
anova(example.1.9)
lm(scale(I(log(sei))) ~ scale(female)+scale(educate)+scale
  (nonwhite)+scale(pasei), data=gss)
```

Example 1.10

```
example.1.10 <- lm(sei~female + educate + nonwhite + pasei +
  educate:pasei, data=gss)
summary(example.1.10)
library(car)
vif(example.1.10)
```

Example 1.11

```
z.edu <- scale(gss$educate, center=TRUE, scale=TRUE)
z.pasei <- scale(gss$pasei, center=TRUE, scale=TRUE)
example.1.11 <- lm(sei~female + z.edu + nonwhite + z.pasei +
  z.edu:z.pasei, data=gss)
summary(example.1.11)
library(car)
vif(example.1.11)

# The remaining examples and figures in Chapter 1 may be
  replicated by adapting the code already provided.
```

CHAPTER 2

Example 2.1

```
pol.table <- table(gss$polviews)
pol.table
prop.table(pol.table)*100
```

Example 2.2

```
pol.gender <- table(gss$polviews, gss$female)
pol.gender
margin.table(pol.gender, 1)
margin.table(pol.gender, 2)
prop.table(pol.gender, 1)*100
chisq.test(pol.gender)
```

Figure 2.1

```
child.mar <- gss[gss$marital!=5 & gss$childs>0,]
hist(child.mar$childs)
```

Example 2.3

```
example.2.3 <- lm(sei~female+nonwhite+educate+pasei, data=gss)
summary(example.2.3)
anova(example.2.3)

AIC(example.2.3)
BIC(example.2.3)
```

Example 2.4

```
complete.cases(gss)
gsscomp <- gss[complete.cases(gss),]
complete.cases(gsscomp)
```

```
modelfull <- (lm(sei~female+nonwhite+educate+pasei,
  data=gsscomp))
modelred <- lm(sei~female+nonwhite, data=gsscomp)
modelnull <- lm(sei~ 1, data=gsscomp)

anova(modelred, modelfull)

install.packages("survey")
library(survey)
regTermTest(example.2.3, "educate", method="Wald")
regTermTest(example.2.3, "pasei", method="Wald")
```

CHAPTER 3
Example 3.1

```
example.3.1 <- lm(satlife~age, data=depress)
summary(example.3.1)
anova(example.3.1)
```

Figure 3.1

```
plot(fitted(example.3.1), residuals(example.3.1))
```

Figure 3.2

```
qqnorm(rstudent(example.3.1))
abline(0,1)
```

Example 3.2

```
male.sat <- table(depress$male, depress$satlife)
male.sat
margin.table(male.sat, 1)
margin.table(male.sat, 2)
prop.table(male.sat,1)*100
```

Example 3.3

```
example.3.3 <- glm(satlife~male, data=depress, family='binomial')
summary(example.3.3)
confint(example.3.3)
exp(coefficients(example.3.3))
exp(confint(example.3.3))

predict(example.3.3, newdata=data.frame(male=1), type='response')
predict(example.3.3, newdata=data.frame(male=0), type='response')
```

Example 3.4

```
example.3.4 <- glm(satlife ~ male + age + iq, data=depress,
  family='binomial')
summary(example.3.4)
confint(example.3.4)
exp(coefficients(example.3.4))
exp(confint(example.3.4))
```

Example 3.5

```
predict(example.3.4, newdata=data.frame(male=1, age=37.45763,
  iq=91.79091),
  type='response')
predict(example.3.4, newdata=data.frame(male=0,
    age=mean(depress$age),
  iq=mean(na.omit(depress$iq))), type='response')
```

Example 3.6

```
depcomp <- depress[complete.cases(depress),]

modelfull <- glm(satlife~male+age+iq, data=depcomp,
  family='binomial')
modelred <- glm(satlife~male, data=depcomp, family='binomial')
anova(modelred, modelfull, test="Chisq")
```

Example 3.7

```
example.3.7 <- glm(satlife~male+age+iq, data=depcomp, family='bino
  mial'(link="logit"))
dev <- residuals(example.3.7, type = "deviance")
rstand <- rstandard(example.3.7, type= "pearson")
hat <- hatvalues(example.3.7)
pred <- fitted(example.3.7)

plot(density(dev))

plot(pred,dev)

plot(depcomp$age, dev)
abline(lm(dev~depcomp$age), col="red")
lines(lowess(depcomp$age, dev), col="blue")

plot(hat, rstand)
abline(v=3*mean(hat))
```

```
# Delta-deviance was not an option to save in R at the time of
  publication. They could be computed, however, using some
  simple programming steps.
```

Figure 3.5

```
plot(hat, d.dev2)
abline(v=3*mean(hat))
abline(h=4)
```

Example 3.8

```
# Delta-deviance was not an option to save in R at the time of
  publication. They could be computed, however, using some
  simple programming steps.
```

Example 3.9

```
example.3.9 <- glm(satlife~male, data=depress, family='binomial'(l
  ink='probit'))
summary(example.3.9)
confint(example.3.9)
exp(coefficients(example.3.9))
exp(confint(example.3.9))

predict(example.3.9, newdata=data.frame(male=1), type='response')
predict(example.3.9, newdata=data.frame(male=0), type='response')
```

Example 3.10

```
example.3.10 <- glm(satlife~male+sleep1+age+iq, data=depress,
  family='binomial' (link='probit'))
summary(example.3.10)
confint(example.3.10)
exp(coefficients(example.3.10))
exp(confint(example.3.10))

predict(example.3.10, newdata=data.frame(male=0, sleep1=0,
  age=mean(depress$age),
  iq=mean(na.omit(depress$iq))), type='response')
predict(example.3.10, newdata=data.frame(male=1, sleep1=0,
  age=mean(depress$age),
  iq=mean(na.omit(depress$iq))), type='response')
predict(example.3.10, newdata=data.frame(male=0, sleep1=1,
  age=mean(depress$age),
  iq=mean(na.omit(depress$iq))), type='response')
```

```
predict(example.3.10, newdata=data.frame(male=1, sleep1=1,
  age=mean(depress$age),
  iq=mean(na.omit(depress$iq))), type='response')

predict(example.3.10, newdata=data.frame(male=0, sleep1=0, age=30,
  iq=mean(na.omit(depress$iq))), type='response')
predict(example.3.10, newdata=data.frame(male=0, sleep1=0, age=40,
  iq=mean(na.omit(depress$iq))), type='response')
```

Example 3.11

```
# There was no dprobit option in R at the time of publication.
```

Example 3.12

```
# There was no dprobit option in R at the time of publication.
```

Figure 3.7

```
dev <- residuals(example.3.10, type = "deviance")
pred <- fitted(example.3.10)
plot(pred, dev)
```

Example 3.13

```
library(logistf)
example.3.13 <- logistf(satlife ~ male, data=depress)
summary(example.3.13)
```

Example 3.14

```
# R uses the glm command for logistic regression. See
  Example 3.3.
```

Example 3.15

```
library(robustbase)
example.3.15 <- glmrob(satlife~male+age+iq,
    family='poisson'(link='log'), data=depcomp)
summary(example.3.15)
exp(coefficients(example.3.15))
```

CHAPTER 4

Example 4.1

```
spank.table <- table(gss$spanking)
spank.table
```

Figure 4.1

```
plot(gss$educate, gss$spanking)
abline(lm(gss$spanking ~ gss$educate))
```

Example 4.2

```
example.4.2 <- lm(spanking~educate, data=gss)
summary(example.4.2)
anova(example.4.2)
```

Figure 4.2

```
rstdnt.4.2 <- rstudent(example.4.2)
plot(density(rstdnt.4.2))
```

Figure 4.3

```
qqnorm(rstdnt.4.2)
abline(0,1)
```

Example 4.3

```
spank.fem <- table(gss$female, gss$spanking)
spank.fem
margin.table(spank.fem,1)
margin.table(spank.fem,2)
prop.table(spank.fem,1)*100
```

Example 4.4

```
spank.fact <- factor(gss$spanking)
library(MASS)
example.4.4 <- polr(spank.fact ~ female, data = gss, Hess=TRUE)
summary(example.4.4)
exp(coefficients(example.4.4))
```

```
# The polr command in the "MASS" library is used here and
  subsequently; the clm command in the "ordinal" package
  provides an alternative.
```

Example 4.5

```
# There was no omodel or Brant test option available in R at the
  time of publication. However, see the documentation for the R
  "ordinal" package for a similar test
```

Example 4.6

```
gsscomp <- gss[complete.cases(gss$female),]
gsscomp4.1 <- gsscomp[complete.cases(gsscomp$spanking),]
gsscomp4.2 <- gsscomp4.1[complete.cases(gsscomp4.1$educate),]
gsscomp4 <- gsscomp4.2[complete.cases(gsscomp4.2$polviews),]
spank.fact4 <- factor(gsscomp4$spanking)
```

```
spank.fact4 <- factor(gsscomp4$spanking)
library(MASS)
example.4.6 <- polr(spank.fact4 ~ female, data=gsscomp4, Hess=T,
  method="probit")
gsscomp4$pred <- predict(example.4.6, type="p")
aggregate(pred~female, gsscomp4, mean)

1-pnorm(2.606)
pnorm(-.5526)
pnorm(-.5526-0.1297)
pnorm(0.6837)
pnorm(0.6837-0.1297)
pnorm(1.4461)
pnorm(1.4461-0.1297)
```

Example 4.7

```
library(MASS)
example.4.7 <- polr(spank.fact ~ female + educate + polviews,
  data=gss, Hess=T, method="logistic")
summary(example.4.7)
```

Example 4.8

```
example.4.8 <- polr(spank.fact4 ~ female+educate+polviews,
  data=gsscomp4, Hess=T, method="probit")
summary(example.4.8)

gsscomp4$pred <- predict(example.4.8, type="p")
aggregate(pred~female, gsscomp4, mean)
aggregate(pred~(female==0)+(educate==10), gsscomp4, mean)
aggregate(pred~(female==0)+(educate==16), gsscomp4, mean)

AIC(example.4.6)
BIC(example.4.6)
AIC(example.4.8)
BIC(example.4.8)
```

Figure 4.5

```
# Provides a horizontal bar graph
Females <- aggregate(pred ~ (female==1), gsscomp4, mean)
Females$"2" <- NULL
Females$"3" <- NULL

install.packages("reshape2")
```

```
library(reshape2)
Figure4.4 <- melt(Females,id.vars=c("female == 1"), measure.
    vars=c("1", "4"), variable.name="Level", value.name=
    "Predict")
Figure4.4$Sex <- c("Male","Female","Male","Female")
Figure4.4$Agreement <- c("Strongly agree","Strongly
    agree","Strongly disagree","Strongly disagree")

library(lattice)
barchart(Sex~Predict, data=Figure4.4, groups=Agreement, auto.key
    = list(colums = 2))
```

Example 4.9

```
gsscomp4$newspank <- ifelse(gsscomp8.4$spanking <= 1, c(1), c(0))

example.4.9 <- glm(newspank~female+educate+polviews,
    data=gsscomp4, family='binomial'(link="logit"))

dev <- residuals(example.4.9, type = "deviance")
rstand <- rstandard(example.4.9, type= "pearson")
hat <- hatvalues(example.4.9)
pred <- fitted(example.4.9)

# Delta-deviance was not an option to save in R at the time of
    publication. They could be computed, however, using some
    simple programming steps.
```

Figure 4.6

```
plot(pred, dev)
```

Example 4.10

```
# There was no gologit2 option available in R at the time of
    publication. However, the vglm command in the "VGAM" package
    fits proportional and nonproportional models that make
    different assumptions than Stata's gologit2.
```

Example 4.11

```
# See Examples 1.3 and 4.7 for code that may be adapted for this
    example.
```

CHAPTER 5

Example 5.1

```
pol.cat <- table(gsst$polview1)
pol.cat
```

Example 5.2

```
pol.nonw <- table(gss$nonwhite, gss$polview1)
pol.nonw
margin.table(pol.nonw, 1)
margin.table(pol.nonw, 2)
prop.table(pol.nonw, 1)*100
```

Example 5.3

```
install.packages("nnet")
library(nnet)

gsscomp5.1 <- gss[complete.cases(gss$polview1),]
gsscomp5 <- gsscomp5.1[complete.cases(gsscomp5.1$nonwhite),]

gsscomp5$pol.fact1 <- factor(gsscomp5$polview1)

gsscomp5$polview2 <- relevel(gsscomp5$pol.fact1, ref = 2)

example.5.3 <- multinom(polview2 ~ nonwhite, data=gsscomp5)
summary(example.5.3)
exp(coefficients(example.5.3))
z <- summary(example.5.3)$coefficients /summary(example5.3)
   $standard.errors
z
p <- (1 - pnorm(abs(z), 0, 1)) * 2
p

gsscomp5$pred <- fitted(example.5.3)
aggregate(pred~nonwhite, gsscomp5, mean)
```

Example 5.4

```
gsscomp5.2 <- gsscomp5.1[complete.cases(gsscomp5.1$female),]
gsscomp5.3 <- gsscomp5.2[complete.cases(gsscomp5.2$age),]
gsscomp5.4 <- gsscomp5.3[complete.cases(gsscomp5.3$educate),]

gsscomp5.4$pol.fact2 <- factor(gsscomp5.4$polview1)
gsscomp5.4$polview3 <- relevel(gsscomp5.4$pol.fact2, ref = 2)
example.5.4 <- multinom(polview3~female+age+educate,
   data=gsscomp5.4)
summary(example.5.4)
exp(coefficients(example.5.4))
z4 <- summary(example.5.4)$coefficients/summary(example.5.4)
   $standard.errors
z4
```

```
p4 <- (1 - pnorm(abs(z4), 0, 1)) * 2
p4
```

Example 5.5

```
gsscomp5.4$polview4 <- relevel(gsscomp5.4$pol.fact2, ref = 3)
example.5.5 <- multinom(polview4 ~ female + age + educate,
  data=gsscomp5.4)

summary(example.5.5)
exp(coefficients(example.5.5))
z5 <- summary(example.5.5)$coefficients/summary(example.5.5)
  $standard.errors
z5
p5 <- (1 - pnorm(abs(z5), 0, 1)) * 2
p5

gsscomp5.4$pred <- fitted(example5.5)
aggregate(pred~(female==0)+(educate==12), gsscomp5.4, mean)
aggregate(pred~(female==0)+(educate==16), gsscomp5.4, mean)
```

Example 5.6

```
install.packages("MNP")
library(MNP)
example.5.6 <- mnp(polview4 ~ female + age + educate,
  data=gsscomp5.4, base="2")
summary(example.5.6)

pred5.6 <- predict.mnp(example.5.6, type="prob")
gsscomp5.4$pred2 <- pred5.6$p
aggregate(pred2~female, gsscomp5.4, mean)
```

Example 5.7

```
gsscomp5.5 <- gsscomp5.4[which(gsscomp5.4$polview1!='3'),]
gsscomp5.5$libmod <- ifelse(gsscomp5.5$polview1 <= 1, c(1), c(0))

example.5.7 <- glm(libmod~female+age+educate, data=gsscomp5.5, fam
  ily='binomial'(link="logit"))
summary(example.5.7)
exp(coefficients(example.5.7))

gsscomp5.5$dev <- residuals(example.5.7, type = "deviance")
gsscomp5.5$rstand <- rstandard(example.5.7, type= "pearson")
gsscomp5.5$hat <- hatvalues(example.5.7)
gsscomp5.5$pred <- fitted(example.5.7)
```

Delta-deviance was not available as a save option in R at the
  time of publication. They could be computed, however, with a
  few simple programming steps.

Figure 5.3

Delta-deviance was not available as a save option in R at the
  time of publication. They could be computed, however, with a
  few simple programming steps.

Example 5.8

Delta-deviance was not available as a save option in R at the
  time of publication. They could be computed, however, with a
  few simple programming steps.

Example 5.9

There was no mlogtest option in R at the time of publication.

CHAPTER 6

Example 6.1

```
volteer.table <- table(gss$volteer)
volteer.table
prop.table(volteer.table)*100
```

Example 6.2

```
summary(gss$volteer)
sd(gss$volteer)
```

Figure 6.3

```
hist(gss$volteer)
```

Example 6.3

```
install.packages("psych")
library("psych")
describeBy(gss$volteer, gss$female)
```

Example 6.4

```
gsscomp6 <- gss[complete.cases(gss$volteer),]
gsscomp6.1 <- gsscomp6[complete.cases(gsscomp6$female),]

example.6.4 <- glm(volteer ~ female, data=gsscomp6.1, family =
    "poisson")
summary(example.6.4)
gsscomp6.1$pred <- fitted(example.6.4)
aggregate(pred~female, gsscomp6.1, mean)
```

Example 6.5

```
gsscomp6.2 <- gsscomp6.1[complete.cases(gsscomp6.1$nonwhite),]
gsscomp6.3 <- gsscomp6.2[complete.cases(gsscomp6.2$educate),]
gsscomp6.4 <- gsscomp6.3[complete.cases(gsscomp6.3$income),]

example.6.5 <- glm(volteer ~ female + nonwhite + educate +
    income, data=gsscomp6.4, family = "poisson")
summary(example.6.5)
exp(coefficients(example.6.5))

gsscomp6.5$pred <- fitted(example.6.5)

aggregate(pred~female+(nonwhite==0)+(educate==12)+(income==5),
  gsscomp6.4, mean)
```

Figure 6.5

```
plot(gsscomp6.4$educate, gsscomp6.4$pred)
abline(glm(gsscomp6.4$pred~gsscomp6.4$educate), col="red")
lines(lowess(gsscomp6.4$educate, gsscomp6.4$pred), col="blue")
```

Example 6.6

```
# R uses the glm command for Poisson regression. See the R code
  in Example 6.5.

gsscomp6.4$dev <- residuals(example.6.5, type = "deviance")
gsscomp6.4$hat <- hatvalues(example.6.5)
gsscomp6.4$cookd <- cooks.distance(example.6.5)
gsscomp6.4$adjdev <- gsscomp6.4$dev + (1/(6*sqrt(gsscomp6.4$pred)))
```

Figure 6.6

```
plot(gsscomp6.4$hat, gsscomp6.4$dev)
abline(h=4)
abline(v=0.0077)
```

Figure 6.7

```
x <- seq(-4,4,length=200)
y <- dnorm(x,mean=0, sd=1)
plot(density(gsscomp6.4$adjdev))
lines(x,y, col="blue")
```

Example 6.7

```
example.6.7 <- glm(volteer ~ female+nonwhite+educate+income,
  data=gsscomp6.4, family = "quasipoisson")
summary(example.6.7)
```

This extradispersed Poisson regression model in R uses the
  Pearson-based dispersion (2.24) rather than the deviance-based
  dispersion (1.27) to adjust the standard errors; this results
  in larger standard errors for the coefficients. See the cmp
  command in the R package "COMPoissonReg" for an alternative
  model that considers extradispersion.

Example 6.8

```
install.packages("MASS")
library(MASS)
example.6.8 <- glm.nb(volteer~female+nonwhite+educate+income,
  data=gsscomp6.4)
exp(coefficients(example.6.8))

gsscomp6.4$pred.nb <- fitted(example.6.8)
aggregate(pred.nb~female+(nonwhite==0)+(educate==12)+(income==5),
  gsscomp6.4, mean)
```

Example 6.9

```
# R uses the glm command for negative binomial regression. See
  Example 6.8.

gsscomp6.4$dev.nb <- residuals(example.6.8, type = "deviance")
gsscomp6.4$hat.nb <- hatvalues(example.6.8)
gsscomp6.4$cookd.nb <- cooks.distance(example.6.8)
gsscomp6.4$adjdev.nb <- gsscomp6.4$dev.nb + (1/
    (6*sqrt(gsscomp6.4$pred.nb)))
```

Figure 6.9

```
plot(gsscomp6.4$hat.nb, gsscomp6.4$dev.nb)
abline(h=4)
abline(v=0.0077)
```

Figure 6.10

```
gss.vol <- gss[which(gss$volteer!=0), ]
hist(gss.vol$volteer)
```

Example 6.10

```
install.packages("pscl")
library(pscl)
example.6.10 <- zeroinfl(volteer~female+nonwhite+educate +income |
  educate, data=gsscomp6.4)
summary(example.6.10)
exp(coefficients(example.6.10))

vuong(example.6.10, example.6.5)
```

Example 6.11

```
example.6.11 <- zeroinfl(volteer~female+nonwhite+educate +income |
  educate, data=gsscomp6.4, dist="negbin")
summary(example.6.11)
exp(coefficients(example.6.11))

vuong(example.6.11, example.6.8)
```

Example 6.12

```
gsscomp6.4$pred.zeroc <- predict(example.6.10, type="count")
aggregate(pred.zeroc~female+(nonwhite==0)+(educate==12)+(inc
  ome==5), gsscomp6.4, mean)

gsscomp6.4$pred.zerop <- predict(example.6.10, type="prob")
aggregate(pred.zerop~female+(nonwhite==0), gsscomp6.4,
  mean)
```

Example 6.13

```
gsscomp6.4$resid <- residuals(example.6.10)
gsscomp6.4$zresid <- scale(gsscomp6.4$resid, center=TRUE,
  scale=TRUE)
gsscomp6.4$rate <- predict(example.6.10, type="response")

gsscomp6.4$anscombe <- (1.5*((gsscomp6.4$volteer^(2/3)) -
  (gsscomp6.4$rate^(2/3)))) / (gsscomp6.4$rate^(1/6))
```

Figure 6.12

```
gss6.graph <- gsscomp6.4[gsscomp6.4$volteer!= 0,]
plot(gss6.graph$rate, gss6.graph$zresid)
```

Figure 6.13

```
plot(gss6.graph$rate, gss6.graph$anscombe)
```

Figure 6.14

```
plot(density(gss6.graph$anscombe))
```

CHAPTER 7

Example 7.1

```
event1[10:20,]
```

Example 7.2

```
library(survival)
event1$mar.surv <- with(event1, Surv(marriage, marriage>0))
event1$mar.surv
```

Figure 7.2

```
example.7.2 <-survfit(mar.surv~1, data=event1)
plot(example.7.2)
```

Example 7.3

```
summary(example.7.2)
```

Example 7.4

```
example.7.2
mean(mar.surv[,1])
rate <- 1/mean(mar.surv[,1])
rate
```

Figure 7.3

```
figure.7.3 <- survfit(mar.surv~cohab, data=event1)
plot(figure.7.3, conf.int=FALSE, lty=c(1,3), lwd=3, col=c("black",
  "blue"))
legend(600, 1, c("Cohab = 0", "Cohab=1"), lty=c(1,3), lwd=3,
  col=c("black", "blue"))
```

Example 7.5

```
survdiff(Surv(marriage, marriage>0) ~ cohab, data=event1, rho=1)
```

Example 7.6

```
event2$mar2.surv <- with(event2, Surv(marriage, evermarr==1))
event2$mar2.surv
```

Example 7.7

```
example.7.7 <- survreg(Surv(marriage, evermarr==1) ~ educate +
  age1sex + attend14 + cohab + race, data=event2,
  dist="lognormal")
summary(example.7.7)
exp(coefficients(example.7.7))
```

Example 7.8

```
example.7.8 <- survreg(Surv(marriage, evermarr==1) ~ educate +
  age1sex + attend14 + cohab + race, data=event2,
  dist="exponential")
summary(example.7.8)
exp((coefficients(example.7.8)*(-1)*(1/example.7.8$scale)))
```

Example 7.9

```
example.7.9 <- survreg(Surv(marriage, evermarr==1) ~ educate +
   age1sex + attend14 + cohab + race, data=event2, dist="weibull")
summary(example.7.9)
exp((coefficients(example.7.9)*(-1)*(1/example.7.9$scale)))
```

Example 7.10

```
# There was no option to calculate Cox-Snell residuals with the
   lognormal distribution in R at the time of publication.
```

Figure 7.4

```
# There was no option to calculate Cox-Snell residuals with the
   lognormal distribution in R at the time of publication.
```

Figure 7.6

```
event2$deviance <- resid(example.7.7, type="deviance")
plot(event2$marriage, event2$deviance)
```

Example 7.11

```
example.7.11 <- coxph(Surv(marriage, evermarr == 1) ~ educate +
   age1sex + attend14 + cohab + race, data=event2, ties = "efron")
summary(example.7.11)

cox.zph(example.7.11, transform = 'rank')
```

Example 7.12

```
example.7.12 <- coxph(Surv(marriage, evermarr == 1) ~ educate +
   age1sex + attend14 + cohab + cohab:marriage, data=event2,
   ties="efron")
summary(example.7.12)
```

Figure 7.7

```
figure7.7 <- (Surv(event2$marriage, event2$evermarr==1))
plot(survfit(figure7.7 ~ event2$cohab), col=c("black", "red"),
   fun="cloglog")
```

Example 7.13

```
# The Surv(start.time, stop.time, status) syntax can be used in
   R for time-dependent covariates. However, the data need to
   contain the start and stop times for each participant, which
   this data set does not. Different ways to work around this and
   rearrange data can be found in the documentation for the R
   package "survival."
```

Example 7.14

```
event3[121:132,1:5]
```

Example 7.15

```
event3$year.fact <- factor(event3$year)
example.7.15 <- glm(delinq1 ~ stress+cohes+year.fact, data=event3,
    family='binomial'(link="logit"))
summary(example.7.15)
exp(example.7.15$coefficients)
```

Example 7.16

```
event3$year.s <- (event3$year-1)

event3$year.start[event3$year.s== 0] <- 0
event3$year.start[event3$year.s== 1] <- 365
event3$year.start[event3$year.s== 2] <- 730
event3$year.start[event3$year.s== 3] <- 1095

event3$year.end[event3$year== 1] <- 365
event3$year.end[event3$year== 2] <- 730
event3$year.end[event3$year== 3] <- 1095
event3$year.end[event3$year== 3] <- 1460

example.7.16 <- coxph(Surv(year.start, year.end, delinq1==1) ~
    stress+cohes, data=event3, ties="efron")
summary(example.7.16)
cox.zph(example.7.16, transform = 'rank')
```

Example 7.17

```
event4$year.fact <- factor(event4$year)
example.7.17 <- glm(delinq ~ stress + cohes + year.fact,
    data=event4, family='binomial'(link="logit"))
summary(example.7.17)
exp(example.7.17$coefficients)
```

Example 7.18

```
example.7.18 <- coxph(Surv(year.start, year.end, delinq==1) ~
    stress+cohes, data=event4, ties="exact")
summary(example.7.18)
```

CHAPTER 8

Example 8.1

```
event4[121:132,1:6]
```

Example 8.2

```
install.packages("plm")
library(plm)
example.8.2 <- plm(esteem~ stress+cohes+male+nonwhite,
   data=esteem, index=c("newid", "year"), model="within")
summary(example.8.2)
```

Example 8.3

```
install.packages("plm")
library(plm)
example.8.3 <- plm(esteem~ stress+cohes+male+nonwhite,
   data=esteem, index=c("newid", "year"), model="random")
summary(example.8.3)

phtest(example.8.2, example.8.3)
```

Example 8.5

```
install.packages("geepack")
library(geepack)
example.8.5 <- geeglm(delinq ~ stress + cohes, data=event4,
   id=newid, family=binomial("logit"), corstr="independence")
summary(example.8.5)
exp(example.8.5$coefficients)
```

Example 8.6

```
library(geepack)
example.8.6 <- geeglm(delinq ~ stress+cohes, data=event4,
   id=newid, family=binomial("logit"), corstr="exchangeable")
summary(example.8.6)
exp(example.8.6$coefficients)
```

Example 8.7

```
library(geepack)
example.8.7 <- geeglm(delinq~stress+cohes, data=event4, id=newid,
   family=binomial("logit"), corstr="ar1")
summary(example.876)
exp(example.8.7$coefficients)
```

Example 8.8

```
library(geepack)
example.8.8 <- geeglm(delinq~stress+cohes, data=event4, id=newid,
   family=binomial("logit"), corstr="unstructured")
summary(example.8.8)
exp(example.8.8$coefficients)
```

Example 8.9

```
example.8.9 <- geeglm(bi_depress~male, data=esteem, id=newid,
    family=binomial("logit"), constr="exchangeable")
summary(example.8.9)
exp(example.8.9$coefficients)
```

Example 8.10

```
install.packages("pglm")
library(pglm)
example.8.10 <- pglm(bi_depress ~ male, data=esteem, family =
    binomial('logit'), model = "random")
summary(example.8.10)
exp(example.8.9$estimate)
```

Example 8.11

```
library(pglm)
example.8.11 <- pglm(bi_depress ~ fstruct, data=esteem, family =
    binomial('logit'), model = "random")
summary(example.8.11)
exp(example.8.11$estimate)
```

CHAPTER 9

Example 9.1

```
example.9.1 <- lm(income ~ male + married, data=multilevel_
    data12)
summary(example.9.1)
anova(example.9.1)
```

Example 9.2

```
cl <- function(dat,fm, cluster){
  require(sandwich, quietly = TRUE)
  require(lmtest, quietly = TRUE)
  M <- length(unique(cluster))
  N <- length(cluster)
  K <- fm$rank
  dfc <- (M/(M-1))*((N-1)/(N-K))
  uj <- apply(estfun(fm),2, function(x) tapply(x, cluster, sum));
  vcovCL <- dfc*sandwich(fm, meat=crossprod(uj)/N)
  coeftest(fm, vcovCL)  }

example.9.2 <- lm(income~male+married, data=multilevel_data12)
cl(multilevel_data12, example.9.2, multilevel_data12$idcomm)
```

Example 9.3

```
install.packages("psych")
library(psych)
describe.by(multilevel_data12$income, multilevel_
   data12$idcomm)
```

Example 9.4

```
install.packages("lme4")
library(lme4)
example.9.4 <- lmer(income ~ (1|idcomm), data=multilevel_data12)
summary(example.9.4)
```

Figure 9.1

```
plot(multilevel_data12$male,multilevel_data12$income)
abline(lm(income[multilevel_data12$idcomm==700]~male[multilevel_
   data12$idcomm==700], data=multilevel_data12), col="red")
abline(lm(income[multilevel_data12$idcomm==1500]~male[multilevel_
   data12$idcomm==1500], data=multilevel_data12),
   col="blue")
abline(lm(income[multilevel_data12$idcomm==2075]~male
   [multilevel_data12$idcomm==2075], data=multilevel_data12),
   col="green")
abline(lm(income[multilevel_data12$idcomm==4480]~male
   [multilevel_data12$idcomm==4480], data=multilevel_data12),
   col="purple")
```

Example 9.5

```
library(lme4)
example.9.5 <- lmer(income ~ male+(1|idcomm), data=multilevel_
   data12)
summary(example.9.5)
```

Example 9.6

```
install.packages("plm")
library(plm)
example.9.6 <- plm(income ~ male, data=multilevel_data12,
   index=c("idcomm"), model="random")
summary(example.9.6)
```

Example 9.7

```
library(lme4)
example.9.7 <- lmer(income ~ male + (1+male|idcomm),
   data=multilevel_data12)
summary(example.9.7)
```

```
multilevel_data12$pred <- predict(example.9.7)

beta <- coef(example.9.7)$idcomm
colnames(beta) <- c("Intercept", "Slope")
```

Figure 9.2

```
multilevel_data12$rstandard <- residuals(example.9.7, scale=TRUE,
    type="pearson")
plot(density(multilevel_data12$rstandard))
```

Figure 9.3

```
plot(beta$Intercept, beta$Slope)
text(beta$Intercept, beta$Slope, labels=row.names(beta))
```

Example 9.8

```
example.9.8 <- lmer(income ~ male + pop2000 + disadvantage +
    (1+male|idcomm), data=multilevel_data12)
summary(example.9.8)
```

Example 9.9

```
example.9.9 <- lmer(income ~ age + disadvantage +
    age:disadvantage + (1+age|idcomm), data=multilevel_data12)
summary(example.9.9)

multilevel_data12$pred.9.9 <- predict(example.9.9)
```

Figure 9.4

```
multilevel_highdis <- multilevel_data12[which(multilevel_
    data12$disadvantage > 0.57),]
multilevel_lowdis <- multilevel _ data12[which(multilevel_
    data12$disadvantage < -0.57),]

plot(multilevel_data12$age, multilevel_data12$pred8.9)
abline(lm(pred9.9~age, data=multilevel_highdis), col="red",
    lwd=5)
abline(lm(pred9.9~age, data=multilevel_lowdis), col="green",
    lwd=5)
```

Example 9.10

```
library(lme4)
example.9.10 <- glmer(trust ~ age + educate + (1|idcomm), data =
    multilevel_data12, family = binomial("logit"))
summary(example.9.10)
```

```
exp(-3.68750)
exp(0.03019)
exp(0.06070)
multilevel_data12$pred8.10 <- predict(example.8.10)

multilevel_data12$anscombe <- (1.5 * ((multilevel_
  data12$trust^(2/3)) - (multilevel_data12$pred9.10^(2/3)))) /
  (multilevel_data12$pred9.10^(1/6))
```

Figure 9.6

```
plot(density(na.omit(multilevel_data12$anscombe)))
```

Figure 9.7

```
esteem_14 <- esteem[which(esteem$newage == 14),]
esteem_23 <- esteem[which(esteem$newage == 23),]
esteem_34 <- esteem[which(esteem$newage == 34),]
esteem_53 <- esteem[which(esteem$newage == 53),]

plot(esteem$age,esteem$esteem)

abline(lm(esteem_14$esteem~esteem_14$age), col="red", lwd=5)
abline(lm(esteem~age, data=esteem_23), col="green", lwd=5)
abline(lm(esteem~age, data=esteem_34), col="blue", lwd=5)
abline(lm(esteem~age, data=esteem_53), col="purple", lwd=5)
```

Example 9.11

```
example.9.11 <- lmer(esteem ~ (1|newid), data=esteem)
summary(example.9.11)
```

Example 9.12

```
example.9.12_RI <- lmer(esteem ~ age+(1|newid), data=esteem)
summary(example.9.12_RI)

example.9.12_RS <- lmer(esteem ~ age+(1+age|newid), data=esteem)
summary(example.9.12_RS)

estimates.RI <- c(23.2, 24.7)
estimates.RS <- c(196.661, 0.852, 20.494)

install.packages("lmtest")
library(lmtest)
lrtest(example.9.12_RS, example.9.12_RI)
anova(example.9.12_RS, example.9.12_RI)
```

Example 9.13

```
example.9.13 <- lmer(esteem ~ age + stress + cohes + male + (1 +
  age + stress + cohes|newid), data=esteem)
summary(example.9.13)
```

Example 9.14

```
example.9.14 <- lmer(esteem ~ age+stress+cohes+male+age:male+
  (1+age+stress+cohes|newid), data=esteem)
summary(example.9.14)
```

Figure 9.8

```
install.packages("lattice")
library(lattice)
esteem$pred9.14 <- predict(example9.14)
xyplot(pred9.14~age,data=esteem, group=male, col=c('red','blue'),
  type=c('p','r'), lwd=4, lty=c(3,2), cex=.5, pch=c(1,2))
```

Example 9.15

```
# At the time of publication, there was no unstructured option
  for the lmer command in R. However, the lme command in the
  "nlme" package offers several options that may be useful for
  this type of problem.
```

Figure 9.9

```
# At the time of publication, there was no unstructured option
  for the lmer command in R. However, the lme command in the
  "nlme" package offers several options that may be useful for
  this type of problem.
```

Figure 9.10

```
# At the time of publication, there was no unstructured option
  for the lmer command in R. However, the lme command in the
  "nlme" package offers several options that may be useful for
  this type of problem.
```

Example 9.16

```
library(lme4)
example.9.16 <- lmer(esteem ~ age + stress + cohes + male +
  nonwhite + (1+age+stress+cohes|newid), data=esteem)

beta <- coef(example.9.16)$newid
colnames(beta) <- c("Intercept", "Slope_Age", "Slope_Stress",
  "Slope_Cohes", "Slope_Male", "Slope_Nonwhite")
```

Example 9.17

```
library(lme4)
example.9.17 <- glmer(stress ~ age + (1|newid), data = event4,
   family = poisson)
summary(example.9.17)
exp(1.8510)
exp(-0.0874)
```

CHAPTER 10

Example 10.1

```
summary(factor1)
factorcomp <- factor1[complete.cases(factor1),]
```

Example 10.2

```
install.packages("FactoMineR")
library(FactoMineR)
example.10.2 <- PCA(factorcomp[,2:14], graph=FALSE)
example.10.2$eig
head(example.10.2$var$coord)
```

Example 10.3

```
install.packages("psych")
library(psych)
example.10.3 <- fa(factorcomp[1:731,2:14], nfactors = 5, rotate =
   "none", fm = "pa")
example.10.3
example.10.3$values

varimax.10.3 <- fa(factorcomp[1:731,2:14], nfactors = 2, rotate =
   "varimax", fm = "pa")
varimax.10.3
varimax.10.3$loadings
install.packages("GPArotation")
library(GPArotation)
library(psych)

oblimin.10.3 <- fa(factorcomp[1:731,2:14], nfactors = 2, rotate =
   "oblimin", fm = "pa")
oblimin.10.3
oblimin.10.3$loadings

pred <- predict(oblimin.10.3, factorcomp[1:731,2:14])
colnames(pred) <- c("Depress", "SelfEsteem")
```

Example 10.4

```
install.packages("polycor")
library(polycor)
corr.10.4 <- polychoric(factorcomp[,2:5])
corr.10.4

example.10.4 <- fa(factorcomp[1:731,2:14], nfactors = 2, rotate =
  "none", fm = "pa", cor="poly")
example.10.4
```

Example 9.5

```
install.packages("polycor")
library(polycor)
corr.10.5 <- polychoric(factorcomp[,2:14])
corr.10.5

example.10.5 <- principal(r = corr.10.5$rho, nfactors = 2, rotate =
  "none")
example.10.5
```

Example 10.6

```
library(polycor)
esteemcorr <- polychoric(factorcomp[c("worthy", "capable",
  "goodqual", "attitude")])
example.esteem <- principal(r = esteemcorr$rho, nfactors = 1,
  rotate = "none")
example.esteem
example.esteem$values

library(polycor)
depresscorr <- polychoric(factorcomp[,4:12])
example.depress <- principal(r = depresscorr$rho, nfactors = 1,
  rotate = "none")
example.depress
example.depress$values
```

Example 10.7

```
pred.esteem <- predict(example.esteem, factorcomp[c("worthy",
  "capable", "goodqual", "attitude")])
pred.depress <- predict(example.depress, factorcomp[1:731,4:12])

summary(pred.esteem)
summary(pred.depress)
cor(pred.esteem, pred.depress)
```

Example 10.8

```
factor1comp <- factor1[complete.cases(factor1),]
factor1comp$id <- NULL

install.packages("lavaan")
library(lavaan)

model.10.8 <- `
  Esteem =~ worthy + capable + goodqual + attitude
  Depress1 =~ bother + noeat + blues + depress + failure + lonely
    + crying + sad + tired
  `
sem.10.8 <- cfa(model.10.8, data=factor1comp)
summary(sem.10.8, fit.measures = TRUE, standardized = TRUE)
```

Example 10.9

```
# Note: the results differ due to different estimation
  techniques used by Stata and the R package lavaan

library(lavaan)

# Note: estimate model.10.8 first
sem.10.9 <- sem(model.10.8, data=factor1comp, ordered =c('worthy',
  'capable', 'goodqual', 'attitude', 'bother', 'noeat', 'blues',
  'depress', 'failure', 'lonely', 'crying', 'sad', 'tired'))
summary(sem.10.9, fit.measures = TRUE, standardized = TRUE)
```

Example 10.10

```
library(lavaan)
model.10.10 <- `
  General =~ mood + argues + tense + complains + fearful +
    getsalong
  Delinq =~ propdamage + shoplift + stole50 + stole49
  Delinq ~ General
  Delinq ~ male
  `
sem.10.10 <- sem(model.10.10, data=nlsy_delinq)
summary(sem.10.10, fit.measures=TRUE, standardized=TRUE)
```

Using Simulations to Examine Assumptions of OLS Regression

Many books on linear models, applied statistics, econometrics, and biostatistics furnish proofs of the central limit theorem (CLT) and the Gauss–Markov theorem, as well as several other postulates, lemmas, and theorems that are foundational for regression models. As discussed in Chapters 1 and 2, the favorable aspects of ordinary least squares (OLS) and maximum-likelihood estimation (MLE) regression depend largely on a set of assumptions. In Chapter 1, we learned in some detail about these assumptions, how some are critical for estimation, and how several are related. Moreover, it is important to emphasize that often, when these assumptions are not met, it may be because we have not thought through the process that generates the model conceptually. For instance, although linear regression models are very useful tools for studying statistical relationships among sets of variables, perhaps we should focus more—conceptually and empirically—on nonlinear models. After all, there are far more ways that sets of variables can be related in a nonlinear fashion than in a linear fashion. Regardless, the main point is that when an assumption is not met, our first goal should be to figure out why rather than to immediately look for a quick analytical fix. Why does a partial residual plot show that a linear fit is not satisfactory? Why is there a heteroscedastic pattern evident in the residuals? Where did that outlier come from?

These are important issues and, if we are careful about answering them and others like them, it is unlikely that the linear regression model will become obsolete anytime soon. Therefore, it is important that we have a firm grasp of its assumptions. If one has sufficient experience with probability and statistical theory, with a good grounding in calculus and linear algebra, then the assumptions likely make sense. If one is proficient with proofs, then all the better. However, many students and researchers do not have this type of a background and, thus, the role of these assumptions can be opaque. Fortunately, there is another way to get a better handle on many of them: computing power now allows us to use simulations to demonstrate important aspects of linear regression models and what happens to the estimates when assumptions are violated.

Briefly, a simulation involves (a) choosing a specific topic of interest, such as examining the behavior of an estimator in a particular probability distribution; (b) defining a population and setting up a *data generating process* (DGP); (c) randomly selecting several independent data sets that represent samples from the population; (d) estimating a model on each data set; and (e) compiling the results of the model and determining how the estimator—or other feature of the model—behaves. The DGP is critical since it defines, in a sense, the "true" model in the population and its specific characteristics. For example, the DGP might be a population regression model with fixed slopes and intercepts. A key advantage of simulations is that we may generate different scenarios, so we do not have to wait for data sets that have distinct characteristics, or gather population data from which to sample. To give a flavor of why this is useful, think about the Gauss–Markov theorem mentioned in Chapter 1. This theorem is foundational to the linear model, for it demonstrates that the OLS estimators are the best linear unbiased estimators when certain assumptions are met. Its proof is very persuasive (Fox, 2016). Along with the CLT, it has given a firm base to the OLS regression model and also led to the development of alternative models that are suitable when the assumptions break down (e.g., robust regression models). Yet these assumptions are based on what some claim is an elusive target: we assume what happens when we repeatedly sample from a population. Fortunately, we may use simulations to sample from a type of population that generates a specific DGP, one that the researcher has control over.

The most common type of simulation uses Monte Carlo methods, so researchers often use the phrase *Monte Carlo simulation* (Carsey and Harden, 2013). How can this help us understand the characteristics and assumptions of the OLS regression model better? Think about one of its main characteristics: $E[\hat{\beta}] = \beta$. In words, this equality asserts that the expected value of the sample slope equals the slope in the population. But how can we know this? Some argue we cannot because the definition of a population is so slippery. Others rely on proofs that rest on postulates and are developed using statistical theory. The major difficulty is that we rarely draw repeated samples from populations; this is impractical on a large scale (although it does depend on how populations are defined). But we can use simulations to get better information about this presumed equality. In a simulation, researchers can set up different scenarios or DGPs and then examine what happens to the sample slopes when, say, the OLS assumptions are met, when one or more is violated, when different sampling distributions are used, and so forth.

Most statistical software includes methods to simulate various scenarios. In Stata, the `boot-strap`, `simulate`, and `postfile` commands are useful for conducting simulations (Adkins and Gade, 2012). However, even though we extolled the virtues of Stata early in this book, the statistical software R is actually bit more user-friendly and efficient for conducting simulations.[1]

The following provides some simulations in R to examine a few of the assumptions and characteristics of OLS regression analysis. One could use an actual population data set and sample from it, but it is more common to define some variables and a DGP, followed by the repeated sampling and estimation steps. The analyst may set the size and other characteristics of the samples, although the following examples present merely an elementary set of simulations (see Carsey and Harden, 2013, for detailed examples).

The first simulation is a simple OLS regression model that is designed to test whether $E[\hat{\beta}] = \beta$. Some of its characteristics include a sample size of 1,000 and 500 simulations. These, of course, may be varied to change the size of the population, the size of the samples, and the number of samples. We also define two explanatory variables, X1, which is drawn from a normal distribution, and X2, which is correlated with X1. We then define the DGP as the following "true" model: $y = 1 + (0.5 \times X1) + \varepsilon(0,1)$. Thus, the model has parameters $\alpha = 1$ and $\beta = 0.5$, with the errors following a standard normal distribution. The relevant R code is

```
set.seed(945793)        # Set the seed for reproducible results
sims <- 500             # Set the number of simulations (try
                          changing the number)
n <- 1000               # Sample size (try varying it)
```

---

1. As mentioned in the Chapter 1, R has a rather steep learning curve, but it is free and very rich in its ability to estimate numerous statistical models, construct graphs, and allow analysts to develop and test new techniques. If one is so inclined, it is worth learning. The environment software *R Studio* simplifies many R tasks.

```
B.1 <- numeric(sims)        # Create an empty vector for storing
                              simulated slopes, X1
B.2 <- numeric(sims)        # Create an empty vector for storing
                              simulated slopes, X2
a <- 1                      # True value of the intercept
b1 <- 0.5                   # True value of slope 1
b2 <- 0.75                  # True value of slope 2
X1 <- rnorm(n,12.45,3.35)   # Create X1, which has 1,000 observations,
                              with mean=12.45 and sd=3.35

    # Note that X1 is outside the loop because Xs should be fixed in
    repeated samples. X1 is drawn from a normal distribution, but it
    could also be drawn from others (e.g., uniform)

X2 <- X1*3 + rnorm(n,0,3)   # X2 is constructed to be based on X1 plus
                              some normal error

# the correlation between X1 and X2 is about 0.96
```

The next step in the simulation is to set up the GDP within the actual simulation. This is accomplished with a loop, as shown in the following R code:

Example B.1

```
# A simple OLS regression model
for(i in 1:sims){              # Begin the loop
    Y1 <- a + b1*X1 + rnorm(n,0,1) # The "true" DGP, with standard
                                     normal error
    model1 <- lm(Y1 ~ X1)      # Estimate OLS model
    B.1[i] <- model1$coef[2]   # Place the slope estimates in the
                                 vector B.1
}                              # End the loop
```

The program then conducts 500 repetitions of the regression model and places each estimated value of X1 in a data vector. Once this is completed, we may examine the estimated slopes to see how they behave. Since we have met the assumption of normally distributed and constant variance among the errors, we should find that the mean of the sample slopes equals (with a bit of sampling error) the population slope, 0.5. First, what does a kernel density plot show?

```
plot(density(B.1), main="Distribution of OLS slope - normal errors")
abline(v=0.5, col="black")
```

Note that there is a vertical line at 0.5, the population or "true" slope (see figure B.1). The distribution of sample slopes approximates it quite well. The mean (type mean(B.1)) is 0.4997 with a standard deviation of 0.009 and confidence intervals {0.4989, 0.5005}. If we increase the number of simulations, which serves a proxy for increasing the number of samples from the population, the estimate of the slope is even more precise.

But what happens when the errors do not follow a normal distribution? Recall that normally distributed errors is an assumption (though not as critical as others) of the OLS regression model.

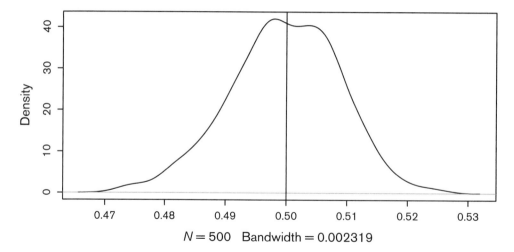

FIGURE B.1

Will the CLT[2] save the model? How precise is the estimate? Change the error distribution to log-normal and find out.

Example B.2

```
# Test of the normality assumption
for(i in 1:sims){
    Y2 <- a + b1*X1 + rlnorm(n,1,2) # DGP with lognormal error
    model2 <- lm(Y2 ~ X1)
    B.1[i] <- model2$coef[2]
}
```

The kernel density plot is shown in figure B.2:

Although the spread of the slopes appears much more peaked than in Example B.1, consider the x-axis scale. Notice that it is much wider than the scale in the first graph. This demonstrates that there is a loss in efficiency when the errors do not follow a normal distribution. From this simulation, the mean of the slopes is 0.4988, with SD = 1.11 and CIs = {0.401, 0.597}. Nevertheless, as predicted by the CLT, the slopes do seem to follow roughly a normal distribution. As an exercise, vary the mean and standard deviation of the error distribution in the DGP. The farther away from the normal distribution, the greater the loss in efficiency.

One of the most consequential assumptions of the OLS regression model is homoscedastic errors. Example B.3 shows what happens when this assumption is violated.

---

2. As a refresher, the CLT states: For random variables with finite variance, the sampling distribution of the standardized sample means approaches the standard normal distribution as the sample size approaches infinity. We should also see the sample slopes approaching a normal distribution with repeated sampling.

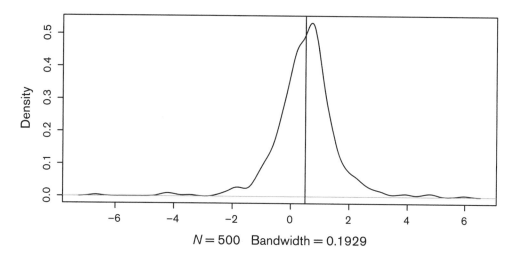

N = 500   Bandwidth = 0.1929

FIGURE B.2

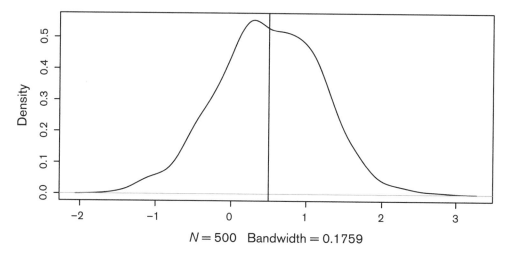

N = 500   Bandwidth = 0.1759

FIGURE B.3

Example B.3

```
# Test of the homoscedasticity assumption
for(i in 1:sims){
  Y3 <- a + b1*X1 + rnorm(n,0,sd=1:30)    # DGP, with heteroscedastic
                                            error
  model3 <- lm(Y3 ~ X1)
  B.1[i] <- model3$coef[2]
}
```

Once again, there is some loss of efficiency with heteroscedastic errors, though the slope appears unbiased (see figure B.3). The mean of the slopes is 0.492, with SD = 0.164 and CIs = {0.477, 0.506}. If we vary the degree of heteroscedasticity, the precision of the estimate changes also.

The next two simulations demonstrate that the slope may be biased when certain assumptions of the model are not met. First, recall that one of the assumptions is that the explanatory variables are not systematically related to the errors. Omitted variable bias is one way that this assumption is not met. Recall that we designed X2 so that it is correlated with X1. What happens when it is included in the DGP, but is left out of the OLS regression model?[3]

Example B.4

```
# Test of omitted variable bias
for(i in 1:sims){
    Y4 <- a + b1*X1 + X2 + rnorm(n,0,1)    # X2 included in the DGP
    model4 <- lm(Y4 ~ X1)                  # X2 not included in the OLS
                                           # model

    B.1[i] <- model4$coef[2]
}
```

The mean of the sample slopes from this model is 3.47, which is far from the population slope. However, if we include X2 multiplied by its "true" slope (b2 = 0.75) in the model, then the mean of the slopes for X1 is very close to 0.5, with only a slightly larger standard deviation and wider CIs than in Example B.1.

The final example involves the linearity assumption. Recall that violating this assumption is a form of misspecification bias. In Example B.5, observe what happens when the population association between X1 and Y is quadratic.

Example B.5

```
# Test of the linearity assumption
for(i in 1:sims){
    Y5 <- a + b1*(X1^2) + rnorm(n,0,1)     # X1-squared in DGP
    model5 <- lm(Y5 ~ X1)
    B.1[i] <- model5$coef[2]
}
```

The results of this model demonstrate well what happens when the linearity assumption is violated and the wrong functional form is used. The sample slopes have a mean of 12.65. Moreover, a component-plus-residual plot (see Chapter 1, figure 1.8) shows the evident curvature as the residuals are plotted against X1 (in R this plot—executed with `crPlots`(*model name*)—is part of the `car` library).

Other assumptions, such as those involving collinearity and independence of observations or errors, may also be examined with simulations. Moreover, the properties of MLE models, including linear, logistic, Poisson, and others, can be studied with simulations (see Carsey and Harden, 2013). One of the limitations with the examples provided here is that they did not consider what happens when the sample size varies. But it should not be surprising that with larger random samples, as the sample size approaches the population size, the estimates become more precise.

---

3. This could also be considered a test of measurement error if we assume that the "true score" of X1 is a combination of observed X1 and unobserved X2, which are correlated (see Chapter 1, Assumption 8).

Working with Missing Data

Most data sets that are based on surveys of individuals or organizations include some missing data. Data can be missing for several reasons: people may drop out of a study, they may not feel comfortable answering certain questions, the response options may not include one that fits their situation, or they may simply skip over a question. Some data are missing at random and other data are missing systematically. For example, many people, perhaps those who earn more money, may feel uncomfortable answering questions about their personal income. If it is true that those who earn more money are less likely to answer questions about income, then income data are not missing at random. Books and review articles on missing data furnish much more detail about patterns of, as they say, *missingness* (e.g., Allison, 2001; Graham, 2009). This appendix provides only a general overview of some things to consider when confronted with missing data.

A portion of the *USData* data set has been modified so there are now a few missing observations. We will use this artificial data set (*usdata_missing.dta*) to evaluate some missing data issues. Open the data set and examine its characteristics. Example C.1 provides a brief section of the data. There are some dots rather than numbers that appear in this example. Dots (or periods) are a commonly used representation of missing values (R uses NA; in spreadsheet software, the cells are often blank). But oftentimes missing values are coded with untenable values such as −99 for age or a 9 for race/ethnicity when the other choices are 1 = Caucasian, 2 = African-American, 3 = Latino, and 4 = other race/ethnicity. There are many other possibilities; if available, check the codebook accompanying the data for information about how missing values are treated. The `misstable summarize` command in Stata is useful for examining patterns of missing data in a data set. It also offers other features for exploring missing data (type `help misstable`).

Example C.1

```
list in 1/10
```

	state	robbrate	density	unemprat	mig_rate	perinc
1.	Alabama	185.75	84.15	5.1	1967.001	19.086
2.	Alaska	155.13	1.05	7.8	−1163.601	21.98
3.	Arizona	173.76	37.9	5.5	6802.762	20.068
4.	Arkansas	125.68	47.63	.	3185.381	17.935
5.	California	331.16	201.78	7.2	−4918.604	23.901

6.	Colorado	96.18	36.04	4.2	6553.892	24.29
7.	Connecticut	163.21	673.23	5.7	-4568.013	32.073
8.	Delaware	198.74	367.62	8.5	.	25.666
9.	Florida	299.91	262.9	5.1	4739.002	22.665
10.	Georgia	205.21	124.11	4.6	4966.259	21.689

misstable summarize

			Obs<.			
Variable	Obs=.	Obs>.	Obs<.	Unique values	Min	Max
density	4		46	45	1.05	1073.22
unemprat	4		46	32	2.9	8.5
mig_rate	2		48	48	-5803.268	15703.08

Regardless of the coding scheme for missing data, before using a variable it is important to decide how you are going to handle missing values. As you do this, though, first consider their source. Why do missing values appear in the data set? The reasons will likely vary depending on the nature of the variable. An important issue to consider is whether the data entry person forgot to place a value in a specific cell of the data set. Perhaps, in Example C.1, we need to go back to the data source and find the unemployment rate for Arkansas or the migration rate for Delaware. However, in some cases these data may not exist. Moreover, when using secondary survey data, it may be impossible to get the data if they were collected from an unidentifiable person some years earlier.

A common source of missing values in survey data involves skip patterns. Skip patterns in surveys are a convenient way to make responding more efficient. For instance, suppose we ask a sample of adolescents to fill out a questionnaire about their use of cigarettes. Our questions ask details such as how often they smoke, how many cigarettes they smoke in a day or a week, from where do they obtain cigarettes, and so forth. A majority of adolescents don't smoke at all, so it is common to find initial questions such as the following:

Q.1. Have you ever smoked cigarettes? *(circle one answer only)*

    (a) No   [Go to question Q.16]

    (b) Yes   [Go to question Q.2]

Q.2. When was the last time you smoked a cigarette? *(circle one answer only)*

    (a) Today

    (b) In the last week

    (c) More than a week ago but less than a month ago

    (d) A month ago or longer

In contemporary surveys these types of questions are often programmed into a computer so that those responding "no" to the first question are automatically asked question 16 next. Suppose we wish to analyze the second question either as an outcome or as an explanatory variable (perhaps by creating dummy variables). Would we only want to analyze data from those who actually answered the question or would we want to also include those who never smoked in the statistical analysis? This is an essential issue to address since those who answered "no" to the first question will be missing on the second question. If we decide to include those who never smoked in the analysis, then it is a simple matter to create a new variable that includes a code for "never smoked" along with others for those who smoked in the last day, week, etc.

It is also not uncommon to find missing data due to refusals or "don't know" responses (Question: "When was the last time you smoked a cigarette?" Response: "I don't know."). As mentioned earlier, many people don't like to answer questions about personal or family income. A perusal of various survey data shows that missing values on income variables are frequent, with usually more than 10% of respondents refusing to answer. Yet many researchers are interested in the association between income and a host of potential explanatory or outcome variables. Is it worth it to lose 10–20% of your sample so that you can include income in the analysis? Are people who refuse to answer income questions (or other types of questions) different from those who do answer? If they are systematically different, then the sample used in an analysis that omits them may no longer be a representative sample from the target population.[1]

Making decisions about missing values is a crucial part of most statistical exercises. There are several widely used solutions to missing data problems. First, some researchers simply ignore them by omitting them from the analysis. They argue that a few missing cases do not bias the results of their models too much. This may be true in some cases, but verification checks are still required. However, be very careful not to leave them with codes (such as −9 or 999) that will be included in the analysis. Imagine, for instance, if you left a missing code of 999 in an analysis of years of education! Now that would be an influential observation.

The most commonly used technique for omitting missing cases is known as *listwise deletion* (this is the default in most regression models estimated with Stata, SAS, and SPSS). It is also known as *complete case analysis*. Suppose we analyze the association between robberies and the migration rate in the *usdata_missing* data set shown earlier with a correlation or a linear regression model. In this situation, listwise deletion removes both Ohio and Delaware from the analysis. Now imagine if we estimate a multiple linear regression model with, say, five explanatory variables. Suppose further that 10 different observations were each missing only one value, but from 5 different explanatory variables. Looking over the frequencies, we would find 48 observations for each explanatory variable, but once we place all of them in the regression model and use listwise deletion, our sample size goes from 50 to 40; we lose one-fifth of the sample. This can be costly.

A problem that is not infrequent—and is discussed briefly in Chapter 1—involves the use of multiple nested regression models with different patterns of missing data. Say we estimate a model with robberies per 100,000 predicted by per capita income. Since there are no missing data on either variable, the sample size is 50. We then add the migration rate, which is missing two observations. A third model adds population density and the unemployment rate, each of which has four missing observations. We therefore have 10 missing observations. It is common to claim that the first two models are nested within the third, thus allowing statistical comparisons. However, they are not truly nested because they have different numbers of observations (n): 50 in the first model, 48 in the second model, and 40 in the third model. Not only do nested F-tests break down when this occurs, but also the regression coefficients should not be compared across models. It is therefore imperative that missing data be handled before estimating any of the regression models so that each has the same number of observations. Moreover, if you do wish to compare models that have only complete data, the e(sample) command is useful (see Chapter 1).

There are a variety of techniques for handling missing values in addition to listwise deletion. These include *mean substitution, regression substitution, raking* techniques, *hot deck* techniques,

---

1. It is also important to consider that some people may not have an income, such as those who are not in the workforce.

creating a code for missing values, maximum-likelihood (ML) estimation with the *expectation–maximization (EM) algorithm*, and *multiple imputation*. It is important to note that all of these assume that the data are missing at random, rather than systematically. If data are missing systematically, then a model that estimates what contributes to their *missingness* should be considered.

A common, but no longer acceptable, approach (according to most experts) is mean substitution. In this technique, one replaces missing values of a variable with the mean of that variable. For example, for the 48 cases that do have data, the mean of the migration rate per 100,000 is 1,212.8. We could then replace the missing value in Stata using the following code:

```
recode mig_rate (. = 1212.8)
```

A problem with mean substitution is that it can lead to biased slope coefficients and standard errors, especially as a higher proportion of observations is missing. Unfortunately, there is no rule of thumb about the proportion of missing values that lead to biased regression results when using mean substitution. In addition, mean substitution is obviously not appropriate with categorical variables. Although some variant, such as mode substitution, may look attractive, it is also not recommended.

Regression substitution replaces missing values with values that are predicted from a regression equation. For instance, suppose we wish to use this method to get complete data for the migration rate. We have complete data for robberies and per capita income. If a regression equation does a good job of predicting the valid migration rate values with these two variables, then the missing values may be predicted by using the coefficients from the following linear regression model:

$$mig\_rate_i = \alpha + \beta_1 (robbrate_i) + \beta_2 (perinc_i).$$

An easy technique is to save the predicted values from the model and ask the program to substitute the predicted values for migration rate when it is missing. An example of this is provided in the following set of Stata commands:

```
regress mig_rate robbrate perinc
predict missmigrate, xb
replace mig_rate=missmigrate if mig_rate == .
```

As with mean substitution, regression substitution can also lead to biased estimates. Note that its utility assumes we have a good prediction equation. Moreover, it is not a good option for categorical variables. Some analysts use some form of logistic or probit regression in this situation.

An alternative that is simpler to implement in Stata is to use the impute command: `impute mig_rate robbrate perinc, gen(newmig_impute)`. This creates a new variable, named *newmig_impute*, which has imputed values for the migration rate based on values for robberies and per capita income. Again, though, there should be a good prediction model in order to reduce bias in estimates based on the imputed variable. Moreover, the three imputation techniques discussed thus far result in imputed variables that have smaller standard deviations than the original variable. In other words, there is less variation when these approaches are used.

Raking is an iterative procedure that is based on using values of adjacent observations to come up with an estimated value for the observation that is missing. Hot deck involves partitioning the data into groups that share similar characteristics and then randomly assigning a value based on the values of this group. Both require specialized software to implement (type `help hotdeck`).

Creating a code for a missing value is also a common procedure. Suppose that we have a problem with a personal income variable: 10% of the observations are missing. We could create a new dummy variable that is coded as 0 = non-missing on income and 1 = missing on income, and include this variable in the regression model, along with other income dummy variables. This is useful mainly if the variable is transformed into a set of dummy variables, though. So income should be categorized into a number of discrete groups (e.g., $0–$10,000, $10,001–$20,000, etc.) that would then be the basis for a set of dummy variables. Hence, it may be more suitable for categorical variables than the other approaches discussed previously. However, there are limitations to this approach (Chen et al., 2008).

It is rarely, if ever, a good idea to use these techniques to replace missing values for outcome variables. Some analysts argue that raking and hot deck techniques are useful if there are only a few missing values in the outcome variable. However, if this is not the case, full information maximum likelihood (FIML), ML with the EM algorithm, and multiple imputation are better approaches. The FIML approach uses marginal probabilities based on the non-missing observations to estimate a likelihood function for the full data set, whereas the EM algorithm approach uses a two-step process: first, missing values are filled by predicted values from a series of regression equations and, second, ML estimates are estimated based on the values from the first step.

Multiple imputation has become a widely used technique throughout the regression world when there are missing values. It involves three steps:

1. Impute, or fill in, the missing values of the data set, not once, but q times (as often as 25 times in many studies). The specific imputation method varies. The imputed values are drawn from a particular probability distribution, which depends on the nature of the variable (e.g., continuous, binary). This step results is q complete data sets.

2. Analyze each of the q data sets. This results in q analyses.

3. Combine the results of the q analyses into a final result. There are rules that are followed to combine them, usually by taking some average of the coefficients and standard errors if a regression model is used (Graham, 2009; Schafer, 1999).

Beginning with version 11, Stata includes a suite of procedures for multiple imputation. Analysts may impute continuous or categorical variables using these procedures. The command `mi` (type `help mi`) offers the initial procedure for multiple imputation in Stata.[2]

The software that is available often drives the choice between choosing ML or multiple imputation. If one relies on Stata, for example, then multiple imputation is the most likely option. However, some experts prefer using FIML because, among other things, it is more efficient and estimates the *missingness* model and the analysis model simultaneously. Structural equation modeling in SAS and Stata allows FIML for missing data. These programs also have the ability to use the EM algorithm to estimate models with missing data. For a general review of procedures for missing data, including some approaches not covered here and excellent advice regarding which type of model to use, see Allison (2001), Graham (2009), or van Buuren (2012).

Example C.2 provides an illustration of multiple imputation in Stata. Using the *usdata_missing* data set, we first estimate a linear regression model with robberies per 100,000 as the outcome variable and the variables with missing data as the explanatory variables. Note that we lose 10 observations out of 50 because of missing values.

---

2. Multiple imputation procedures are also available in SPSS (*multiple imputation* command), SAS (*Proc MI* and *Proc MIANALYZE*), and R (e.g., package *amelia*, *mi*, or *mitools*).

Example C.2

```
regress robbrate density unemprat mig_rate
```

Source	SS	df	MS			
				Number of obs	=	40
				F(3, 36)	=	3.22
Model	90696.9364	3	30232.3121	Prob > F	=	0.0341
Residual	338257.94	36	9396.05388	R-squared	=	0.2114
				Adj R-squared	=	0.1457
Total	428954.876	39	10998.843	Root MSE	=	96.933

robbrate	Coef.	Std. Err.	t	P>|t|	[95% Conf. Interval]	
density	.1462075	.0708553	2.06	0.046	.0025063	.2899088
unemprat	26.18443	12.93278	2.02	0.050	-.0444562	52.41332
mig_rate	.0019356	.0044171	0.44	0.664	-.0070227	.0108938
_cons	-2.546117	70.73095	-0.04	0.971	-145.9951	140.9029

Next, we set up the multiple imputation model. This model uses multivariate normal regression (mvn) to impute the values. If we were conducting a careful analysis, it would be wise to check the characteristics of the missingness process to determine if this is the best approach to take. Here we simply wish to evaluate an elementary example.

Example C.2 (Continued)

```
set seed 5544                          * set a seed so we can replicate
mi set wide                            * tells Stata we have missing data
mi register imputed density unemprat mig_rate
                                       * identifies the variables with missing data
mi impute mvn density unemprat mig_rate =
    robbrate perinc, add(5)            * requests five imputed data sets
```

```
Multivariate imputation              Imputations =        5
Multivariate normal regression            added =        5
Imputed: m=1 through m=5                 updated =        0

Prior: uniform                        Iterations =      500
                                         burn-in =      100
                                         between =      100
```

	Observations per m			
Variable	Complete	Incomplete	Imputed	Total
density	46	4	4	50
unemprat	46	4	4	50
mig_rate	48	2	2	50

(complete + incomplete = total; imputed is the minimum across m
 of the number of filled-in observations.)

mi estimate: regress robbrate density unemprat mig_rate

* reestimate the linear regression model

```
Multiple-imputation estimates        Imputations      =           5
Linear regression                    Number of obs    =          50
                                     Average RVI      =      0.0656
                                     Largest FMI      =      0.1553
                                     Complete DF      =          46
DF adjustment:    Small sample       DF:      min     =       31.94
                                              avg     =       36.71
                                              max     =       41.25
Model F test:          Equal FMI     F(   3,    41.6) =        3.87
Within VCE type:             OLS     Prob > F         =      0.0158
```

robbrate	Coef.	Std. Err.	t	P>|t|	[95% Conf.	Interval]
density	.1562048	.0647055	2.41	0.020	.0254948	.2869147
unemprat	23.2527	11.14398	2.09	0.045	.5513575	45.95405
mig_rate	.0031498	.0039171	0.80	0.426	-.0047596	.0110591
_cons	.3726648	60.32033	0.01	0.995	-122.3496	123.0949

Comparing the two regression models, we see that the coefficients differ by a modest margin. For example, the coefficient for the unemployment rate is about four units higher in the model that excludes the missing data. Moreover, not surprisingly since there are more data, the standard errors in the multiply imputed model are smaller than in the original model. The *p*-values are smaller and the confidence intervals narrower in the multiply imputed model. This is normally the case. However, we should always evaluate whether the multiple imputation model used is the correct one for our particular data (e.g., are the missing data for the migration rate missing at random or are they systematically missing?). There are also some diagnostics that are available after the data are imputed that should be examined (van Buuren, 2012).

REFERENCES

Acock, Alan C. 2013. *Discovering Structural Equation Modeling Using Stata*, 2nd ed. College Station, TX: Stata Press.

Adkins, Lee C., and Mary N. Gade. 2012. "Monte Carlo Simulations Using Stata: A Primer with Examples." LearnEconometrics.com. Accessed September 15, 2015. http://www .learneconometrics.com/pdf/MCstata/MCstata.pdf.

Agresti, Alan. 2010. *Analysis of Ordinal Categorical Data*, 2nd ed. New York: John Wiley & Sons.

Agresti, Alan. 2013. *Categorical Data Analysis*, 3rd ed. New York: John Wiley & Sons.

Allison, Paul. 1995. *Survival Analysis using SAS*. Cary, NC: SAS Institute.

Allison, Paul. 2001. *Missing Data*. Thousand Oaks, CA: Sage Publications.

Allison, Paul. 2009. *Fixed-Effects Regression Models*. Newbury Park, CA: Sage Publications.

Allison, Paul. 2012. "Logistic Regression for Rare Events." Statistical Horizons. Accessed September 15, 2015. http://statisticalhorizons.com/logistic-regression-for-rare-events.

Angrist, Joshua D., and Jörn-Steffen Pischke. 2008. *Mostly Harmless Econometrics: An Empiricist's Companion*. Princeton, NJ: Princeton University Press.

Arceneaux, Kevin, and David W. Nickerson. 2009. "Modeling Certainty with Clustered Data: A Comparison of Methods." *Political Analysis* 17(2): 177–190.

Baldwin, Scott A., and John P. Hoffmann. 2002. "The Dynamics of Self-Esteem: A Growth-Curve Analysis." *Journal of Youth and Adolescence* 31(2): 101–113.

Bijleveld, Catrien C. J. H., and J. Theo van der Kamp. 1998. *Longitudinal Data Analysis: Designs, Models and Methods*. Newbury Park, CA: Sage Publications.

Blossfeld, Hans-Peter, Katrin Golsch, and Gotz Rohwer. 2009. *Event History Analysis with Stata*. New York: Psychology Press.

Bollen, Kenneth A. 1989. *Structural Equations with Latent Variables*. New York: Wiley.

Box-Steffensmeier, Janet M., and Bradford S. Jones. 2004. *Event History Modeling: A Guide for Social Scientists*. New York: Cambridge University Press.

Breusch, T. S., and A. R. Pagan. 1979. "A Simple Test for Heteroscedasticity and Random Coefficient Variation." *Econometrica* 47(5): 1287–1294.

Carsey, Thomas M., and Jeffrey J. Harden. 2013. *Monte Carlo Simulation and Resampling Methods for Social Science*. Los Angeles, CA: Sage Publications.

Casacci, Sara, and Adriano Pareto. 2015. "Methods for Quantifying Ordinal Variables: A Comparative Study." *Quality and Quantity* 49(5): 1859–1872.

Chatterjee, Samprit, and Ali S. Hadi. 2006. *Regression Analysis by Example*, 4th ed. New York: Wiley.

Chen, Feinian, Patrick J. Curran, Kenneth A. Bollen, James Kirby, and Pamela Paxton. 2008. "An Empirical Evaluation of the Use of Fixed Cutoff Points in RMSEA Test Statistic in Structural Equation Models." *Sociological Methods & Research* 36(4): 462–494.

Chen, Jarvis T., Afamia Kaddour, and Nancy Krieger. 2008. "Implications of Missing Income Data." *Public Health Reports* 123(3): 260.

Congdon, Peter. 2014. *Applied Bayesian Modeling*, 2nd ed. New York: Springer.

Cox, Nicholas J. 2004. "Speaking Stata: Graphing Categorical and Compositional Data." *Stata Journal* 4(2): 190–215.

Czepiel, Scott A. n.d. "Maximum Likelihood Estimation of Logistic Regression Models: Theory and Implementation." Accessed September 15, 2015. http://czep.net/stat/mlelr.pdf.

Daigle, Leah E., Francis T. Cullen, and John Paul Wright. 2007. "Gender Differences in the Predictors of Juvenile Delinquency Assessing the Generality-Specificity Debate." *Youth Violence and Juvenile Justice* 5(3): 254–286.

Dobson, Annette J., and Adrian Barnett. 2011. *An Introduction to Generalized Linear Models*, 3rd ed. Boca Raton, FL: CRC Press.

Duncan, Terry E., Susan C. Duncan, and Lisa A. Strycker. 2006. *An Introduction to Latent Variable Growth Curve Modeling: Concepts, Issues, and Application*, 2nd ed. New York: Routledge.

Fan, Xitao, and Stephen A. Sivo. 2007. "Sensitivity of Fit Indices to Model Misspecification and Model Types." *Multivariate Behavioral Research* 42(3): 509–529.

Finkel, Steven E. 2004. "Change Scores." In *Encyclopedia of Social Science Research Methods*, edited by Michael S. Lewis-Beck, Alan Bryman, and Tim Futing Liao, 120–121. Thousand Oaks, CA: Sage Publications.

Fox, John. 2010. *Maximum Likelihood Estimation: Basic Ideas*. Accessed September 15, 2015. http://socserv.socsci.mcmaster.ca/jfox/Courses/SPIDA/mle-mini-lecture-notes.pdf.

Fox, John. 2016. *Applied Regression Analysis and Generalized Linear Models*, 3rd ed. Los Angeles, CA: Sage Publications.

Gelman, Andrew. 2015. "Working Through Some Issues." *Significance* 12(3): 33–35.

Gelman, Andrew, and Jennifer Hill. 2007. *Data Analysis Using Regression and Multilevel/Hierarchical Models*. New York: Cambridge University Press.

Glejser, Herbert. 1969. "A New Test for Heteroskedasticity." *Journal of the American Statistical Association* 64(325): 316–323.

Gould, William, Jeffrey Pitblado, and Brian Poi. 2010. *Maximum Likelihood Estimation with Stata*, 4th ed. College Station, TX: Stata Press.

Graham, John W. 2009. "Missing Data Analysis: Making It Work in the Real World." *Annual Review of Psychology* 60: 549–576.

Greene, William H. 2000. *Econometric Analysis*. New York: Prentice-Hall.

Guan, Weihua. 2003. "Bootstrapped Standard Errors." *Stata Journal* 3(1): 71–80.

Hardin, James W., and Joseph M. Hilbe. 2012. *Generalized Linear Models and Extensions*, 3rd ed. College Station, TX: Stata Press.

Hardin, James W., and Joseph M. Hilbe. 2013. *Generalized Estimating Equations*, 2nd ed. Boca Raton, FL: CRC Press.

Harrington, Donna. 2008. *Confirmatory Factor Analysis*. New York: Oxford University Press.

Hastie, Trevor, Robert Tibshirani, and Jerome Friedman. 2009. *The Elements of Statistical Learning: Data Mining, Inference, and Prediction*, 2nd ed. New York: Springer.

Hausman, Jerry A. 1978. "Specification Tests in Econometrics." *Econometrica* 46(6): 1251–1271.

Heeringa, Steven G., Brady T. West, and Patricia A. Berglund. 2010. *Applied Survey Data Analysis*. Boca Raton, FL: CRC Press.

Hilbe, Joseph M. 2009. *Logistic Regression Models*. Boca Raton, FL: CRC Press.

Hilbe, Joseph M. 2014. *Modeling Count Data*. New York: Cambridge University Press.

Hoffmann, John P. 2004. *Generalized Linear Models: An Applied Approach*. Boston, MA: Pearson.

Hoffmann, John P., and Kevin Shafer. 2015. *Linear Regression Analysis*. Washington, DC: NASW Press.

Holgado-Tello, Francisco Pablo, Salvador Chacón-Moscoso, Isabel Barbero-García, and Enrique Vila-Abad. 2010. "Polychoric versus Pearson Correlations in Exploratory and Confirmatory Factor Analysis of Ordinal Variables." *Quality & Quantity* 44(1): 153–166.

Hosmer, David W., Stanley Lemeshow, and Rodney X. Sturdivant. 2013. *Applied Logistic Regression*, 3rd ed. New York: Wiley.

Hubbard, Raymond, and R. Murray Lindsay. 2008. "Why *P* Values Are Not a Useful Measure of Evidence in Statistical Significance Testing." *Theory & Psychology* 18(1): 69–88.

Johnson, Dallas E. 1998. *Applied Multivariate Methods for Data Analysts*. Pacific Grove, CA: Duxbury.

Kaplan, David. 2009. *Structural Equation Modeling: Foundations and Extensions*. Newbury Park, CA: Sage Publications.

King, Gary, and Langche Zeng. 2001. "Logistic Regression in Rare Events Data." *Political Analysis* 9(2): 137–163.

Kline, Paul. 1994. *An Easy Guide to Factor Analysis*. New York: Routledge.

Kline, Rex B. 2010. *Principles and Practices of Structural Equation Modeling*, 3rd ed. New York: Guilford.

Knafl, George J., and Margaret Grey. 2007. "Factor Analysis Model Evaluation through Likelihood Cross-Validation." *Statistical Methods in Medical Research* 16(2): 77–102.

Kneib, Thomas. 2013. "Beyond Mean Regression." *Statistical Modelling* 13(4): 275–303.

Kohler, Ulrich, and Frauke Kreuter. 2012. *Data Analysis Using Stata*, 3rd ed. College Station, TX: Stata Press.

Kuha, Jouni. 2004. "AIC and BIC Comparisons of Assumptions and Performance." *Sociological Methods & Research* 33(2): 188–229.

Lai, Mark H. C., and Oi-man Kwok. 2015. "Examining the Rule of Thumb of Not Using Multilevel Modeling: The 'Design Effect Smaller than Two' Rule." *Journal of Experimental Education* 83(3): 423–438.

Leitgöb, Heinz. 2013. "The Problem of Rare Events in Maximum Likelihood Logistic Regression: Assessing Potential Remedies." Paper presented at the 2013 European Survey Research Association Meetings, Reykjavik, Iceland, July 13–17.

Liebetrau, Albert M. 2006. "Proportional Reduction in Error (PRE) Measures of Association." In *Encyclopedia of Statistical Sciences*, edited by Samuel Kotz. New York: Wiley.

Lindgren, Bernard W. 1993. *Statistical Theory*, 4th ed. New York: Chapman & Hall,

Loeys, Tom, Beatrijs Moerkerke, Olivia De Smet, and Ann Buysse. 2012. "The Analysis of Zero-Inflated Count Data: Beyond Zero-Inflated Poisson Regression." *British Journal of Mathematical and Statistical Psychology* 65(1): 163–180.

Long, J. Scott. 2009. *The Workflow of Data Analysis Using Stata*. College Station, TX: Stata Press.

Long, J. Scott, and Jeremy Freese. 2006. *Regression Models for Categorical Dependent Variables Using Stata*, 2nd ed. College Station, TX: Stata Press.

Lumley, Thomas, Paula Diehr, Scott Emerson, and Lu Chen. 2002. "The Importance of the Normality Assumption in Large Public Health Data Sets." *Annual Review of Public Health* 23: 151–169.

McCullagh, Peter, and John A. Nelder. 1989. *Generalized Linear Models*, 2nd ed. Boca Raton, FL: CRC Press.

McFadden, Daniel L. 1973. "Conditional Logit Analysis of Qualitative Choice Behavior." In *Frontiers in Econometrics*, edited by Paul Zarembka, 105–142. New York: Academic Press.

Millar, Russell B. 2011. *Maximum Likelihood Estimation and Inference*. New York: Wiley.

Mulaik, Stanley A. 2009. *Foundations of Factor Analysis*, 2nd ed. Boca Raton, FL: CRC Press.

Myung, In Jae. 2003. "Tutorial on Maximum Likelihood Estimation." *Journal of Mathematical Psychology* 47(1): 90–100.

Newson, Roger. 2014. "RSOURCE: Stata Module to Run R from Inside Stata Using an R Source File." Accessed September 15, 2015. https://ideas.repec.org/c/boc/bocode/s456847.html.

O'Brien, Robert M. 2007. "A Caution Regarding Rules of Thumb for Variance Inflation Factors." *Quality & Quantity* 41(5): 673–690.

Oesterle, Sabrina, Monica Kirkpatrick Johnson, and Jeylan T. Mortimer. 2004. "Volunteerism during the Transition to Adulthood: A Life Course Perspective." *Social Forces* 82(3): 1123–1149.

Osgood, D. Wayne. 2000. "Poisson-Based Regression Analysis of Aggregate Crime Rates." *Journal of Quantitative Criminology* 16(1): 21–43.

Paccagnella, Omar. 2006. "Centering or Not Centering in Multilevel Models? The Role of the Group Mean and the Assessment of Group Effects." *Evaluation Review* 30(1): 66–85.

Poisson, Simeon D. 1837. *Recherches sur la Probabilite' des Jugements en Matie're Criminelle et en Matiere Civile, Prkdedes des Regles Generales du Calcul des Probabilites*. Paris: Bachelier, Imprimeur-Libraire. [In French]

Pregibon, Daryl. 1980. "Goodness of Link Tests for Generalized Linear Models." *Journal of the Royal Statistical Society, Series C (Applied Statistics)* 29(1): 15–24.

Qiao, Xingye. 2015. "Learning Ordinal Data." *WIREs: Computational Statistics* 7(5): 341–346.

Rabe-Hesketh, Sophia, and Anders Skrondal. 2012. *Multilevel and Longitudinal Modeling with Stata*, 3rd ed. College Station, TX: Stata Press.

Ramsey, James B. 1969. "Tests for Specification Errors in Classical Linear Least-Squares Regression Analysis." *Journal of the Royal Statistical Society, Series B (Methodological)* 31(2): 350–371.

Raykov, Tenko, and George A. Marcoulides. 2012. *A First Course in Structural Equation Modeling*. New York: Routledge.

Ross Sheldon M. 2014. *Introduction to Probability Models*, 10th ed. Boston, MA: Elsevier.

Schafer, Joe L. 1999. "Multiple Imputation: A Primer." *Statistical Methods in Medical Research* 8(1): 3–15.

Schwertman, Neil C., and David M. Allen. 1979. "Smoothing an Indefinite Variance-Covariance Matrix." *Journal of Statistical Computation and Simulation* 9(3): 183–194.

Selvin, Steve. 2004. *Statistical Analysis of Epidemiological Data*. New York: Oxford University Press.

Shumway, Robert H., and David S. Stoffer. 2010. *Time Series Analysis and its Applications*. New York: Springer.

Skrondal, Anders, and Sophia Rabe-Hesketh. 2004. *Generalized Latent Variable Modeling: Multilevel, Longitudinal, and Structural Equation Models*. Boca Raton, FL: CRC Press.

Snijders, Tom A. B., and Johannes Berkhof. 2008. "Diagnostic Checks for Multilevel Models." In *Handbook of Multilevel Analysis*, edited by Jan Deleeuw and Erik Meijer, 141–175. New York: Springer.

Snijders, Tom A. B., and Roel J. Bosker. 2012. *Multilevel Analysis: An Introduction to Basic and Advanced Multilevel Modeling*. Thousand Oaks, CA: Sage Publications.

Sowislo, Julia Friederike, and Ulrich Orth. 2013. "Does Low Self-Esteem Predict Depression and Anxiety? A Meta-Analysis of Longitudinal Studies." *Psychological Bulletin* 139(1): 213–240.

Stone, James V. 2013. *Bayes' Rule: A Tutorial Introduction to Bayesian Analysis*. London: JV Stone.

Taylor, Aaron B., Stephen G. West, and Leona S. Aiken. 2006. "Loss of Power in Logistic, Ordinal Logistic, and Probit Regression When an Outcome Variable Is Coarsely Categorized." *Educational and Psychological Measurement* 66(2): 228–239.

Therneau, Terry M., and Patricia M. Grambsch. 2000. *Modeling Survival Data: Extending the Cox Model*. New York: Springer.

Tjur, Tue. 2009. "Coefficients of Determination in Logistic Regression Models—A New Proposal: The Coefficient of Discrimination." *American Statistician* 63(4): 366–372.

van Buuren, Stef. 2012. *Flexible Imputation of Missing Data*. Boca Raton, FL: CRC Press.

von Bortkiewicz, Ladislaus. 1898. *Das Gesetz der Kleinen Zahlen* [*The Law of Small Numbers*]. Leipzig, Germany: B.G. Teubner.

Vuong, Quang H. 1989. "Likelihood Ratio Tests for Model Selection and Non-Nested Hypotheses." *Econometrica* 57(2): 307–333.

Wakefield, Jon. 2013. *Bayesian and Frequentist Regression Methods*. New York: Springer.

Wald, Abraham. 1939. "Contributions to the Theory of Statistical Estimation and Testing Hypotheses." *Annals of Mathematical Statistics* 10(4): 299–326.

Weisberg, Sanford. 2013. *Applied Linear Regression, Fourth Edition*. New York: John Wiley & Sons.

Westland, J. Christopher. 2015. *Structural Equation Models: From Paths to Networks*. New York: Springer.

White, Halbert. 1980. "A Heteroskedasticity-Consistent Covariance Matrix Estimator and a Direct Test for Heteroskedasticity." *Econometrica* 48(4): 817–838.

Williams, Richard. 2006. "Generalized Ordered Logit/Partial Proportional Odds Models for Ordinal Dependent Variables." *Stata Journal* 6(1): 58–82.

Williams, Richard. 2011. "Comparing Logit and Probit Coefficients between Models and across Groups." Accessed September 15, 2015. https://www3.nd.edu/~rwilliam/stats/Oglm.pdf.

Wooldridge, Jeffrey M. 2010. *Econometric Analysis of Cross Section and Panel Data*. Cambridge, MA: MIT Press.

Yamaguchi, Kazuo. 1991. *Event History Analysis*. Newbury Park, CA: Sage Publications.

Yeomans, Keith A., and Paul A. Golder. 1982. "The Guttman-Kaiser Criterion as a Predictor of the Number of Common Factors." *The Statistician* 31(3): 221–229.

Zhou, Mai. 2016. *Empirical Likelihood Method in Survival Analysis*. Boca Raton, FL: CRC Press.

Zorn, Christopher J. W. 1998. "An Analytic and Empirical Examination of Zero-Inflated and Hurdle Poisson Specifications." *Sociological Methods & Research* 26(3): 368–400.

Zou, Guangyong. 2004. "A Modified Poisson Regression Approach to Prospective Studies with Binary Data." *American Journal of Epidemiology* 159(7): 702–706.

INDEX

Note: Page number followed by (*f*) indicates figure.